Polymer Nanocomposite Coatings

Polymer Nanocomposite Coatings

Edited by
Vikas Mittal

CRC Press
Taylor & Francis Group
Boca Raton London New York

CRC Press is an imprint of the
Taylor & Francis Group, an **informa** business

CRC Press
Taylor & Francis Group
6000 Broken Sound Parkway NW, Suite 300
Boca Raton, FL 33487-2742

First issued in paperback 2017

ISBN-13: 978-1-4665-5730-7 (hbk)
ISBN-13: 978-1-138-07498-9 (pbk)

Library of Congress Cataloging-in-Publication Data

Polymer nanocomposite coatings / editor, Vikas Mittal.
    pages cm
  Includes bibliographical references and index.
   ISBN 978-1-4665-5730-7 (hardback)
  1. Plastic coating. 2. Nanocomposites (Materials) 3. Polymeric composites. I. Mittal, Vikas.

TA418.9.C57P65 2013
667'.9--dc23
                                                             2013030831

Visit the Taylor & Francis Web site at
http://www.taylorandfrancis.com

and the CRC Press Web site at
http://www.crcpress.com

Contents

Preface

The main focus of this book is the coatings generated with polymer nano-composite materials. Polymer nanocomposites have been developed continuously for the last two decades. These advancements have led to their applications in many fields like automotive, packaging, etc. Coatings are one such product that are common to many application fields. They are used to enhance the functionality of the substrate in order to gain certain performance parameters. For example, in the case of packaging, these coatings provide a gas barrier to the substrate foil or laminate. On the other hand, in the case of automobiles, they are used to provide properties like antiscratch, anticorrosion, etc.

Generation of coatings has additional requirements like choice of suitable dispersion medium, processing/polymerization conditions, and tuning of the properties according to specific requirements. This may require mixing of preformed polymer matrix with the filler or the formation of polymer in situ in the presence of filler. Formation of filler in the presence of the polymer may also be achieved. These operations also need to be carried out while taking into account the interaction of the coating materials with the substrates, as the adhesion of these two components is of utmost importance in achieving optimal performance.

Chapter 1 provides an overview of the state of the art in the field of silver-polysaccharide systems, with particular attention to the biomedical and food-packaging fields. It also discusses the controversial issue of nano-safety and the potential risk of silver-related toxicity. Chapter 2 describes epoxy-layered silicate nanocomposite coatings, which were achieved by the solution casting method using different amine hardeners as well as different filler surface modifications. It analyzes the effect of system components on the composite microstructure and oxygen permeation performance. Chapter 3 deals with carbon nanotube–polytetrafluoroethylene (CNT–PTFE) nanocomposite coatings. According to results obtained from experiments, CNT–PTFE nanocomposite coating seems to be a promising anti-bio-fouling technique that substantially reduces the requirement for tedious process protocols. Chapter 4 presents an exhaustive literature review on organic–inorganic hybrid coatings with enhanced scratch-resistance properties obtained through a bottom-up approach by the sol–gel process. Chapter 5 summarizes the molecular interactions between chitosan (CS) and magnesium aluminum silicate (MAS) clays in the form of dispersions and films. The characteristics of CS–MAS nanocomposite films and the application of these films for use as a tablet-coating material in the pharmaceutics field are also discussed. In Chapter 6, the authors by viewing from inside to the surface of polymer systems confirm the necessity of multidisciplinary characterization for the description of complex organic–inorganic nanocomposite coatings. The combination of "classical" and "up-to-date" analytical techniques spanning from segmental up to macroscopic level was reported to be the best method. Chapter 7 covers hybrid sol–gel coatings that provide a viable method for improving the mechanical properties of transparent plastics like polycarbonate. Nanocomposite hybrid sol–gel coatings, where nanoparticles can be generated in situ, improve the mechanical properties better than a dispersion of nanoparticles in the hybrid sol–gel matrix. Chapter 8 starts with a discussion on photopolymerizable siloxane-modified acrylic formulation and then describes UV-cured boehmite nanocomposites as potential protective coatings for wood elements. Chapter 9 discusses protective coatings based on silsesquioxane nanocomposite materials. Chapter 10 sheds light on the recent advances of polyhedral oligomeric silsesquioxane (POSS)-based nanocomposites for biomedical use, in particular those that function as a robust coating for medical devices that are implanted in biological systems. Chapter 11 describes the coatings of intrinsically conducting polymers for corrosion protection. Chapter 12 deals with the ultrasound-assisted in situ emulsion polymerization synthesis of polymer–$CaCO_3$ nanocomposites. Applications of polymer–$CaCO_3$ nanocomposites in various fields, for example, coatings, are also discussed.

Vikas Mittal
Abu Dhabi, United Arab Emirates

Editor

Vikas Mittal is an assistant professor in the Chemical Engineering Department of The Petroleum Institute, Abu Dhabi. He obtained his PhD in 2006 in polymer and materials engineering from the Swiss Federal Institute of Technology in Zurich, Switzerland. Later, he worked as materials scientist in the Active and Intelligent Coatings section of SunChemical in London, UK, and as polymer engineer at BASF Polymer Research in Ludwigshafen, Germany. His research interests include polymer nanocomposites, novel filler surface modifications, thermal stability enhancements, and polymer latexes with functionalized surfaces. He has authored over 50 scientific publications, book chapters, and patents on these subjects.

Contributors

Magdalena Aflori
Department of Polymer Materials
    Physics
"Petru Poni" Institute of
    Macromolecular Chemistry
Iasi, Romania

Bharat A. Bhanvase
Department of Chemical
    Engineering
Vishwakarma Institute of
    Technology
Pune, India

Irina-Elena Bordianu
Department of Polymer Materials
    Physics
"Petru Poni" Institute of
    Macromolecular Chemistry
Iasi, Romania

Massimiliano Borgogna
Department of Life Sciences
University of Trieste
Trieste, Italy

Jiří Brus
Department of NMR Spectroscopy
Institute of Macromolecular
    Chemistry
Academy of Sciences of the Czech
    Republic
Prague, Czech Republic

Carola Esposito Corcione
Department of Engineering for
    Innovation
University of Salento
Lecce, Italy

Corneliu Cotofana
Department of Polymer Materials
    Physics
"Petru Poni" Institute of
    Macromolecular Chemistry
Iasi, Romania

Ivan Donati
Department of Life Sciences
University of Trieste
Trieste, Italy

Florica Doroftei
Department of Polymer Materials
    Physics
"Petru Poni" Institute of
    Macromolecular Chemistry
Iasi, Romania

Yasmin Farhatnia
Division of Surgery and
    Interventional Science
Centre for Nanotechnology and
    Regenerative Medicine
University College London
London, United Kingdom

Mariaenrica Frigione
Department of Engineering for
    Innovation
University of Salento
Lecce, Italy

Soojin Jun
Department of Human Nutrition,
    Food and Animal Sciences
University of Hawaii at Manoa
Honolulu, Hawaii

Kirill L. Levine
Department of General and
    Technical Physics
National Mineral Resources
    University "Mining"
St. Petersburg, Russia

Eleonora Marsich
Department of Medical, Surgical,
    and Health Sciences
University of Trieste
Trieste, Italy

Massimo Messori
Department of Engineering
    'Enzo Ferrari'
University of Modena and Reggio
    Emilia
Modena, Italy

Vikas Mittal
Department of Chemical
    Engineering
The Petroleum Institute
Abu Dhabi, United Arab Emirates

Mihaela Olaru
Department of Polymer Materials
    Physics
"Petru Poni" Institute of
    Macromolecular Chemistry
Iasi, Romania

Sergio Paoletti
Department of Life Sciences
University of Trieste
Trieste, Italy

Thaned Pongjanyakul
Division of Pharmaceutical
    Technology
Faculty of Pharmaceutical Sciences
Khon Kaen University
Khon Kaen, Thailand

Natthakan Rungraeng
Department of Molecular
    Biosciences and Bioengineering
University of Hawaii at Manoa
Honolulu, Hawaii

Marco Sangermano
Department of Applied Science and
    Technology
Polytechnic of Turin
Torino, Italy

Alexander M. Seifalian
Division of Surgery and
    Interventional Science
Centre for Nanotechnology and
    Regenerative Medicine
University College London
and
Royal Free London
NHS Foundation Trust
London, United Kingdom

Bogdana Simionescu
Department of Polymer Materials
    Physics
"Petru Poni" Institute of
    Macromolecular Chemistry
Iasi, Romania

and

NMR Department "Costin D.
    Nenitescu" Centre of Organic
    Chemistry
Bucharest, Romania

Shirish H. Sonawane
Department of Chemical
    Engineering
National Institute of Technology
Warangal, India

Milena Špírková
Department of Nanostructured
  Polymers and Composites
Institute of Macromolecular
  Chemistry
Academy of Sciences of the Czech
  Republic
Prague, Czech Republic

Raghavan Subasri
Center for Sol-Gel coatings
International Advanced Research
  Centre for Powder Metallurgy
  and New Materials (ARCI)
Hyderabad, India

Aaron Tan
Division of Surgery and
  Interventional Science
Centre for Nanotechnology and
  Regenerative Medicine
and
UCL Medical School
University College London
London, United Kingdom

Andrea Travan
Department of Life Sciences
University of Trieste
Trieste, Italy

1

Silver–Polysaccharide Nanocomposite Antimicrobial Coatings

**Andrea Travan, Eleonora Marsich, Ivan Donati,
Massimiliano Borgogna, and Sergio Paoletti**

CONTENTS

1.1 Introduction

Recently, research on nanoparticle synthesis and use has been gaining considerable attention in the area of medicine, biology, and materials engineering owing to their chemical, biological, and physical properties, which depends on particle chemistry, structure, and dimensions.

Nanocomposites are by definition materials that contain domains or inclusions in the nanometer size scale. Nowadays, there is an increasing interest toward nanocomposite materials that exploit the antimicrobial properties of silver at the nanoscale for various aims, especially in the biomedical and food industry. For these fields, the regulation of harmful effects (toxicity) is of primary importance, and in this perspective, the preparation and stabilization of nanoparticles must be carried out using biocompatible polymers; to this end, natural polysaccharides are emerging as the most appropriate choice acting both as nontoxic reducing agents and as cell-friendly matrix.

These bioactive systems are particularly important when the antimicrobial activity has to be exerted by direct contact with the material surface; in these cases, silver–polysaccharide nanocomposites can be constructed in the form of coatings, films, or sheets.

This chapter aims at providing an overview of the state of art in the field of silver–polysaccharide systems, with particular attention to biomedical and food-packaging fields. Following a brief introduction and background concepts of antimicrobial surfaces, the antimicrobial activity of silver nanoparticles (AgNPs) is discussed. Then, details on the main routes of preparation of silver–polysaccharide coatings and films are reported, and some relevant applications and significant results over the last few years are presented. Subsequently, a section is devoted to discuss the controversial issue of nano-safety and the potential risk of silver-related toxicity. Finally, the future of this promising class of materials is tentatively devised.

1.2 Antimicrobial Surfaces: Background Concepts

Bacterial adhesion to material surfaces and interfaces is the first step in bacterial colonization, which can lead to a mature biofilm. The insertion of a biomaterial into a body causes an immediate reaction that leads to the deposition of organic matter on its surface, indicated as "conditioning film," composed essentially of proteins and platelets. This intermediate layer connects the substratum and the adhering microorganisms [1]. Bacterial adhesion is a complex interplay of different phenomena such as Brownian motion, gravitation, diffusion, convection, and the intrinsic motility of microorganisms. The overall process of bacterial adhesion can be described by means of the so-called model. This is composed of an initial reversible interaction between the bacterial membrane and the material surface, followed by a second stage that includes specific and nonspecific interactions between adhesion proteins expressed on the bacterial surface and binding molecules on the material surface. The second step is slowly reversible, although it is often referred to as irreversible [2].

There are several aspects to take into account when designing an antimicrobial surface. Roughness, stiffness, charge, degree of hydrophobicity, Lewis acid–base character, hydrogen bonding capacity [3], van der Waals forces, and specific receptor–adhesion interactions [4] play an important role on bacterial adhesion. Several authors have attempted to theoretically predict the adhesion of microbes on surfaces by means of a physical–chemical approach indicated as the DLVO (Derjaguin, Landau, Verwey, Overbeek) theory [5–7] (for an extended discussion, see [8]). The DLVO approach is based on a thermodynamic treatment of the bacterial adhesion, which includes in the description of the free energy of microbial adhesion (ΔG^{adh}), a balance between Lifshitz–van der Waals forces (ΔG^{LW}, generally attractive) and electrostatic forces generated from the overlap of the electrical double layer of the microbial cell and the substratum (ΔG^{EL}, generally repulsive). By means of this thermodynamic approach, predictions on the physical–chemical properties of the surface to be sought for discouraging bacterial adhesion have been successfully attained. It was found that the increase in hydrophilicity of the surface led to a decrease in microbial adhesion, as predicted on the of the DLVO theory. Along this line, the modification of the surface with poly(ethyleneoxide) was reported to reduce the bacterial adhesion on polyurethane [4]. The increase in hydrophilicity of poly(vinyl chloride) by oxygen plasma was found to decrease the adhesion of *Pseudomonas aeruginosa* strains, although bacterial biofilm formation is not completely impeded [9].

Microbial adhesion is also influenced by the physical–chemical characteristics of bacteria. The hydrophobicity of the bacterial cell wall, determined basically by its chemical composition in terms of biomolecules synthesized, has a lower effect when compared to the hydrophobicity of the substratum. As a general consideration, it can be stated that for the same substratum, the higher the hydrophobicity of the bacteria the higher their adhesion, and the biofilm is formed. The more hydrophilic *Staphylococcus aureus* was found to adhere to a lower extent than the more hydrophobic *Escherichia coli* [10]. In addition, different strains of the same bacterium show marked differences in the hydrophobicity/hydrophilicity properties [11,12]. For reasons resumed earlier, predictions of the bacterial adhesion on a substratum based on the DLVO approach shows several limitations. Indeed, the colonization of bacteria at the implant interface strongly depends on many factors that often are microorganism dependent (temperature, flow conditions, and concentration of glucose and oxygen). Consequently, even for a single strain and material surface, environmental stimuli can affect the relative importance of both adhesion mechanisms and material surface characteristics. In addition, the use of medium containing proteins, due to the adsorption of the latter on the surface of materials, complicates the analysis. Moreover, bacterial cells show peculiar features that could affect their adhesion to surfaces. In fact, microbial cell surfaces are heterogeneous in composition and are with a high level of complexity, which is normally not taken into account when considering a physical–chemical modeling at the molecular level. Simple concepts like, for example, distance—partially lose their meaning considering that appendages on microbial cell surfaces can become as long as 1 μm [8]. The importance of surface roughness on the adhesion of bacteria was recently reviewed [13,14]. It is generally agreed that irregularities on the polymeric surface promote bacterial adhesion, while ultrasmooth ones prevent biofilm deposition [15–17]. It has been reported that roughening of the surface increased bacterial adhesion and biofilm formation [18], although a direct correlation has not been found [19]. This is because the initial bacterial adhesion probably locates where the cells are sheltered from the shear forces of the body fluids. In addition, the higher the roughness of the surface, the higher the total surface available for adhesion [10]. Among the different topographic modifications, the presence of well-distributed etched pits on the material surface was reported to bring about an effect on bacterial adhesion, which depends on the dimension of the pits and of the bacterial species considered [20,21].

As to the influence of the stiffness of the surface of the material, it has been reported that this physical parameter modulates mechano-selective adhesion of *Staphylococcus epidermidis* and *E. coli*. In particular, the increase in Young's modulus of the polyelectrolyte layers on the substrate correlates positively with the adhesion of the bacterial strains in the range of 1–100 MPa [22]. However, an extended analysis on this line is still partially lacking.

Prevention or limitation of bacterial colonization on the surface of biomaterials has been generally pursued by means of three strategies (Figure 1.1).

In the first strategy, the surface of the biomaterial is selected or modified to show hydrophilic or ultrahydrophobic properties [4,23–25]. This passive substrate aims at preventing the bacteria to achieve the two adhesion stages based on a physical–chemical incompatibility between the material surface and the bacterial cell wall. However, different bacterial species have different physical–chemical features, and the "passive surface" approach has a limited spectrum activity that hampers its efficacy. Polymeric materials engineered at the nano- and microstructures play a fundamental role in the design and creation of surfaces capable of influencing the attachment of both prokaryotic and eukaryotic cells [26,27]. In the study of antimicrobial coatings, charged polymers or oligomers are being employed in the development of antibacterial surfaces; in fact, the presence of highly hydrophilic cationic groups has been generally correlated with potent antimicrobial effects in various studies [28]. A convenient route to bind charged polymers like polyelectrolytes to biomaterials is represented by the surface adsorption driven by electrostatic interactions into "poly electrolyte multilayers"; in the field of biomaterials, this type of surface modification offers numerous opportunities for studying and developing materials whose surfaces are capable of inhibiting or promoting cell adhesion and proliferation.

The second strategy is based on the release of a biocide from the surface of the biomaterial over time. The latter can be designed to release antibiotics [29,30], peptides [31], and metal ions or nanoparticles, such as silver in a controlled fashion [24,32]. Active localized administration of antimicrobial drugs limits the phenomenon of "resistance" related to the pathogens associated

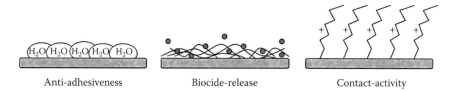

Anti-adhesiveness   Biocide-release   Contact-activity

FIGURE 1.1
Schematic representation of the three main strategies employed to prevent bacteria colonization of surfaces.

with implant infections and prevents undesired side effects associated with systemic administration [33,34]. The short-term release of the biocide from the "active" substrate accounts for a reduction in the surface colonization and prevention of biofilm formation in the postimplantation period, which is the most critical one for infections. In addition, a long-term release of antimicrobial drugs limits, to some extent, the formation of protective fibrous capsule [35]. However, the risk of toxicity for eukaryotic cells holds for these surfaces. Moreover, over time, the surface, which acts as a reservoir, is depleted of the biocide and loses its bioactivity. Over recent years, the incorporation of enzymes on the surface of the material is gaining increasing interest. Enzymes are selected to damage proteins or glycoproteins used by bacterial species for adhesion [36] or to impair the development of the bacterial cells once they are attached to the surface [37]. Moreover, enzymes can also locally produce biocides like hydrogen peroxide or hypohalogenic acid (see [21] for extended discussion).

The third strategy is based on the killing of the bacteria as a consequence of a direct contact with the surface of the biomaterial. In this case, biocides like antimicrobial peptides [38], quaternary ammonium compounds [28], guanidine polymers [39], phosphonium salts [40], and polycations [28] are chemically anchored on the surface of the biomaterial. The concern over the loss of mobility of the drugs, and hence of their antimicrobial activity, has been overcome by tethering them to polymer chains like polyethylene glycol. In this case, flexible polymer spacers allow a rapid and free orientation of the antimicrobial drug at implant–medium interface. As an example, antimicrobial peptides can be immobilized onto the surface of a biomaterial via chemical bonding (for a recent review, see [41]), which overcomes biocide leaching and short-term availability. However, in order to exploit the antimicrobial activity of the drug, effects of tethering parameters like drug surface concentration and orientation, spacer length, and spacer flexibility have to be optimized.

1.3 Silver as Antimicrobial Agent

The use of silver as antimicrobial agent is nowadays well established, being its biocidal properties known also to the ancient civilizations [42]. The recent diffusion of the nanotechnological approach has led to the development of consumer and medical products based on silver-engineered nanomaterials [42,43], due also to the need of new strategies for the treatment of infections caused by antibiotic-resistant bacteria and to reduce drug adverse side effects [44,45]. Silver-based (nano) materials have shown their activity against bacteria (also multidrug-resistant strains), fungi, viruses, and algae [43,46]. The elucidation of the antimicrobial mechanisms of such materials represents

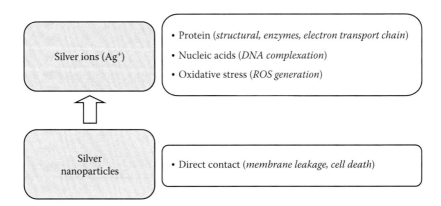

FIGURE 1.2
Biocidal mechanisms and main target sites of the silver-based materials as reported in the literature. (Adapted from Travan, A. et al., Silver nanocomposites and their biomedical applications, In: Kumar, C. (ed.), *Nanocomposites*, pp. 81–137, Wiley, New York, 2010.)

a challenging field in the biomaterial science, since they are still not completely uncovered [43,47,48]. Moreover, the reasonable concerns related to the environmental diffusion of (nano-sized) silver-based products underline the need of a deep knowledge on the risk of toxicity and in particular of undesired human effects [42,43]. These issues will be tackled in Section 1.6.

Figure 1.2 shows a summary of the biocidal mechanisms and reports the main target sites of silver. It is believed that a combination of phenomena occurs. Furthermore, it is still not clear whether the activity of AgNPs is based only on the release of silver ions (Ag⁺) or whether it involves different mechanisms when compared to bulk or ionic silver [43,49,50]. Indeed, nanomaterials present unique properties due to the high surface area/volume ratio, and several metal nanoparticles resulted effective in the development of antibacterial strategies [44].

Some of the suggested mechanisms focus on the interaction of silver ions with biological macromolecules [50]. Proteins represent important Ag⁺-sensitive sites, whose alteration leads to cell disruption, due to structural and metabolic damages. Silver ions inhibit various enzymatic activities, by reacting with electron donor groups, such as carboxylates, phosphates, hydroxyls, amines, imidazoles, indoles, and, in particular, sulfhydryl groups [50–54].

Several authors have proposed the role of the Na⁺-NQR (Na⁺-translocating NADH: ubiquinone oxidoreductase) as primary target for silver ions. This factor is a component of the respiratory chain of some bacteria and induces a transmembrane electrochemical Na⁺ potential [55,56]. Moreover, it has been suggested that the biocidal action of silver ions is related to the interaction with ribosomal subunits and also with the inhibition of the factors involved in the ATP production [57].

Silver ions can induce DNA condensation by forming complexes with the DNA bases. This inhibits the replication process, which is effective only when DNA molecules are in a relaxed state [47,48,52]. The morphological and structural study after treatment with Ag$^+$ ions reported by Feng and coworkers [52] highlighted the detachment of the cytoplasm membrane from the cell, and the presence of electron-dense granules both around the cell wall and inside the cell, probably formed by a combination of silver and proteins. Moreover, the authors proposed that the Ag$^+$ ions can cause the formation of a (somewhat undefined) low-molecular-weight region in the center of the cell, a defense structure used by the bacteria to protect the DNA from the presence of toxic compounds.

Other mechanisms have been described as responsible for the biocide action of silver-based materials. For instance, the production of high amounts of free radicals such as reactive oxygen species (ROS) induces oxidative stress in the cells, which results in alterations of the membrane lipids, breakdown of membrane, and DNA damage [43]. The increase in free-radical content is recognized as a potential antimicrobial mechanism of silver-based materials, including AgNPs, and can be related to different modes of actions [43,58,59]. Metals can catalyze the generation of ROS in the presence of dissolved oxygen. Moreover, silver ions could induce ROS production by damaging the respiratory chain enzymes and the superoxide radical scavenging enzymes (e.g., superoxide dismutases), via a direct interaction with their thiol groups [43]. Le Pape and coworkers suggested a biocidal activity based on the combination of the effects of silver ions, superoxides ions, and hydrogen peroxide [43,60].

Su and coworkers explored free-radical burst and ROS production in bacteria treated with AgNPs/clay in the presence of an indicator of intracellular ROS levels (2,7-dichlorofluorescin-diacetate) [49]. The study demonstrated the involvement of the generation of ROS in the killing mechanism, and also the role of such nanocomposites in the alteration of cytoskeleton motor functions and in the prevention of cytokinesis [50].

In another work, the kinetics of bacterial (*E. coli*) inactivation induced by a nanocrystalline silver-supported carbon matrix has been studied [61]. The ROS scavengers employed in the study showed inhibitory effects on the bactericidal activity only in the first hour: these results suggest that the generation of ROS is responsible for bacterial inactivation only at the beginning of the contact.

Several mechanisms have been suggested according to the morphological and structural changes observed in studies on bacteria. Silver could interact with the bacterial membrane, either by direct nanoparticle–membrane contact, causing a direct transfer (solvent-free) of silver ions, or by the release of silver ions into the medium. Reasonably, the combination of the two mechanisms is possible [50].

It has been suggested that when bacteria are in direct contact with AgNPs in the medium, they encounter locally high concentrations of silver ions, which induce the cell death. This is enhanced by the very high surface-to-volume ratio of the nanosized materials, which enables the sustained local supply of Ag^+ ions at the material–bacterium interface, thus preventing bacterial adhesion and biofilm formation [62].

As already mentioned, the possibility of an antimicrobial activity of the AgNPs independent from the silver ion release is extensively debated. In the study of Su and coworkers [49] on the antimicrobial mechanism of AgNPs/ silicate clay nanocomposites, the constructs exerted a biocidal effect by direct contact with the nanoparticles, and not through the release of silver ions. In the system taken as a model, a simple contact with silver nanoparticulate was enough to trigger membrane leakage and cell death.

Bacterial membrane represents an important target site for AgNPs: the induction of morphological changes leads to membrane permeability increase and to changes in transport mechanisms through the plasma membrane [63,64]. In a recent study [64], a colloidal system based on the combination of an engineered lactose-modified chitosan and AgNPs showed clear effects on membrane potential and permeability. The study of cell membrane depolarization was carried out by employing the fluorescent probe *bis*-(1,3-dibarbituric acid)-trimethine oxanol (DiBAC4(3)) as a selective label, which has the peculiarity to enter only into the bacteria characterized by a collapsed membrane potential. The fluorescence intensity measured in bacteria after a 10 min treatment with the silver-based colloidal solution was higher than that in control cells. A relevant depolarization was observed (97.5% of cells after 10 min and 99.6% after 30 min), which means that a strong interaction between silver and the bacterial membrane occurred. The loss of membrane integrity was also evaluated and confirmed by using propidium iodide as fluorescent probe.

In the study of Morones and coworkers [47], the effect of AgNPs was determined on four types of gram-negative bacteria (*E. coli, V. cholera, P. aeruginosa,* and *Salmonella typhus*). The nanoparticles detected were on the cell membrane and also into the bacterial cytoplasm. In another study, Guzman and colleagues [46] tested the effect of AgNPs of different sizes on four bacterial strains (*S. aureus* and *S. aureus* MRSA, gram positive; *E. coli* and *P. aeruginosa,* gram negative) and found a correlation between the reduction of the size and the stronger bactericidal effect. The shape of the nanomaterials can also affect the biocidal activity, as pointed out by some studies [50,65].

As for other biocidal agents, for silver-based materials also, the risk of the onset of resistance mechanisms generates growing concerns. As it has been recently reviewed by Marambio-Jones and coworkers [43], the resistance to silver ions, and in general to heavy metals, can be encoded by both plasmid and chromosomal genes. Relevant examples are the *Salmonella* plasmid pMG101, whose gene products are involved in efflux systems combined to periplasmic binding proteins, and the chromosomal Ag^+ resistance genes (*sil*) found in some *E. coli* strains (K-12 and O157:H7). However, the studies and the evidences of resistance to AgNPs are still limited. Recently, the involvement of mechanisms related to the anaerobic respiration have been hypothesized [66].

1.4 Preparation and Characterization of Silver–Polysaccharide Nanocomposites

In recent years, many efforts have been addressed to find convenient and reproducible methods to obtain polymer-based silver nanocomposites. One of the main issues in the preparation of homogeneous nanocomposite materials containing AgNPs is the tendency of the particles to aggregate [67], which causes a nonhomogeneous structure of the construct and a reduction in the effective surface area and brings about a reduced efficacy against bacteria.

One way to obtain silver nanocomposite coatings is to mix preformed nanoparticles with the polymer matrix; this *ex situ* method is considered more likely to cause nonhomogeneous structures. At variance, when AgNPs are synthesized within the polymer matrix (*in situ* method), the dispersion of the particles is typically more homogeneous because the particles are not added subsequently as an exogenous phase; this approach is particularly suitable in the case of polysaccharides as a polymer matrix [64].

The most popular method to obtain silver–polysaccharide nanocomposites is the wet chemical synthesis in the presence of a reducing agent and a stabilizing agent. This approach is based on the reduction of silver ions (typically from a silver salt) to zeroth-valent metal in an aqueous solution in the presence of a stabilizing agent, which hinders the aggregation of the nanoparticles. The wet chemical synthesis is based on a two-step mechanism: (1) nucleation and (2) growth of the particle. In step 1, part of the metal ions is reduced, and these atoms act as nucleation centers catalyzing the reduction of the remaining ions in the solution; in step 2, the atomic coalescence leads to the formation of metal clusters whose dimensions can be controlled by a stabilizing agent like ligands, surfactants, and polymers [68]. This synthesis typically involves the reduction and stabilization of silver ions in aqueous solution; the use of nontoxic components for this purpose is fundamental for biomedical and food-related applications.

Currently, sustainable approaches based on green chemistry are becoming of utmost importance to lower the impact and protect the global environment; in fact, the use of nontoxic solvents and various natural components instead of toxic chemicals for the reduction and stabilization of metallic nanoparticles is receiving considerable attention. Biomacromolecules have been utilized in the high-yield production of metal nanoparticles under relatively mild conditions, and the "green" aspects of such syntheses are particularly relevant for potential *in vitro* or *in vivo* applications [69]. Several natural compounds like proteins, plant products, and especially polysaccharides represent excellent scaffolds for this purpose.

Natural polysaccharides are the most abundant natural polymers in the biosphere and are excellent candidates for the preparation of metal nanoparticles, and they find already many applications in medicine and food fields [70]. In this perspective, the chemical structure of polysaccharide plays a fundamental role; in fact, it possesses hydroxyl groups, hemiacetal reducing ends, and other functionalities that can play important roles in both reduction and stabilization of metallic nanoparticles [71]. Moreover, the presence of side groups like amino, carboxyl, and sulfate, and their persistence length and the length of the polymeric chain (number of saccharide units) are all important factors that influence the interaction of polysaccharides with water and proteins [72].

Several polysaccharides are particularly suited for the preparation of silver nanocomposites, and when the antimicrobial activity of the nanoparticles is combined with the bioactivity of the polysaccharide, the resulting nanocomposite system can exploit both of their properties in a synergistic manner.

Plant polysaccharides (starch, cellulose, alginate, and dextran) or polysaccharides from shellfish (chitosan) are all considered biocompatible polymers, and some of them are also able to exert bioactive properties (Figure 1.3); these carbohydrates are widely used in the synthesis of metal nanoparticles, as briefly summarized in the following paragraphs.

Heparin is a polysaccharide commonly extracted from animal tissue consisting of repeating, 1,4-linked hexosamine, and uronic acid residues; it has a variety of biological functions (e.g., blood anticoagulation, anti-inflammation, promotion of cell adhesion, cell migration, and mitogenesis) mostly resulting from its selective binding to different groups of physiologically important proteins. Heparin has been used for the preparation of AgNPs [73,74].

Hyaluronic acid (HA) is a polysaccharide composed of repeating disaccharide units of glucuronic acid and N-acetyl glucosamine; it is one of the main components of the extracellular matrix, and it contributes significantly to cell proliferation and migration. It has been employed in thermal synthesis of AgNPs as both reducing and stabilizing agents [75].

Chitosan is a polysaccharide derived from chitin, the main component of the exoskeleton of crustaceans; chitin is a homopolymer composed of $\beta(1,4)$-linked N-acetyl-D-glucosamine residues, and the removal of N-acetyl groups yields chitosan with improved solubility in water. It exhibits properties of technological interest, which include excellent film-forming characteristics, biodegradability, and antimicrobial activity [64]; the primary amine endows this polysaccharide with bacteriostatic properties [76], and it effectively supports the stabilization of metallic nanoparticles [77,78].

Cellulose is a polysaccharide consisting of a linear chain of $\beta(1,4)$-linked D-glucose (Figure 1.3), and it is a major component of plant cell walls. It has been used for the preparation of AgNPs for potential applications as antimicrobial surfaces [79], especially in the form of its derivative carboxymethylcellulose (MC) [80]. The synthesis of AgNPs in combination with bacterial cellulose obtained by immersing the polysaccharide in silver nitrate solution using various reducing agents was recently reported [81,82].

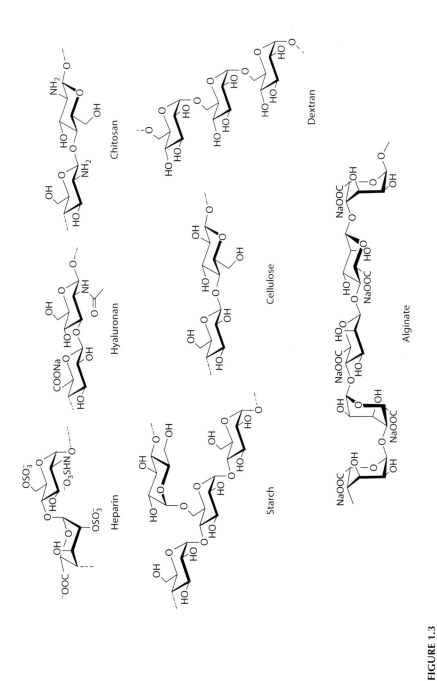

FIGURE 1.3
Structures of the polysaccharides mostly used for the synthesis and stabilization of silver nanoparticles.

Starch is a carbohydrate synthesized in plants as a mixture of α-amylose and amylopectin, which finds applications in AgNPs synthesis acting as both reducing and stabilizing agents [83].

Dextran is a branched polysaccharide made of many glucose molecules composing chains of varying lengths; it has antithrombotic properties and is capable of reducing blood viscosity. Dextran has been used as both a reductive and protective reagent of AgNPs for biosensor applications [84].

Alginate is a polysaccharide widely distributed in the cell walls of brown algae. In a typical alginate chain, homopolymeric blocks of (1,4)-linked β-D-mannuronic acid (M) and α-L-glucuronic acid (G), and alternating (MG) sequences are present. This polysaccharide is used in several applications including pharmaceutical, biomedical, and food industries [85]. Alginate has been successfully used in the photochemical synthesis of AgNPs [86] and in their stabilization in a biocompatible matrix [87–89].

In this context, one of the main advantages of these polysaccharides is that the presence of several functional groups accounts for the potential (in proper conditions) to reduce silver ions, thus avoiding the use of hazardous reducing agents. In fact, polyol synthesis of metal nanoparticles is being widely investigated, and it is particularly suited in the case of AgNPs [90].

Common exogenous reducing agents for a green synthesis of AgNPs include ascorbic acid [77,91], sodium citrate [92,93], alcohols, and aldehyde groups of sugars [94,95]; reducing saccharides are indeed used as environmental-friendly agents typically exploiting the Tollens reaction that involves the reduction of a silver ammonia solution with sugar (glucose, maltose, xylose, …) [96].

Various methodologies have been employed for the preparation of silver–polysaccharide nanocomposites; typically these procedures are based on polymer dispersion in aqueous solution followed by the reduction of silver ions to form metallic nanoparticles. When these nanocomposites are developed in the form of coatings and films, the technique employed should ensure an even distribution of the nanoparticles within the matrix. Several techniques are being studied to produce silver-based polysaccharide coatings and films, as reported in the following paragraphs.

Electrophoretic deposition (EPD) is a process in which charged colloidal particles suspended in a liquid medium migrate under the influence of an electric field and are deposited onto an electrode (Figure 1.4); this coating technique enables good control over the thickness, morphology, crystallinity, and stoichiometry of the deposits.

The interest in EPD for biomedical applications stems from the high purity of the deposited material and from the possibility to form uniform deposits on substrates of complex shapes, like surgical instruments, prosthesis, and stents. This technique was employed by Pang et al. [97] to develop a nanocomposite coating based on silver and hydroxyapatite (HA–Ag) embedded in a chitosan matrix on various conductive substrates; films were deposited

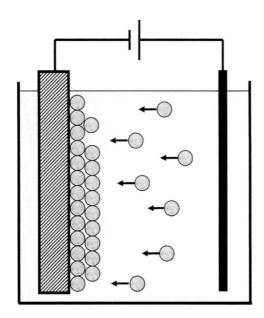

FIGURE 1.4
Schematic representation of the electrophoretic deposition technique.

as HA–Ag–chitosan monolayers or multilayers containing individual HA–chitosan and Ag–chitosan layers with a thickness in the range of 0–20 μm. SEM analysis revealed that the use of the polysaccharide enabled the formation of uniform and crack-free layers whose thickness depended on the deposition time. The release rate of silver ions upon the immersion of coating in water could be tuned by controlling the multilayer construction (Figure 1.5), which represents a desired feature for biocompatible antimicrobial coatings.

Another promising technique to obtain polysaccharide-based films is the dip-coating process in which a substrate is dipped into a liquid coating solution to form the desired layer (Figure 1.6).

This technique has been recently applied to form silver-based coatings for food-packaging applications; Costa et al. [98] described the preparation of alginate coatings loaded with silver–montmorillonite nanoparticles to prolong the shelf-life of fresh-cut carrots. The coatings were prepared by dipping the substrate into alginate solution containing silver–montmorillonite nanoparticles followed by the immersion in $CaCl_2$ solution to promote the gel-forming process of the polysaccharide matrix. In the same perspective, Gammariello et al. [99] prepared a bio-based coating containing silver–montmorillonite nanoparticles in alginate solution combined with packaging material to prevent the microbial and sensory quality decay of cheese.

The dip-coating technique is particularly employed in the construction of polyelectrolyte multilayers: in this case, the alternate deposition of

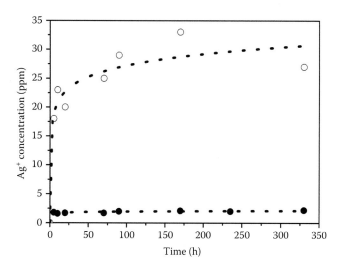

FIGURE 1.5
Silver release profile as a function of the immersion time from Ag–chitosan monolayer coating on stainless steel (white circles) and from HA–chitosan/Ag–chitosan bilayer coating on stainless steel. (Adapted from Pang, X. and Zhitomirsky, I., *Surf. Coat. Technol.*, 202(16), 3815, 2008.)

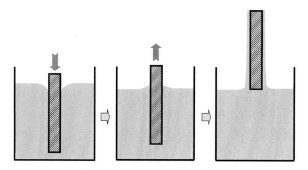

FIGURE 1.6
Schematic representation of the dip-coating process in which a substrate is dipped into liquid solution to form desired coating layer.

polyelectrolytes with opposite charges lead to the construction of a coating with the desired number of layers and thickness. This method enables the polymers to self-assemble on a surface, the film growth depending both on the intrinsic properties of the polyelectrolytes and on the experimental conditions. In each layer, it is possible to incorporate bioactive agents like silver.

With this approach, Malcher et al. [45] developed an antibacterial coating made of poly(L-lysine) ploycation (PLL)/HA multilayer films and liposome aggregates loaded with silver ions. The buildup of PLL/HA multilayer films by the layer-by-layer technique consisted of sequential dipping in PLL and in

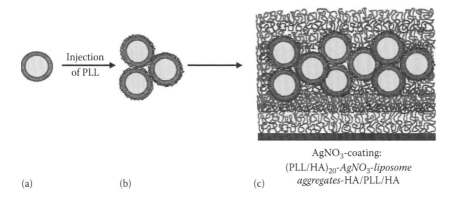

AgNO$_3$-coating:
(PLL/HA)$_{20}$-*AgNO$_3$-liposome
aggregates*-HA/PLL/HA

(a)　　　　　　　(b)　　　　　　　(c)

FIGURE 1.7
Construction of a nanocomposite multilayer obtained through the dip-coating technique. Silver–liposome aggregates are prepared (a→b) and embedded (c) in alternate layers of poly(L-lysine) (PLL) and hyaluronic acid (HA). (Reproduced from Malcher, M. et al., *Langmuir*, 24(18), 10209, 2008. With permission.)

the polyanion (HA) solutions, while the silver–liposomes were incorporated in the multilayer structure (Figure 1.7).

It is interesting to notice that with this technique, various bactericidal agents like antibiotics and peptides could in principle be encapsulated to enhance the bioactivity of the coating, thus pointing out the versatility of this approach.

On the other hand, the main challenge of this technique is to obtain a coating with sufficient stability at the interface with the substrate material; for this reason, the dip-coating approach is usually employed when there is a chemical and/or physical interaction between the substrate material and the coating material, otherwise the dipping procedure would not allow the formation of a stable and consistent layer. In some cases, it is possible to modify the surface of the substrate material in order to promote a strong interaction with the coating material. This strategy has been successfully followed to realize a silver–polysaccharide coating on methacrylate thermosets, a substrate material very widely used in dentistry and orthopedic applications. In these studies, the methacrylate substrate (bisphenol A glycidylmethacrylate/triethyleneglycol dimethacrylate) underwent a chemical modification through a hydrolysis reaction of esters, in order to introduce negative charges on the surface of the material; then, the surface-activated substrate was immersed in a colloidal solution composed of AgNPs synthesized within a lactose-modified chitosan, which is able to coordinate the nanoparticles and possess positive charges due to protonated amino groups [100–102] (Figure 1.8a).

In this case, the dip-coating technique enables to obtain a nanocomposite layer with a thickness of several micrometers in which the nanoparticles are evenly dispersed within the polysaccharide matrix (Figure 1.8b). This

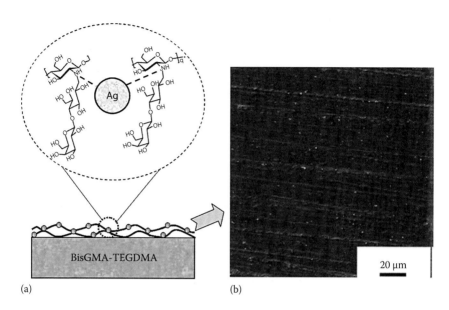

(a)   (b)

FIGURE 1.8

(a) Schematic representation of a nanocomposite antimicrobial coating for a methacrylate substrate (BisGMA-TEGDMA) composed of silver nanoparticles dispersed in a lactose-modified chitosan matrix (chitlac–nAg). (Reproduced from Marsich, E. et al., *Acta Biomaterialia*, In press, 2012. with permission.) (b) SEM image of the chitlac–nAg nanocomposite coating: the bright spots indicate the clusters of silver nanoparticles homogeneously dispersed in the polysaccharide layer. (Reproduced from Travan, A. et al., *Acta Biomater.*, 7(1), 337, 2011. With permission.)

coating preparation and technique appears to be particularly suited for biomaterials applications, since it can be applied to any geometry of the substrate after surface activation.

Dip coating on PET substrates was described by Bai-Liang Wang et al. [104] to produce films of chitosan/poly(vinylpyrrolidone) (PVP) doped with AgNPs. The nanoparticles were obtained through thermal reduction of the silver ions initially coordinated by hydroxyl and amino groups of the polysaccharide, which acted as reductive agents. The nanocomposite films with spherical particles (diameter range 10–50 nm) were shown to be stable after being immersed in phosphate-buffered saline for 35 days.

A similar approach was followed by Bal et al. [105], who developed a system for silver particle delivery by coating the micro-spongy structure of the vegetable *Luffa cylindrica* with chitosan. The coating material was prepared by immersing the *L. cylindrica* fibers in a sodium hydroxide solution followed by immersion in chitosan and in AgNO$_3$ solutions. SEM analyses confirmed the presence of AgNPs formed on the surface of the fibers with a size ranging from 20 to 150 nm.

Nanotechnologies enable the preparation of nanoparticles embedded into polymeric supports such as membranes and thin films: recently, many researches focused on the development of silver–polysaccharide constructs

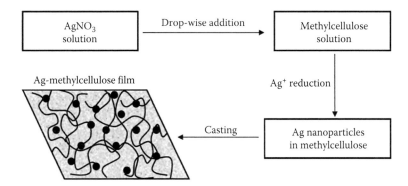

FIGURE 1.9
Schematic representation of the casting technique used to obtain silver–polysaccharide films and membranes. (Adapted from Maity, D. et al., *Carbohydr. Polym.*, 90(4), 1818, 2012.)

prepared in such forms for antimicrobial applications. Maity et al. [106] reported the preparation of MC–silver nanocomposite films by mixing MC with silver nitrate solution; the modified polysaccharide was employed for both Ag+ reduction and stabilization of the nanoparticles. The silver–MC colloidal solutions were cast in the form of nanocomposite films (Figure 1.9), and after the evaporation of solvent, yellow transparent nanocomposite films were obtained with an average thickness of 30 μm.

FTIR analyses of such films pointed out the interaction between the AgNPs and the MC matrix, while on the macroscopic scale, the nanocomposite films exhibited enhanced mechanical properties with respect to the films devoid of nanoparticles.

In a similar study, Pinto et al. [107] prepared colloidal AgNPs using exogenous reducing agents in aqueous solution; such particles were investigated as fillers in chitosan and chitosan derivative matrices to obtain thin films by a casting technique with a thickness of 9–19 μm. AFM analysis pointed out the presence of nanoparticle domains in agglomerates dispersed in the biopolymer matrix on the surface of the film, while the optical transparency of the films was dependent on the amount of silver and average particle size. These chitosan/silver composites displayed good potentials in applications requiring flexible films with tuned optical properties and antimicrobial activity.

The casting method was followed by Yoksan and Chirachanchai [108], who reported the fabrication of AgNPs by γ-ray irradiation reduction of silver nitrate in chitosan solution; the colloidal solution was then mixed with rice starch to obtain antimicrobial nanocomposite films by casting and drying the solution. The incorporation of AgNPs led also to a slight improvement of the tensile and oxygen gas barrier properties, although water vapor/moisture barrier properties were decreased.

Regardless of the technique employed to obtain the nanocomposite coatings and films, polysaccharides are widely explored for the synthesis of AgNPs, which typically represents the first step for preparing the final material. In the literature, there are several examples of the use of polysaccharides employed solely for the synthesis of silver nanoparticles. Recently, Li et al. [70] have described the preparation of AgNPs by using a triple helical polysaccharide (lentinan) dissolved in water as a matrix; the electronegativity of the hydroxyl groups of lentinan allowed capturing of the metal ions in a noncovalent interaction between the polysaccharide and silver, which led to a good dispersion of AgNPs with mean size of 6 nm in water. The system was studied by means of UV–Vis, TEM, and DLS, which revealed that the colloidal solution was stable for 9 months. Interestingly, upon the addition of NaOH, the conformation transition of the polysaccharide from the triple helix to random coil in the solution caused the aggregation of the denatured polysaccharide chains, but it did not influence the shape and size of the AgNPs (Figure 1.10).

Shukla et al. [109] recently described a green synthesis of a silver nanocomposite material based on agar extracted from the red alga *Gracilaria dura*; the thermal treatment of the polysaccharide–AgNO$_3$ solution enabled the preparation of nanoparticles homogeneously dispersed with a small average size (6 nm).

Another thermal method was introduced for the synthesis of stable AgNPs: soluble starch was used as both a reducing and a stabilizing agent of nanoparticles obtained by an autoclaving process that led to the formation of particles entrapped inside the helical amylose structure, as confirmed by iodometric titration analyses [83].

Venkatpurwar et al. [110] described a green route for the synthesis of AgNPs through sulfated polysaccharides isolated from marine red algae (*Porphyra vietnamensis*); the involvement of sulfate group in the synthesis of AgNPs (average size 13 nm) was pointed out by FTIR analysis. These composites showed good stability at a wide range of pH (2–10) and at a high electrolyte concentration (up to 10^{-2} M of NaCl).

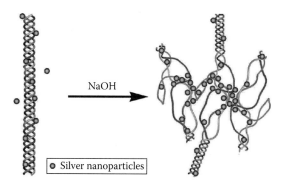

NaOH

Silver nanoparticles

FIGURE 1.10
Schematic representation of the binding of silver nanoparticles to the triple helix of the polysaccharide lentinan and to its denatured structure upon the addition of NaOH. (Reproduced from Li, S. et al., *Biomacromolecules*, 12(8), 2864, 2011. With permission.)

1.5 Applications in the Biomedical and Food-Packaging Fields

The previous paragraph described the main routes of preparation of silver–polysaccharide coatings and films. Since the integration of polysaccharide and nanomaterials is emerging as an exciting research opportunity for various purposes, this section provides an overview of how these nanocomposite materials are being studied and employed in biomedical and food research.

There are three main fields of applications of silver–polysaccharide antimicrobial coatings and films (Figure 1.11): (1) biomaterials for topical applications (e.g., wound healing), (2) implantable biomaterials, and (3) food packaging.

Biomaterials are materials designed to interact in a proper way with biological systems, and most specifically with human body; they represent one of the main targets to exploit the specific properties of polysaccharide–silver nanocomposites. In fact, the simultaneous presence of cell-friendly polymers and antimicrobial nanoparticles offers wide solutions for the development of biocompatible and bioactive medical devices. In the case of biomaterials for topical applications, wound dressings should act as barriers against exogenous bacteria and prevent wound contamination or infection; antiinfective agents incorporated in these membranes include antibiotics, antiseptics, and disinfectants [111]. In these conditions, an ideal biomaterial should possess a wide-spectrum antimicrobial activity and hygroscopic properties in order to adsorb exudates from the lesion site; these features can be achieved by using polysaccharides like alginate, chitosan, and HA as a matrix for AgNPs. Silver nanocomposites in the form of films, membranes, and coatings for topical applications are designed to be in contact with skin infections, lesions, and ulcers; some solutions are already present in clinical practice.

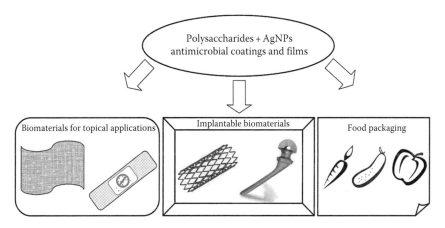

FIGURE 1.11
Main fields of applications of silver–polysaccharide antimicrobial coatings and films.

To evaluate the effects of commercial silver-based dressings on wound healing in the treatment of contaminated and infected acute or chronic wounds in randomized controlled trials, Vermeulen et al. [111] found a greater reduction of ulcer size with silver-containing membranes and observed a reduced leakage occurrence in patients with leg ulcers and chronic wounds treated with silver dressing. Nevertheless, clinical data is still limited, and some commercial materials did not significantly increase complete ulcer healing as compared with standard foam dressings. Since the use of natural polymers appears particularly appropriate for such applications, novel silver–polysaccharide solutions are continuously sought by researches, as evidenced by the studies mentioned in the following paragraphs. Alginate fibers were prepared by wet-spinning in a $CaCl_2$ and chemically cross-linked to obtain a biodegradable hydrogel; the fibers were loaded with AgNPs for potential applications as wound healing membranes [112]. In a similar study, AgNPs were synthesized by γ-irradiation in the presence of sodium alginate as stabilizers and incorporated in a chitin matrix to obtain a nanocomposite antimicrobial wound dressing; biological *in vitro* results demonstrated a dose-dependent antimicrobial effect against wound related microorganisms like *S. aureus* and *P. aeruginosa* [113]. Shukla et al. [109] described the antibacterial effect of agar-based films with AgNPs as a potential candidate for wound dressing use: the material was studied against *Bacillus pumilus* (a gram positive bacillus commonly found in soil) by disk diffusion and bacterial killing kinetics assay, and the results showed that the silver-based system could retard the bacterial growth and kill bacteria at 99.93% of reduction. In another study, the micro-spongy structure of the vegetable *L. cylindrica* has been proposed for wound healing applications due to the ability of this material to promote fluid absorption, which is particularly important in the case of exudates [105]. In order to endow this substrate with antimicrobial properties, the material was coated with a chitosan–silver layer; this polysaccharide is considered particularly useful in wound healing due to the presence of metal chelating groups (i.e., amine groups). The nanocomposite biomaterial was shown to exhibit antibacterial activity against *S. aureus* and *E. coli* with an inhibitory effect for 18 days (Figure 1.12).

In the biomedical field, another difficult challenge is to prevent infections occurring on materials inserted into the body. Bacterial colonization of implantable biomaterials represents a major threat, which is receiving increasing attention by the scientific community; in fact, infections at the site of implanted medical devices can lead to chronic microbial infection, inflammation, tissue necrosis, septicemia, and eventually to death [114]. For example, *S. epidermidis*, *E. coli*, and *S. aureus* are common bacterial strains that are able to colonize intravascular catheters and metallic implants, and their progressive accumulation on the material surface can ultimately cause the formation of a bacterial biofilm. As explained in Section 1.2, biofilms are aggregates of microorganisms embedded within a self-produced matrix of extracellular polymeric substances that adhere to the substrate and are

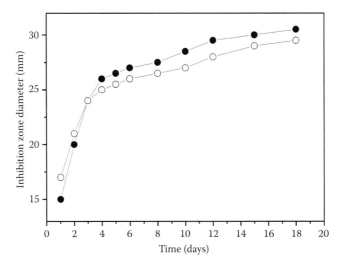

FIGURE 1.12
Evolution of the inhibition zone diameters upon contact of the nanocomposite material based on chitosan and silver with *S. aureus* (black circles) and *E. coli*.(white circles). (Adapted from Bal, K.E. et al., *Mater. Lett.*, 79, 238, 2012.)

very difficult to eradicate [115]. In this scenario, an ideal "active" coating should be able to deliver the antimicrobial agent at the site of implantation at high local concentration during the critical short-term postimplantation period (several hours) without reaching the systemic toxicity level of the agent. To this end, a smart solution would be to develop a coating composed of a highly biocompatible matrix (e.g., a polysaccharide) that embeds a wide spectrum antimicrobial agent (e.g., silver). When a silver–polysaccharide system is employed for coating a medical device to be inserted into the body, the implantable biomaterial undergoes a surface functionalization to endow the substrate with the specific properties of the nanocomposite interface layer. Several studies have explored this possibility for medical use. For bone prosthesis implants, the main goal is to achieve a strong integration between biomaterial and bone; for this reason, many studies consider the use of synthetic hydroxyapatite as a coating material since it is the major inorganic component of natural bones and it has the ability to form strong bonds with bone, although this bioceramic does not possess any antimicrobial activity. The possibility to prevent the implant-related infections using the antimicrobial properties of silver has generated interest in the development of silver-doped hydroxyapatite coatings. Pang et al. [97] have recently developed nanocomposite coatings based on hydroxyapatite–silver–chitosan by means of the electrodeposition technique; this layered nanocomposite coating could be tuned to control the release rate of Ag^+, which represents a key issue to develop systems that possess simultaneously low cytotoxicity and high antibacterial activity. Moreover, electrochemical impedance spectroscopy and potentiodynamic polarization studies showed that these multilayered coatings provide corrosion protection for stainless steel, a widely used material for implants, in the simulated physiological environment.

Recently, several studies were focused on the preparation of an antimicrobial coating for methacrylate-based bone implants [100–103]. Such polymeric material is widely used for dentistry applications and is being studied to replace metallic implants in the orthopedic field owing to the possibility to tailor the mechanical performance (e.g., in association with reinforcing fibers) for a closer match with that of bone tissue. In order to endow the surface of this material with bioactive properties, a nanocomposite coating was developed with the aim of preventing bacterial biofilm formation while providing a favorable substrate for the proliferation of osteoblasts. To this end, a lactose-modified chitosan (chitlac) was employed for the synthesis and stabilization of AgNPs [64,90], and the resulting nanocomposite system was grafted on the methacrylate surface by exploiting the electrostatic interactions. Such chitlac–AgNPs coating proved to be highly effective in reducing bacteria proliferation and in limiting biofilm formation, as shown by the *in vitro* results summed in Figure 1.13.

At the same time, this silver–polysaccharide coating was shown to stimulate osteoblasts and stem cell proliferation *in vitro* and to allow for a good osteointegration *in vivo* (Figure 1.14), pointing out the excellent biocompatibility of the nanocomposite material.

Further studies showed that this coating was stable against degradation by highly concentrated lysozyme or H_2O_2; thus, a long-term stability of the coating may be expected for implant applications, regardless of inflammatory reactions occurring during the healing process [116]. Given these results, the nanocomposite layer formed by the polysaccharide chitlac with AgNPs has

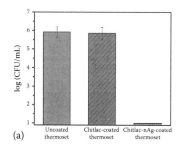
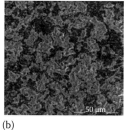
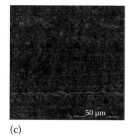

FIGURE 1.13

(a) Antimicrobial efficacy tests on *S. aureus* in direct contact with silver–polysaccharide coated and uncoated thermosets after 3 h of incubation. Biofilm formation of *S. aureus* on BisGMA/TEDGMA uncoated thermosets, (b) or with the silver–polysaccharide coating (c). (Reproduced from Travan, A. et al., *Acta Biomater.*, 7(1), 337, 2011; Marsich, E. et al., *Acta Biomaterialia*, In press, 2012. With permission.)

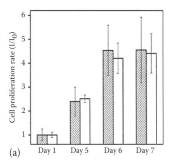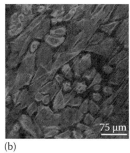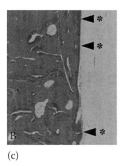

(a)   Day 1   Day 5   Day 6   Day 7   (b)   (c)

FIGURE 1.14

(a) Alamar blue proliferation assay on human adipose-derived stem cells on chitlac–nAg coated (striped bars) and uncoated (white bars) thermosets. (b) Confocal images of Saos-2 osteoblasts distribution over the chitlac–nAg coating. (c) Light micrographs of bone–implant interfaces after 8 weeks of implantation in cortical bone stained by van Gieson method of chitlac–nAg-coated implant; arrows. Areas of direct BIC are shown with arrowheads. Asterisks indicate newly formed woven bone contact. (Reproduced from Travan, A. et al., *Acta Biomater.*, 7(1), 337, 2011; Marsich, E. et al., *Acta Biomaterialia*, In press, 2012. With permission.)

shown considerable potential as a multifunctional surface coating of biomaterials, showing both antibacterial activity and cell biocompatibility, which paves the way toward future clinical studies involving this type of solutions for implantable biomaterials.

Implants like catheters and vascular stents are typical targets for bacterial colonization; in these cases, a bioactive coating should prevent biofilm formation while being biocompatible for blood and endothelial cells. In this context, Bai-Liang Wang et al. [104] studied the antibacterial properties and biocompatibility of chitosan-Ag/PVP nanocomposite films prepared by the dip-coating method; a fast antibacterial activity was demonstrated against *S. aureus* and *E. coli*, while the biocompatibility on human umbilical vein endothelial cells was shown to be dependent on silver concentration, with a threshold value of 0.25 mM for the silver nitrate concentration.

Another emerging field of interest for silver–polysaccharide systems regards food-related materials like films, sheets, membranes, and coatings. The use of coatings for a wide range of food products like fresh fruit and vegetables has received increasing attention because coatings are supposed to keep intact product characteristics like texture and hydration, and may act as carriers for a wide range of food additives, including antibrowning or antimicrobial agents. In this field, toxicological features are controlled by EU safety regulations, which limit silver amount to 0.05 mg Ag/kg of food [117]. For food-related products, coatings made of polysaccharides, proteins, and lipids can provide a semipermeable barrier to carbon dioxide, oxygen, moisture, and solute movement, thus reducing water loss, respiration, and oxidation reaction rate [98,118,119].

In this frame, the application of AgNPs for antimicrobial food packaging is being explored: some efforts have been made to incorporate AgNPs into synthetic polymers (e.g., poly(vinylalcohol) [120]) as well as into polysaccharides like starch [121] and chitosan [122]. The combination of polysaccharides and silver appears as an effective solution as pointed out by the promising results described in the recent literature. For instance, calcium alginate coatings loaded with silver–montmorillonite nanoparticles were developed by Costa et al. [98] to prolong the shelf-life of fresh-cut carrots; these coatings were tested with mesophilic and psychrotrophic bacteria (*Enterobacteriaceae* spp., *Pseudomonas* spp.) yeasts and molds, and the results highlighted that bacteria loads remained below the selected microbial threshold. Microbiological tests showed that the combined use of proper packaging and Ag-alginate-based coating maintained a good preservation of carrots prolonging the shelf-life to about 70 days with respect to the uncoated samples (about 4 days).

Chitosan/AgNPs have also been investigated for food applications [119]. Pinto et al. [107] studied the antibacterial activity of optically transparent nanocomposite films based on chitosan (and chitosan derivatives) and AgNPs; both qualitative (halo tests) and quantitative assessments pointed out a strong antimicrobial activity against *S. aureus, Klebsiella pneumoniae*, and *E. coli*, with a drop of several colony-forming units after 24 h of contact with the material. This effect was observed already in pure chitosan-derivative films, and it was further increased by the introduction of AgNPs.

MC–silver films synthesized by a casting method were recently developed for antimicrobial applications [106]. The material exhibited antibacterial activity against *B. subtilis, B. cereus, P. aeruginosa, S. aureus*, and *E. coli* as demonstrated by inhibition zone tests, while the introduction of silver led to both strengthening and toughening of MC matrix, as shown by the increase in Young's modulus and tensile strength in the presence of the metal (Figure 1.15).

In a similar study, Yoksan and Chirachanchai [108] prepared nanocomposite antimicrobial films by casting technique; to this end, AgNPs were dispersed in a chitosan–starch matrix, and these films exhibited bactericidal activity against *E. coli, S. aureus*, and *B. cereus*, as demonstrated by the inhibitory zone surrounding the films. Moreover, the incorporation of AgNPs into the polysaccharide matrix led to a slight improvement of tensile and gas barrier properties (oxygen) of the films.

Overall, this paragraph provided an overview of the main applications of silver–polysaccharide coatings and films for biomedical and food applications; the increasing number of papers in the recent literature stresses the attention on the versatile use of these nanocomposite materials and on the importance of an interdisciplinary characterization that comprises chemical, mechanical, and biological studies to assess stability, efficacy, and biocompatibility of the proposed solutions.

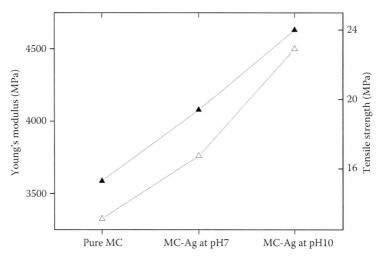

FIGURE 1.15
Young's modulus (white triangles) and tensile strength (black triangles) of methylcellulose (MC) and methylcellulose–silver (MC–Ag) nanocomposite films prepared at different pH. (Adapted from Maity, D. et al., *Carbohydr. Polym.*, 90(4), 1818, 2012. With permission.)

1.6 Silver Toxicity and Nano-Safety

Nanomaterials are expected to become the cornerstone of a number of sectors not only in healthcare but also in microelectronics, textiles, energy, and cosmetic goods. Nanoparticles are being introduced rapidly into the consumer market, but there is still little understanding on their potential consequences for human and environmental health. The most widespread metal-based nanoparticles in the world are certainly AgNPs, due to their large presence within a number of diverse products that primarily exploit their antibacterial behavior. Among more than 800 commercial products that contain nanostructures, roughly 30% are claimed to contain silver particles. At present, silver ions and AgNPs are largely available in the market both in medical areas such as wound dressing and surgical instruments, and in daily life like detergents, soaps, room sprays, water purifiers, textiles, personal care products, handles, and furniture loaded with silver-based products. In the last few years, internal prosthesis and devices (e.g., bone cement and catheters) containing silver are used in the medical field. The flip side of silver's desired toxicity toward microbes is that it might have toxic effects for eukaryotic cells as well, and this fed a debate about the safety of nanosilver products. Hundreds of nanosilver-containing products are in the market, but there are no official government indications and regulations for the commercialization of these products. Recently, a public discussion is emerging and rapidly increasing about the potential impact of nanomaterials on health. Very contradictory opinions and scientific evidences are reported by the scientific community and by the regulatory agencies. In the last decades, the widespread diffusion of silver ion-based products is partially justified by the fact that silver in ionic form is considered to have low toxicity in the human body and then that minimal risks are expected following exposure in different routes, such as inhalation, ingestion, dermal application, and hematogenous route. Silver is not known as a systemic toxic for humans, except at extreme doses. Silver in the form of ions into the human body is rapidly bound by key metal-binding proteins such as metallothioneins to form complexes that are eliminated by the liver and kidneys. These protein complexes can mitigate the cellular toxicity of silver [123,124].

The most evident effects in humans ingesting silver for long-term are argyria and argyrosis, a medically benign but permanent bluish-gray discoloration of the skin and eyes or other organs due to the deposition of inert precipitates of silver selenide and sulfide when metal absorption exceeds the capacity of liver and kidney to eliminate it. Although the deposition of silver is permanent, it is still not associated with any adverse health effects. No pathologic changes or inflammatory reactions have been seen in the patients from silver deposition [125]. Only transitory changes of renal and hepatic enzymes in blood have been described [126,127].

In 1991, the U.S. Environmental Protection Agency (EPA) established an oral reference dose, or daily intake limit, of 0.005 mg/kg/day of silver, which is "an estimate of the daily exposure to human population that is likely to be without an appreciable risk of deleterious effects during a lifetime." In 1993, EPA in an official regulatory document (EPA-738-F-93-005), stated the potential risk for humans, animals, and environment following extensive and acute exposure to pesticides containing silver compounds, they declared that "only the use of currently registered pesticide products containing silver in accordance with approved labeling will not pose unreasonable risks or adverse effects to humans or the environment."

In the last 10 years, the wide development and diffusion of new materials containing AgNPs led to a corresponding increase in human and ecosystem exposures, renewing the scientific and medical debates on the safety of AgNP-containing materials. A lot of the discussion is centered on the asserted assumption that nanoparticles are something fundamentally "new" and thus cannot be compared to conventional chemicals or bulk materials. Nanoparticles show specific physical and chemical properties that can influence and modify their biological actions often unpredictably. The potential harmfulness of AgNPs is accentuated by the fact that these particles are easily internalized by clathrin- or caveolae-mediated endocytosis and phagocytosis, thus delivering a huge quantity of silver inside the cells, which release metal ions in proximity of sensible targets—organelles and proteins [128–133]. The accumulation of AgNPs by cells is significantly lowered in the presence of endocytosis inhibitors such as chloroquine or amiloride [130]. Several studies report oxidative, inflammatory, and genotoxic consequences associated with silver particulate exposure of different cell types [134–140]. It is demonstrated that AgNPs in large part mediate their toxicity through induction of oxidative stress with increase in ROS levels and following activation of the apoptotic mitochondrial pathway. AgNPs induce ROS generation, suppression of reduced glutathione, mitochondrial membrane perturbation, and reduction of mitochondrial function in BRL 3A rat liver cells [140], rat alveolar macrophages [131], human THP-1 monocytic cells [141], lung cancer keratinocyte and epithelial adenocarcinoma cell lines [134,137], human liver cells [138], and normal bronchial epithelial (BEAS-2B) cells [135]. ROS generated by AgNPs resulted in damage to various cellular components, DNA breaks, lipid membrane peroxidation, and protein carbonylation. Upon AgNP exposure, cell viability decreased due to apoptosis, as demonstrated by the formation of apoptotic bodies, sub-G(1) hypodiploid cells, and DNA fragmentation [138]. AgNPs induce a mitochondria-dependent apoptotic pathway via modulation of Bax and Bcl-2 expressions, resulting in the disruption of mitochondrial membrane potential. Loss of mitochondrial membrane potential is followed by cytochrome c release from the mitochondria, resulting in the activation of caspases. The apoptotic effect of AgNPs appears to be exerted via the activation of c-Jun NH(2)-terminal kinase (JNK) and p53 genes [138,142]. Similarly, treatments with uncoated and polysaccharide-coated AgNPs of mouse embryonic stem cells and of embryonic fibroblast have shown p53 upregulation in both cell lines and p53 phosphorylation in embryonic stem cells. Moreover, the high induced expression of the antiapoptotic Bcl-2 protein in human colon cancer HCT116 cells protects these cells from apoptosis induced by nanosilver [142]. Many biochemical and molecular changes related to ROS-induced genotoxicity are promoted by AgNPs in cultured cells [143]. ROS, whose increase is associated to oxidative stress and subsequence redox imbalance, react with many biological macromolecules, including DNA, enzymes, and lipids. Mutation of DNA, induced by genotoxic materials, leads to carcinogenesis and has a deep impact on the biology of reproductive cells. Micronuclei formation and DNA breakage are observed in many cell lines treated with AgNPs at different concentrations. In normal bronchial epithelial (BEAS-2B) cells, AgNPs induce DNA breakage and micronuclei formation in a dose-dependent manner. The genotoxic effects of AgNPs can be partially blocked by ROS scavengers such as superoxide dismutase [135]. Micronuclei formations are also observed in HepG2 human hepatoma cells [144], in normal human lung fibroblast cells, and in human glioblastoma cells [132]. In mouse embryonic stem cells and mouse embryonic fibroblasts, AgNPs upregulate the expression of the cell cycle checkpoint protein p53 and the DNA damage repair protein Rad51, a protein involved in DNA double-strand breaks. Moreover, they induce the formation of phosphorylated-H2AX, a histone that is rapidly phosphorylated in the chromatin microenvironment surrounding a DNA

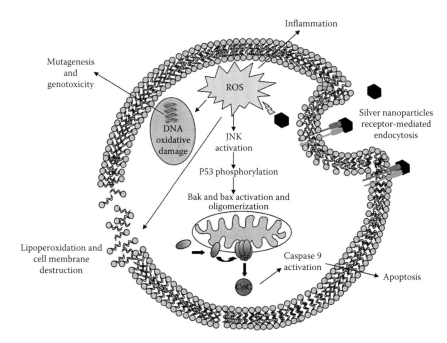

FIGURE 1.16
Cytotoxic effects of silver nanoparticles on eukaryotic cells.

double-strand break [128]. Various chromosomal aberrations such as aneu-ploidy, chromatid lesions, centromere spreading, chromatid exchanges, and micronuclei formation are observed in fish cells [145]. These effects are sum-marized in Figure 1.16.

Most of the biological effects induced by AgNPs can be ascribed to the action of Ag^+ ions, supporting the hypothesis that the toxic potentiality of AgNPs is mainly due to the Ag^+ release from them [133,146]. However, some *in vitro* studies show that toxic effects cannot be solely explained by the release of metal ions in solution, but also that an intrinsic, still unknown, toxic mechanism mediated by nanoparticles should be considered. As an example, Kawata et al. report that in hepatocarcinoma cells, the DNA dam-ages induced by toxic doses of AgNPs are only partially abolished by cyste-ine, a strong ionic Ag^+ ligand, indicating that ionic Ag^+ derived from AgNPs could not fully explain these biological actions [144]. Similarly, Chae et al. investigated the changes in the expression of stress-related genes induced by AgNPs and by soluble silver ions in the Japanese Medaka fish. The results suggest that the two silver forms have distinguishable toxic fingerprints and that ionic silver results in a lower overall stress response when compared with the AgNPs [147]. Although all *in vitro* studies clearly evidence that AgNPs can be highly toxic for different cell types, the real potential harmfulness of these compounds can be fully demonstrated only by *in vivo* studies. The toxicity of any material or molecule in an organism is strictly correlated with the amount absorbed into the body, the route of internalization, its metabolism, accumulation, and degradation in target organs, existence and efficiency of detoxification systems, and vulnerability of the different cell types. All these complex biological interactions in a higher-order organism cannot be reproduced by an *in vitro* system.

Absorption of AgNPs into the human body may occur via inhalatory, oral, and dermal routes of exposure [148]. Using rats and mice as animal models, oral administration and intravenous injection of AgNPs in a size range between 10 and 100 nm have shown a dose-dependent silver accumulation in all organs, with the highest levels in liver, spleen, and kidneys [149–155]. Although the metal is cleared up from most organs after some weeks, it is not remarkably removed by brain and testis [150,154]. Silver accumulation in brain appears correlated to the destruction of blood–brain barrier by the alteration of endothelial cell membrane permeability and neurotoxic effects [150,153,156–159]. Alteration of hematologic parameters and metabolic indicators of liver functionality has been described only in animals treated with doses of intravenous AgNPs higher that 20 mg/kg [160,161]. Acute and subchronic inhalation of AgNPs have shown no significant clinical effects or lung function under a concentration of 100 $\mu g/m^3$ [162,163]. Only higher concentrations of nanoparticles induce persistent lung function changes and inflammation in rats and mice [164–166]. Apart from a small increase in blood calcium, no additional effects of systemic exposure of AgNPs on hematology and blood chemistry parameters have been reported after the inhalation exposure [162]. Oral administration of 60 nm AgNPs induced some changes in the red blood cell compartment (increased red blood cell count, hemoglobin, and hematocrit) and on coagulation parameters (decreased active partial prothrombin time) [167].

Several studies have been performed to test the harmfulness of AgNPs on soil and water organisms due to the rapid and high release and accumulation of these particles in the natural environment. AgNPs induce subchronic toxicity in Medaka fish (*Oryzias latipes*) causing oxidative damage with the inhibition of enzyme in the redox system and histological lesions, following bioaccumulation in the liver [168]. In zebra fish embryos, AgNPs cause lethality and morphological malformations in a size-dependent manner and for some concentrations and time points [169,170]. Similarly, dose- and time-related changes in aerobic metabolism, chorionic disruption oxidative stress, morphological changes in embryos including cardiovascular malformations, ischemia, underdeveloped central nervous system, and eyes are described in embryonic Medaka [171,172]. Overall, in sharp contrast to the emphasis on the application of AgNPs, information on the toxicological implication of the use of AgNPs remains still very limited [173]. However, the general feeling is that the toxicological effects induced by nanoparticles on higher organism are not so much alarming as the toxicity studies on cell systems would lead to think. The physiological responses appear mainly dependent on the size of the particles, on their concentrations, and on their routes of administration. Analyzing the different toxicological studies that are reported so far, it appears clear that *in vivo* studies suffer from a very strong limit, which is the lack of uniformity in the experimental conditions used. The toxicity of nanomaterials are influenced by particle concentration, size and size distribution, agglomeration state, shape, chemical, and physical nature of the bulk, physical status of the composite, and finally site and time of exposure. It is a common opinion that in the near future every new nanomaterial entering in the market will need to be screened for toxicity and biopersistence, using low-cost, fast-throughput but scientifically standardized tests.

1.7 Future Trends

Nanotechnology is continuously offering new exciting solutions for the production of multifunctional materials for numerous fields of research, including the biomedical and food industries.

In light of the growing interest of the scientific community in nanobiotechnology, in the next few years, research and applications of nanocomposite materials based on biologically active components will continue to expand at all stages, from fundamental studies to applied technologies.

The requirement of biological activity in the absence of potential hazards for human health will be a central point of this challenge, since nanosafety is a debated issue by the scientific community; research studies are expected to deepen the knowledge of the complex mechanisms of interaction between nanomaterials and the biological environment, which will lead to define the edges of possible applications of these materials in direct contact with human body and will ultimately the guide regulatory organisms. In this perspective, particular attention should be paid to the synthesis of nanoparticles and the production of nanocomposite systems. In fact, a growing need for sustainability initiatives is expected to boost the development of green procedures for the preparation of bioactive nanoparticles (e.g., AgNPs) in order to replace the traditional chemical syntheses: this will decrease the use of chemicals that are toxic to health and environment and will lead to novel engineered nanocomposites specifically designed for biomedical and food use. To this end, the use of environmentally safe materials from renewable sources like natural polysaccharides (e.g., alginate, chitosan, HA) will offer numerous benefits for the preparation of nanocomposite materials since the widespread occurrence of these polymers is pushing their use toward large-scale industrial production. To this end, the results obtained on AgNPs and lactose-modified chitosan seem particularly promising, since these nanocomposite systems display a very effective bactericidal activity toward both gram-positive and gram-negative bacteria; however, the hydrogel does not show any cytotoxic effect toward three different eukaryotic cell lines [64].

Moreover, the possibility of chemically modifying polysaccharides in order to endow them with specific properties can be exploited to obtain materials with additional functionalities and for designing multifunctional architectures. At the same time, the increasing demand for effective antimicrobial solutions is stressing the attention on AgNPs for the development of antimicrobial nanocomposite surfaces for several purposes. In the field of biomaterials for topical applications, nanocomposite films based on silver and natural polysaccharides are expected to play a crucial role in the wound-healing treatment. In fact, the synergistic use of AgNPs within a cell-friendly polymeric matrix can exploit both the antibacterial activity of the metal and the excellent biocompatibility of polysaccharides, which makes this solution ideal for use in medical applications. Another expanding research line in biomaterial science is the possibility to endow implant materials with specific surface properties like antimicrobial activity; to this end, the use of silver–polysaccharide antimicrobial coatings appears to be a very promising solution since important findings in the recent literature demonstrate the possibility to obtain noncytotoxic polysaccharide-based surfaces that are able to prevent or limit bacterial biofilm formation. These studies represent the basis for the development of novel multifunctional materials for clinical use.

Given the encouraging scientific results obtained in biomedical research, the same rationale could be successfully adapted to the food-related field, where the main goal is to prolong the shelf-life of products; the use of silver–polysaccharide coatings and films represents an appealing opportunity for the future development of membranes, films, and coatings for the packaging of fresh food.

Various techniques are being proposed for the production of silver–polysaccharide surfaces; among them, the layer-by-layer deposition bears a great potential in particular for biomaterial applications. In fact, the deposition of polyelectrolytes onto a solid substrate is becoming a popular technology for producing new types of thin coatings with controlled architecture, which in turn enables to control the release profile of the bioactive agent. Owing to their versatility and ease of deposition on various substrate materials, polysaccharide-based films will offer the opportunity to modify the surface of biomaterials while maintaining their bulk properties unmodified.

Overall, in the wide family of polymer nanocomposites, silver–polysaccharide antimicrobial surfaces possess great potential for several applications, and it can be expected that fundamental understanding of these assemblies will guide toward their use in everyday life.

References

1. Busscher HJ, Doornbusch GI, Van der Mei HC. Adhesion of mutants streptococci to glass with and without a salivary coatings as studied in a parallel plate flow chamber. *J Dent Res* 1992;71:491–500.
2. Lichter JA, Van Vliet KJ, Rubner MF. Design of antibacterial surfaces and interfaces: Polyelectrolyte multilayers as a multifunctional platform. *Macromolecules* 2009 October 12;42(22):8573–8586.
3. Abu-Lail NI, Camesano TA. Specific and nonspecific interaction forces between *Escherichia coli* and silicon nitride, determined by Poisson statistical analysis. *Langmuir* 2006 July 18;22(17):7296–7301.
4. Ki DP, Young SK, Dong KH, Young HK, Eun HBL, Hwal S et al. Bacterial adhesion on PEG modified polyurethane surfaces. *Biomaterials* 1998 April 5;19(7–9): 851–859.
5. Derjaguin B, Landau L. Theory of the stability of strongly charged lyophobic sols and of the adhesion of strongly charged particles in solution of electrolytes. *Acta Physicochim URSS* 1941;14:633–662.
6. Rutter PR, Vincent B. The adhesion of microorganisms to surfaces, physico-chemical aspects. In: Berkeley RCW, Lynch JM, Melling J, Rutter PR, Vincent B (eds.). *Microbial Adhesion to Surfaces*. London, U.K.: Ellis Horwood, 1980, pp. 79–91.
7. Verwey EJW, Overbeek JTG. *Theory of the Stability of Lyophobic Colloids*. Amsterdam, the Netherlands: Elsevier, 1948.
8. Bos R, van der Mei HV, Busscher HJ. Physico-chemistry of initial microbial adhesive interactions—Its mechanisms and methods for study. *FEMS Microbiol Rev* 1999;23:179–230.
9. Triandafillu K, Balazs DJ, Aronsson B-O, Descouts P, Tu Quoc P, van Delden C et al. Adhesion of *Pseudomonas aeruginosa* strains to untreated and oxygen-plasma treated poly(vinyl chloride) (PVC) from endotracheal intubation devices. *Biomaterials* 2003 April;24(8):1507–1518.
10. Mauclaire L, Brombacher E, Bunger JD, Zinn M. Factors controlling bacterial attachment and biofilm formation on medium-chain-length polyhydroxyalkanoates (mcl-PHAs). *Colloids Surf B: Biointerfaces* 2010 March 1;76(1):104–111.
11. Jucker BA, Zehnder AJB, Harms H. Quantification of polymer interactions in bacterial adhesion. *Environ Sci Technol* 1998 August 12;32(19):2909–2915.
12. Vacheethasanee K, Temenoff JS, Higashi JM, Gary A, Anderson JM, Bayston R et al. Bacterial surface properties of clinically isolated *Staphylococcus epidermidis* strains determine adhesion on polyethylene. *J Biomed Mater Res* 1998; 42(425):432.
13. An YH, Friedman RJ. Concise review of mechanism of bacterial adhesion to biomaterial surfaces. *J Biomed Mater Res* 1998;43:338–348.
14. Katsikogianni M, Missirlis YF. Concise review of mechanism of bacterial adhesion to biomaterials and of techniques used in estimating bacteria-material interactions. *Eur Cells Mater* 2004;8:37–57.
15. Earl WM, Larry CC, Patrick GB, Richard H, Stephen GK. The role of polymeric surface smoothness of biliary stents in bacterial adherence, biofilm deposition, and stent occlusion. *Gastrointest Endosc* 1993 January 1;39(3):422–425.

16. Scheuerman TR, Camper AK, Hamilton MA. Effects of substratum topography on bacterial adhesion. *J Col Interf Sci* 1998;208:23–33.
17. Speranza G, Gottardi G, Pederzolli C, Lunelli L, Canteri R, Pasquardini L et al. Role of chemical interactions in bacterial adhesion to polymer surfaces. *Biomaterials* 2004 May;25(11):2029–2037.
18. Baker AS, Greenham LW. Release of gentamicin from acrylic bone cement. Elution and diffusion studies. *J Bone Joint Surg Am* 1988 December 1;70(10):1551–1557.
19. Teixeira P, Trindade AC, Godinho MH, Azeredo J, Oliveira R, Fonseca JG. *Staphylococcus epidermidis* adhesion on modified urea/urethane elastomers. *J Biomater Sci Polym Ed* 2006;17(1–2):239–246.
20. Whitehead KA, Colligon J, Verran J. Retention of microbial cells in substratum surface features of micrometer and sub-micrometer dimensions. *Colloids Surf B: Biointerfaces* 2005 March 25;41:129–138.
21. Glinel K, Thebault P, Humblot V, Pradier CM, Jouenne T. Antibacterial surfaces developed from bio-inspired approaches. *Acta Biomater* 2012 May;8(5):1670–1684.
22. Lichter JA, Thompson MT, Delgadillo M, Nishikawa T, Rubner MF, Van Vliet KJ. Substrata mechanical stiffness can regulate adhesion of viable bacteria. *Biomacromolecules* 2008 June;9(6):1571–1578.
23. Genzer J, Efimenko K. Recent developments in superhydrophobic surfaces and their relevance to marine fouling: A review. *Biofouling* 2006;22(5):339–360.
24. Ho CH, Tobis J, Sprich C, Thomann R, Tiller JC. Nanoseparated polymeric network with multiple antimicrobial properties. *Adv Mater* 2004;16(12):957–961.
25. Kang S, Choi H. Effect of surface hydrophobicity on the adhesion of *S. cerevisiae* onto modified surfaces by poly(styrene-ran-sulfonic acid) random copolymers. *Colloids Surf B: Biointerfaces* 2005 December 10;46(2):70–77.
26. Picart C. Polyelectrolyte multilayer films: From physico-chemical properties to the control of cellular processes. *Curr Med Chem* 2008;15(7):685–697.
27. Thompson MT, Berg MC, Tobias IS, Rubner MF, Van Vliet KJ. Tuning compliance of nanoscale polyelectrolyte multilayers to modulate cell adhesion. *Biomaterials* 2005 December;26(34):6836–6845.
28. Lewis K, Klibanov AM. Surpassing nature: Rational design of sterile-surface materials. *Trends Biotechnol* 2005 July;23(7):343–348.
29. Aviv M, Berdicevsky I, Zilberman M. Gentamicin-loaded bioresorbable films for prevention of bacterial infections associated with orthopedic implants. *J Biomed Mater Res A* 2007 October;83(1):10–19.
30. Darouiche RO. Antimicrobial approaches for preventing infections associated with surgical implants. *Clin Infect Dis* 2003 May 15;36(10):1284–1289.
31. Shukla A, Fleming KE, Chuang HF, Chau TM, Loose CR, Stephanopoulos GN et al. Controlling the release of peptide antimicrobial agents from surfaces. *Biomaterials* 2010 March;31(8):2348–2357.
32. Milovic NM, Wang J, Lewis K, Klibanov AM. Immobilized N-alkylated polyethylenimine avidly kills bacteria by rupturing cell membranes with no resistance developed. *Biotechnol Bioeng* 2005 June 20;90(6):715–722.
33. Price JS, Tencer AF, Arm DM, Bohach GA. Controlled release of antibiotics from coated orthopedic implants. *J Biomed Mater Res* 1996 March;30(3):281–286.
34. Stigter M, Bezemer J, de Groot K, Layrolle P. Incorporation of different antibiotics into carbonated hydroxyapatite coatings on titanium implants, release and antibiotic efficacy. *J Control Release* 2004 September 14;99(1):127–137.

35. Anderson JM. Biological responses to materials. *Annu Rev Mater Res* 2001; 31:81–110.
36. Kamino K. Novel barnacle underwater adhesive protein is a charged amino acid-rich protein constituted by a Cys-rich repetitive sequence. *Biochem J* 2001 June 1;356(Pt 2):503–507.
37. Olsen SM, Pedersen LT, Laursen MH, Kiil S, Dam-Johansen K. Enzyme-based antifouling coatings: A review. *Biofouling* 2007;23(5–6):369–383.
38. Toppazzini M, Coslovi A, Boschelle M, Marsich E, Benincasa M, Gennaro R et al. Can the interaction between LL-37 and alginate be exploited for the formulation of new biomaterials with antimicrobial properties? *Carbohydr Pol* 2010;In press.
39. Bromberg L, Hatton TA. Poly(N-vinylguanidine): Characterization, and catalytic and bactericidal properties. *Polymer* 2007 December 13;48(26):7490–7498.
40. Kenawy ER, Worley SD, Broughton R. The chemistry and applications of antimicrobial polymers: A state-of-the-art review. *Biomacromolecules* 2007 April 11;8(5):1359–1384.
41. Onaizi SA, Leong SS. Tethering antimicrobial peptides: Current status and potential challenges. *Biotechnol Adv* 2011 January;29(1):67–74.
42. Wijnhoven SWP, Peijnenburg WJGM, Herberts CA, Hagens WI, Oomen AG, Heugens EHW et al. Nano-silver—A review of available data and knowledge gaps in human and environmental risk assessment. *Nanotoxicology* 2009 January 1;3(2):109–138.
43. Marambio-Jones C, Hoek E. A review of the antibacterial effects of silver nanomaterials and potential implications for human health and the environment. *J Nanopart Res* 2010;12(5):1531–1551.
44. Hajipour MJ, Fromm KM, Ashkarran A, Jimenez de Aberasturi D, Larramendi IRd, Rojo T et al. Antibacterial properties of nanoparticles. *Trends Biotechnol* 2012 October;30(10):499–511.
45. Malcher M, Volodkin D, Heurtault Bü, André P, Schaaf P, Mohwald H et al. Embedded silver ions-containing liposomes in polyelectrolyte multilayers: Cargos films for antibacterial agents. *Langmuir* 2008 August 13;24(18): 10209–10215.
46. Guzman M, Dille J, Godet S. Synthesis and antibacterial activity of silver nanoparticles against gram-positive and gram-negative bacteria. *Nanomed: Nanotechnol, Biol Med* 2012 January;8(1):37–45.
47. Morones JR, Elechiguerra JL, Camacho A, Holt K, Kouri JB, Ramirez JT et al. The bactericidal effect of silver nanoparticles. *Nanotechnology* 2005;16(10):2346–2353.
48. Rai M, Yadav A, Gade A. Silver nanoparticles as a new generation of antimicrobials. *Biotechnol Adv* 2001 January;27(1):76–83.
49. Su HL, Chou CC, Hung DJ, Lin SH, Pao IC, Lin JH et al. The disruption of bacterial membrane integrity through ROS generation induced by nanohybrids of silver and clay. *Biomaterials* 2009 August 20;30(30):5979–5987.
50. Travan A, Marsich E, Donati I, Paoletti S. Silver nanocomposites and their biomedical applications. In: Kumar C (ed). *Nanocomposites*. New York: Wiley, 2010, pp. 81–137.
51. Bragg PD, Rainnie DJ. The effect of silver ions on the respiratory chain of *Escherichia coli*. *Can J Microbiol* 1974 June;20(6):883–889.
52. Feng QL, Wu J, Chen GQ, Cui FZ, Kim TN, Kim JO. A mechanistic study of the antibacterial effect of silver ions on *Escherichia coli* and *Staphylococcus aureus*. *J Biomed Mater Res* 2000 December 15;52(4):662–668.

53. Furr JR, Russell AD, Turner TD, Andrews A. Antibacterial activity of Actisorb Plus, Actisorb and silver nitrate. *J Hosp Infect* 1994 July;27(3):201–208.
54. Gupta A, Matsui K, Lo JF, Silver S. Molecular basis for resistance to silver cations in Salmonella. *Nat Med* 1999 February;5(2):183–188.
55. Hayashi M, Miyoshi T, Sato M, Unemoto T. Properties of respiratory chain-linked Na(+)-independent NADH-quinone reductase in a marine *Vibrio alginolyticus*. *Biochim Biophys Acta* 1992 February 21;1099(2):145–151.
56. Semeykina AL, Skulachev VP. Submicromolar Ag^+ increases passive Na^+ permeability and inhibits the respiration-supported formation of Na^+ gradient in Bacillus FTU vesicles. *FEBS Lett* 1990 August 20;269(1):69–72.
57. Yamanaka M, Hara K, Kudo J. Bactericidal actions of a silver ion solution on *Escherichia coli*, studied by energy-filtering transmission electron microscopy and proteomic analysis. *Appl Environ Microbiol* 2005 November;71(11):7589–7593.
58. Choi O, Hu Z. Size dependent and reactive oxygen species related nanosilver toxicity to nitrifying bacteria. *Environ Sci Technol* 2008 June 15;42(12):4583–4588.
59. Kim JS, Kuk E, Yu KN, Kim JH, Park SJ, Lee HJ et al. Antimicrobial effects of silver nanoparticles. *Nanomed Nanotechnol Biol Med* 2007 March;3(1):95–101.
60. Le Pape H, Solano-Serena F, Contini P, Devillers C, Maftah A, Leprat P. Involvement of reactive oxygen species in the bactericidal activity of activated carbon fibre supporting silver: Bactericidal activity of ACF(Ag) mediated by ROS. *J Inorg Biochem* 2004 June;98(6):1054–1060.
61. Pal S, Tak YK, Joardar J, Kim W, Lee JE, Han MS et al. Nanocrystalline silver supported on activated carbon matrix from hydrosol: Antibacterial mechanism under prolonged incubation conditions. *J Nanosci Nanotechnol* 2009 March;9(3):2092–2103.
62. Stevens KNJ, Crespo-Biel O, van den Bosch EEM, Dias AA, Knetsch MLW, Aldenhoff YBJ et al. The relationship between the antimicrobial effect of catheter coatings containing silver nanoparticles and the coagulation of contacting blood. *Biomaterials* 2009 August;30(22):3682–3690.
63. Sondi I, Salopek-Sondi B. Silver nanoparticles as antimicrobial agent: A case study on *E. coli* as a model for Gram-negative bacteria. *J Colloid Interf Sci* 2004 July 1;275(1):177–182.
64. Travan A, Pelillo C, Donati I, Marsich E, Benincasa M, Scarpa T et al. Non-cytotoxic silver nanoparticle-polysaccharide nanocomposites with antimicrobial activity. *Biomacromolecules* 2009 June 8;10(6):1429–1435.
65. Monteiro DR, Gorup LF, Takamiya AS, Ruvollo-Filho AC, Camargo ERd, Barbosa DB. The growing importance of materials that prevent microbial adhesion: Antimicrobial effect of medical devices containing silver. *Int J Antimicrob Agents* 2009 August;34(2):103–110.
66. Du H, Lo TM, Sitompul J, Chang MW. Systems-level analysis of *Escherichia coli* response to silver nanoparticles: The roles of anaerobic respiration in microbial resistance. *Biochem Biophys Res Commun* 2012 August 10;424(4):657–662.
67. Lok CN, Ho CM, Chen R, He QY, Yu WY, Sun H et al. Silver nanoparticles: Partial oxidation and antibacterial activities. *J Biol Inorg Chem* 2007 May 1;12(4):527–534.
68. Mallick K, Witcomb MJ, Scurrell MS. Polymer stabilized silver nanoparticles: A photochemical synthesis route. *J Mater Sci* 2004 July 1;39(14):4459–4463.
69. Dickerson MB, Sandhage KH, Naik RR. Protein- and peptide-directed syntheses of inorganic materials. *Chem Rev* 2008 October 30;108(11):4935–4978.

70. Li S, Zhang Y, Xu X, Zhang L. Triple helical polysaccharide-induced good dispersion of silver nanoparticles in water. *Biomacromolecules* 2011 May 4;12(8): 2864–2871.

71. Park Y, Hong YN, Weyers A, Kim YS, Linhardt RJ. Polysaccharides and phytochemicals: A natural reservoir for the green synthesis of gold and silver nanoparticles. *IET Nanobiotechnol* 2011 September;5(3):69–78.

72. Crouzier T, Boudou T, Picart C. Polysaccharide-based polyelectrolyte multilayers. *Curr Opin Colloid Interface Sci* 2010 December;15(6):417–426.

73. Philip D. Biosynthesis of Au, Ag and Au-Ag nanoparticles using edible mushroom extract. *Spectrochim Acta A Mol Biomol Spectrosc* 2009 July 15;73(2):374–381.

74. Kemp MM, Kumar A, Clement D, Ajayan P, Mousa S, Linhardt RJ. Hyaluronan- and heparin-reduced silver nanoparticles with antimicrobial properties. *Nanomedicine* (London, U.K.) 2009 June;4(4):421–429.

75. Kemp MM, Kumar A, Mousa S, Park TJ, Ajayan P, Kubotera N et al. Synthesis of gold and silver nanoparticles stabilized with glycosaminoglycans having distinctive biological activities. *Biomacromolecules* 2009 March 9;10(3):589–595.

76. Muzzarelli RAA, Boudrant J, Meyer D, Manno N, DeMarchis M, Paoletti MG. Current views on fungal chitin/chitosan, human chitinases, food preservation, glucans, pectins and inulin: A tribute to Henri Braconnot, precursor of the carbohydrate polymers science, on the chitin bicentennial. *Carbohyd Polym* 2012 January 15;87(2):995–1012.

77. Fu J, Ji J, Fan D, Shen J. Construction of antibacterial multilayer films containing nanosilver via layer-by-layer assembly of heparin and chitosan-silver ions complex. *J Biomed Mater Res A* 2006 December 1;79(3):665–674.

78. Huang H, Yang X. Synthesis of polysaccharide-stabilized gold and silver nanoparticles: A green method. *Carbohyd Res* 2004 October 20;339(15):2627–2631.

79. Cai J, Kimura S, Wada M, Kuga S. Nanoporous cellulose as metal nanoparticles support. *Biomacromolecules* 2008 November 24;10(1):87–94.

80. Chen J, Wang J, Zhang X, Jin Y. Microwave-assisted green synthesis of silver nanoparticles by carboxymethyl cellulose sodium and silver nitrate. *Mater Chem Phys* 2008 April 15;108(2–3):421–424.

81. Maneerung T, Tokura S, Rujiravanit R. Impregnation of silver nanoparticles into bacterial cellulose for antimicrobial wound dressing. *Carbohyd Polym* 2008 April 3;72(1):43–51.

82. de Santa Maria LC, Santos ALC, Oliveira PC, Barud HS, Messaddeq YS, Ribeiro SJL. Synthesis and characterization of silver nanoparticles impregnated into bacterial cellulose. *Mater Lett* 2009 April 15;63:797–799.

83. Vigneshwaran N, Nachane RP, Balasubramanya RH, Varadarajan PV. A novel one-pot green synthesis of stable silver nanoparticles using soluble starch. *Carbohyd Res* 2006 September 4;341(12):2012–2018.

84. Ma Y, Li N, Yang C, Yang X. One-step synthesis of amino-dextran-protected gold and silver nanoparticles and its application in biosensors. *Anal Bioanal Chem* 2005 June;382(4):1044–1048.

85. Donati I, Paoletti S. Material properties of alginates. *Microbiol Monogr* 2009; 13:1–53.

86. Saha S, Pal A, Kundu S, Basu S, Pal T. Photochemical green synthesis of calcium-alginate-stabilized Ag and Au nanoparticles and their catalytic application to 4-nitrophenol reduction. *Langmuir* 2009 December 3;26(4):2885–2893.

87. Torres E, Mata YN, Blazquez ML, Munoz JA, Gonzalez F, Ballester A. Gold and silver uptake and nanoprecipitation on calcium alginate beads. *Langmuir* 2005 July 15;21(17):7951–7958.

88. Wiegand C, Heinze T, Hipler UC. Comparative in vitro study on cytotoxicity, antimicrobial activity, and binding capacity for pathophysiological factors in chronic wounds of alginate and silver-containing alginate. *Wound Repair Regen* 2009 July;17(4):511–521.

89. Yonezawa Y, Sato T, Miyama T, Takami A, Umemura J, Takenaka T. Photoinduced formation of aggregated silver particles from silver salt of polysaccharide. *Surf Rev Lett* 1996;3(1):1109–1112.

90. Donati I, Travan A, Pelillo C, Scarpa T, Coslovi A, Bonifacio A et al. Polyol synthesis of silver nanoparticles: Mechanism of reduction by alditol bearing polysaccharides. *Biomacromolecules* 2009 February 9;10(2):210–213.

91. Sondi I, Goia DV, Matijevic E. Preparation of highly concentrated stable dispersions of uniform silver nanoparticles. *J Colloid Interf Sci* 2003 April 1;260(1):75–81.

92. Bonifacio A, van der Sneppen L, Gooijer C, van der Zwan G. Citrate-reduced silver hydrosol modified with w-mercaptoalkanoic acids self-assembled monolayers as a substrate for surface-enhanced resonance Raman scattering. A study with cytochrome c. *Langmuir* 2004 July 6;20(14):5858–5864.

93. Pillai ZS, Kamat PV. What factors control the size and shape of silver nanoparticles in the citrate ion reduction method? *J Phys Chem B* 2003 December 19;108(3):945–951.

94. Sun Y, Mayers B, Herricks T, Xia Y. Polyol synthesis of uniform silver nanowires: A plausible growth mechanism and the supporting evidence. *Nano Lett* 2003 July;3(7):955–960.

95. Wiley B, Herricks T, Sun Y, Xia Y. Polyol synthesis of silver nanoparticles: Use of chloride and oxygen to promote the formation of single-crystal, truncated cubes and tetrahedrons. *Nano Lett* 2004 September 8;4(9):1733–1739.

96. Panacek A, Kvitek L, Prucek R, Kolar M, Vecerova R, Pizurova N et al. Silver colloid nanoparticles: Synthesis, characterization, and their antibacterial activity. *J Phys Chem B* 2006 August 24;110(33):16248–16253.

97. Pang X, Zhitomirsky I. Electrodeposition of hydroxyapatite-silver-chitosan nanocomposite coatings. *Surf Coat Technol* 2008 May 15;202(16):3815–3821.

98. Costa C, Conte A, Buonocore GG, Lavorgna M, Del Nobile MA. Calcium-alginate coating loaded with silver-montmorillonite nanoparticles to prolong the shelf-life of fresh-cut carrots. *Food Res Int* 2012 August;48(1):164–169.

99. Gammariello D, Conte A, Buonocore GG, Del Nobile MA. Bio-based nanocomposite coating to preserve quality of Fior di latte cheese. *J Dairy Sci* 2011 November;94(11):5298–5304.

100. Travan A, Donati I, Marsich E, Bellomo F, Achanta S, Toppazzini M et al. Surface modification and polysaccharide deposition on BisGMA/TEGDMA thermoset. *Biomacromolecules* 2010 March 8;11(3):583–592.

101. Travan A, Marsich E, Donati I, Benincasa M, Giazzon M, Felisari L et al. Silver-polysaccharide nanocomposite antimicrobial coatings for methacrylic thermosets. *Acta Biomater* 2011 January;7(1):337–346.

102. Travan A, Marsich E, Donati I, Foulc MP, Moritz N, Aro HT et al. Polysaccharide-coated thermosets for orthopedic applications: From material characterization to in vivo tests. *Biomacromolecules* 2012 May 14;13(5):1564–1572.

103. Marsich E, Travan A, Donati I, Turco G, Kulkova J, Moritz N et al. Biological responses of silver-coated thermosets: An in vitro and in vivo study. *Acta Biomaterialia* 2012; In press.

104. Wang BL, Liu XS, Ji Y, Ren KF, Ji J. Fast and long-acting antibacterial properties of chitosan-Ag/polyvinylpyrrolidone nanocomposite films. *Carbohyd Polym* 2012 September 1;90(1):8–15.

105. Bal KE, Bal Y, Cote G, Chagnes A. Morphology and antimicrobial properties of *Luffa cylindrica fibers*/chitosan biomaterial as micro-reservoirs for silver delivery. *Mater Lett* 2012 July 15;79:238–241.

106. Maity D, Mollick M, Mondal D, Bhowmick B, Bain MK, Bankura K et al. Synthesis of methylcellulose-silver nanocomposite and investigation of mechanical and antimicrobial properties. *Carbohydr Polym* 2012 November 6;90(4):1818–1825.

107. Pinto RJB, Fernandes SCM, Freire CSR, Sadocco P, Causio J, Neto CP et al. Antibacterial activity of optically transparent nanocomposite films based on chitosan or its derivatives and silver nanoparticles. *Carbohydr Res* 2012 February 1;348:77–83.

108. Yoksan R, Chirachanchai S. Silver nanoparticle-loaded chitosan-starch based films: Fabrication and evaluation of tensile, barrier and antimicrobial properties. *Mater Sci Eng C* 2010 July 20;30(6):891–897.

109. Shukla MK, Singh RP, Reddy CRK, Jha B. Synthesis and characterization of agar-based silver nanoparticles and nanocomposite film with antibacterial applications. *Biores Technol* 2012 March;107:295–300.

110. Venkatpurwar V, Pokharkar V. Green synthesis of silver nanoparticles using marine polysaccharide: Study of in-vitro antibacterial activity. *Mater Lett* 2011 March 31;65(6):999–1002.

111. Vermeulen H, van Hattem JM, Storm-Versloot MN, Ubbink DT. Topical silver for treating infected wounds. *Cochrane Database Syst Rev* 2007;(1):CD005486.

112. Neibert K, Gopishetty V, Grigoryev A, Tokarev I, Al-Hajaj N, Vorstenbosch J et al. Wound-healing with mechanically robust and biodegradable hydrogel fibers loaded with silver nanoparticles. *Adv Healthcare Mater* 2012 September; 1(5):621–630.

113. Singh R, Singh D. Chitin membranes containing silver nanoparticles for wound dressing application. *Int Wound J* 2012; In press.

114. Rupp ME, Archer GL. Coagulase-negative staphylococci: Pathogens associated with medical progress. *Clin Infect Dis* 1994 August 1;19(2):231–245.

115. Marsich E, Travan A, Donati I, Turco G, Bellomo F, Paoletti S. *Tissue-Implant Antimicrobial Interfaces. Antimicrobial Polymers*. New York: John Wiley & Sons, Inc., 2011, pp. 379–428.

116. Nganga S, Travan A, Donati I, Crosera M, Paoletti S, Vallittu PK. Degradation of silver-polysaccharide nanocomposite in solution and as coating on fiber-reinforced composites by lysozyme and hydrogen peroxide. *Biomacromolecules* 2012 August 13;13(8):2605–2608.

117. Fernandez A, Soriano E, Loez-Carballo G, Picouet P, Lloret E, Gavara R et al. Preservation of aseptic conditions in absorbent pads by using silver nanotechnology. *Food Res Int* 2009 October;42(8):1105–1112.

118. Mastromatteo M, Conte A, Del Nobile MA. Packaging strategies to prolong the shelf life of fresh carrots (*Daucus carota* L.). *Innovat Food Sci Emerg Techn* 2012 January;13:215–220.

119. Devlieghere F, Vermeulen A, Debevere J. Chitosan: Antimicrobial activity, interactions with food components and applicability as a coating on fruit and vegetables. *Food Microbiol* 2004 December;21(6):703–714.

120. Bernabò M, Pucci A, Galembeck F, Leite CA, Ruggeri G. Thermal- and sun-promoted generation of silver nanoparticles embedded into poly(vinyl alcohol) films. *Macromol Mater Eng* 2009;294(4):256–264.

121. Bozanic DK, Djokovic V, Blanusa J, Nair PS, Georges MK, Radhakrishnan T. Preparation and properties of nano-sized Ag and Ag_2S particles in biopolymer matrix. *Eur Phys J E Soft Matter* 2007;22(1):51–59.

122. Zhitomirsky I, Hashambhoy A. Chitosan-mediated electrosynthesis of organic-inorganic nanocomposites. *J Mater Process Technol* 2007 August 1; 191(1–3):68–72.

123. Luther EM, Schmidt MM, Diendorf J, Epple M, Dringen R. Upregulation of metallothioneins after exposure of cultured primary astrocytes to silver nanoparticles. *Neurochem Res* 2012 August;37(8):1639–1648.

124. Shinogi M, Maeizumi S. Effect of preinduction of metallothionein on tissue distribution of silver and hepatic lipid peroxidation. *Biol Pharm Bull* 1993 April;16(4):372–374.

125. Lansdown AB. A pharmacological and toxicological profile of silver as an antimicrobial agent in medical devices. *Adv Pharmacol Sci* 2010;2010:910686.

126. Coombs CJ, Wan AT, Masterton JP, Conyers RA, Pedersen J, Chia YT. Do burn patients have a silver lining? *Burns* 1992 June;18(3):179–184.

127. Trop M, Novak M, Rodl S, Hellbom B, Kroell W, Goessler W. Silver-coated dressing acticoat caused raised liver enzymes and argyria-like symptoms in burn patient. *J Trauma* 2006 March;60(3):648–652.

128. Ahamed M, Karns M, Goodson M, Rowe J, Hussain SM, Schlager JJ et al. DNA damage response to different surface chemistry of silver nanoparticles in mammalian cells. *Toxicol Appl Pharmacol* 2008 December 15;233(3):404–410.

129. Kim S, Choi IH. Phagocytosis and endocytosis of silver nanoparticles induce interleukin-8 production in human macrophages. *Yonsei Med J* 2012 May;53(3):654–657.

130. Luther EM, Koehler Y, Diendorf J, Epple M, Dringen R. Accumulation of silver nanoparticles by cultured primary brain astrocytes. *Nanotechnology* 2011 September 16;22(37):375101.

131. Carlson C, Hussain SM, Schrand AM, Braydich-Stolle LK, Hess KL, Jones RL et al. Unique cellular interaction of silver nanoparticles: Size-dependent generation of reactive oxygen species. *J Phys Chem B* 2008 October 30;112(43): 13608–13619.

132. Asharani PV, Hande MP, Valiyaveettil S. Anti-proliferative activity of silver nanoparticles. *BMC Cell Biol* 2009;10:65.

133. Singh RP, Ramarao P. Cellular uptake, intracellular trafficking and cytotoxicity of silver nanoparticles. *Toxicol Lett* 2012 September 3;213(2):249–259.

134. Foldbjerg R, Dang DA, Autrup H. Cytotoxicity and genotoxicity of silver nanoparticles in the human lung cancer cell line, A549. *Arch Toxicol* 2011 July;85(7):743–750.

135. Kim HR, Kim MJ, Lee SY, Oh SM, Chung KH. Genotoxic effects of silver nanoparticles stimulated by oxidative stress in human normal bronchial epithelial (BEAS-2B) cells. *Mutat Res* 2011 December 24;726(2):129–135.

136. Mei N, Zhang Y, Chen Y, Guo X, Ding W, Ali SF et al. Silver nanoparticle-induced mutations and oxidative stress in mouse lymphoma cells. *Environ Mol Mutagen* 2012 July;53(6):409–419.

137. Mukherjee SG, O'Claonadh N, Casey A, Chambers G. Comparative in vitro cytotoxicity study of silver nanoparticle on two mammalian cell lines. *Toxicol In Vitro* 2012 March;26(2):238–251.

138. Piao MJ, Kang KA, Lee IK, Kim HS, Kim S, Choi JY et al. Silver nanoparticles induce oxidative cell damage in human liver cells through inhibition of reduced glutathione and induction of mitochondria-involved apoptosis. *Toxicol Lett* 2011 February 25;201(1):92–100.

139. Srivastava M, Singh S, Self WT. Exposure to silver nanoparticles inhibits selenoprotein synthesis and the activity of thioredoxin reductase. *Environ Health Perspect* 2012 January;120(1):56–61.

140. Hussain SM, Hess KL, Gearhart JM, Geiss KT, Schlager JJ. In vitro toxicity of nanoparticles in BRL 3A rat liver cells. *Toxicol In Vitro* 2005 October;19(7):975–983.

141. Foldbjerg R, Olesen P, Hougaard M, Dang DA, Hoffmann HJ, Autrup H. PVP-coated silver nanoparticles and silver ions induce reactive oxygen species, apoptosis and necrosis in THP-1 monocytes. *Toxicol Lett* 2009 July 14;190(2):156–162.

142. Hsin YH, Chen CF, Huang S, Shih TS, Lai PS, Chueh PJ. The apoptotic effect of nanosilver is mediated by a ROS- and JNK-dependent mechanism involving the mitochondrial pathway in NIH3T3 cells. *Toxicol Lett* 2008 July 10;179(3):130–139.

143. Nymark P, Catalan J, Suhonen S, Jarventaus H, Birkedal R, Clausen PA et al. Genotoxicity of polyvinylpyrrolidone-coated silver nanoparticles in BEAS 2B cells. *Toxicology* 2012 November 8.

144. Kawata K, Osawa M, Okabe S. In vitro toxicity of silver nanoparticles at noncytotoxic doses to HepG2 human hepatoma cells. *Environ Sci Technol* 2009 August 1;43(15):6046–6051.

145. Wise JP Sr., Goodale BC, Wise SS, Craig GA, Pongan AF, Walter RB et al. Silver nanospheres are cytotoxic and genotoxic to fish cells. *Aquat Toxicol* 2010 April 1;97(1):34–41.

146. Lubick N. Nanosilver toxicity: Ions, nanoparticles—Or both? *Environ Sci Technol* 2008 December 1;42(23):8617.

147. Chae YJ, Pham CH, Lee J, Bae E, Yi J, Gu MB. Evaluation of the toxic impact of silver nanoparticles on Japanese medaka (*Oryzias latipes*). *Aquat Toxicol* 2009 October 4;94(4):320–327.

148. Johnston HJ, Hutchison G, Christensen FM, Peters S, Hankin S, Stone V. A review of the in vivo and in vitro toxicity of silver and gold particulates: Particle attributes and biological mechanisms responsible for the observed toxicity. *Crit Rev Toxicol* 2010 April;40(4):328–346.

149. Dziendzikowska K, Gromadzka-Ostrowska J, Lankoff A, Oczkowski M, Krawczynska A, Chwastowska J et al. Time-dependent biodistribution and excretion of silver nanoparticles in male Wistar rats. *J Appl Toxicol* 2012 November;32(11):920–928.

150. Hadrup N, Loeschner K, Mortensen A, Sharma AK, Qvortrup K, Larsen EH et al. The similar neurotoxic effects of nanoparticulate and ionic silver in vivo and in vitro. *Neurotoxicology* 2012 June;33(3):416–423.

151. Kim WY, Kim J, Park JD, Ryu HY, Yu IJ. Histological study of gender differences in accumulation of silver nanoparticles in kidneys of Fischer 344 rats. *J Toxicol Environ Health A* 2009;72(21–22):1279–1284.

152. Lankveld DP, Oomen AG, Krystek P, Neigh A, Troost-de JA, Noorlander CW et al. The kinetics of the tissue distribution of silver nanoparticles of different sizes. *Biomaterials* 2010 November;31(32):8350–8361.

153. Tang J, Xiong L, Wang S, Wang J, Liu L, Li J et al. Distribution, translocation and accumulation of silver nanoparticles in rats. *J Nanosci Nanotechnol* 2009 August;9(8):4924–4932.

154. van der ZM, Vandebriel RJ, Van DE, Kramer E, Herrera RZ, Serrano-Rojero CS et al. Distribution, elimination, and toxicity of silver nanoparticles and silver ions in rats after 28-day oral exposure. *ACS Nano* 2012 August 28;6(8):7427–7442.

155. Loeschner K, Hadrup N, Qvortrup K, Larsen A, Gao X, Vogel U et al. Distribution of silver in rats following 28 days of repeated oral exposure to silver nanoparticles or silver acetate. *Part Fibre Toxicol* 2011;8:18.

156. Liu Y, Guan W, Ren G, Yang Z. The possible mechanism of silver nanoparticle impact on hippocampal synaptic plasticity and spatial cognition in rats. *Toxicol Lett* 2012 March 25;209(3):227–231.

157. Sharma HS, Ali SF, Hussain SM, Schlager JJ, Sharma A. Influence of engineered nanoparticles from metals on the blood-brain barrier permeability, cerebral blood flow, brain edema and neurotoxicity. An experimental study in the rat and mice using biochemical and morphological approaches. *J Nanosci Nanotechnol* 2009 August;9(8):5055–5072.

158. Cha K, Hong HW, Choi YG, Lee MJ, Park JH, Chae HK et al. Comparison of acute responses of mice livers to short-term exposure to nano-sized or micro-sized silver particles. *Biotechnol Lett* 2008 November;30(11):1893–1899.

159. Rahman MF, Wang J, Patterson TA, Saini UT, Robinson BL, Newport GD et al. Expression of genes related to oxidative stress in the mouse brain after exposure to silver-25 nanoparticles. *Toxicol Lett* 2009 May 22;187(1):15–21.

160. Kim YS, Song MY, Park JD, Song KS, Ryu HR, Chung YH et al. Subchronic oral toxicity of silver nanoparticles. *Part Fibre Toxicol* 2010;7:20.

161. Tiwari DK, Jin T, Behari J. Dose-dependent in-vivo toxicity assessment of silver nanoparticle in Wistar rats. *Toxicol Mech Methods* 2011 January;21(1):13–24.

162. Ji JH, Jung JH, Kim SS, Yoon JU, Park JD, Choi BS et al. Twenty-eight-day inhalation toxicity study of silver nanoparticles in Sprague-Dawley rats. *Inhal Toxicol* 2007 August;19(10):857–871.

163. Sung JH, Ji JH, Song KS, Lee JH, Choi KH, Lee SH et al. Acute inhalation toxicity of silver nanoparticles. *Toxicol Ind Health* 2011 March;27(2):149–154.

164. Song KS, Sung JH, Ji JH, Lee JH, Lee JS, Ryu HR et al. Recovery from silver-nanoparticle-exposure-induced lung inflammation and lung function changes in Sprague Dawley rats. *Nanotoxicology* 2012 January 20.

165. Stebounova LV, Adamcakova-Dodd A, Kim JS, Park H, O'Shaughnessy PT, Grassian VH et al. Nanosilver induces minimal lung toxicity or inflammation in a subacute murine inhalation model. *Part Fibre Toxicol* 2011;8(1):5.

166. Sung JH, Ji JH, Park JD, Yoon JU, Kim DS, Jeon KS et al. Subchronic inhalation toxicity of silver nanoparticles. *Toxicol Sci* 2009 April;108(2):452–461.

167. Kim YS, Kim JS, Cho HS, Rha DS, Kim JM, Park JD et al. Twenty-eight-day oral toxicity, genotoxicity, and gender-related tissue distribution of silver nanoparticles in Sprague-Dawley rats. *Inhal Toxicol* 2008 April;20(6):575–583.

168. Wu Y, Zhou Q. Subchronic toxicity of Ag nanoparticles to medaka silver nanoparticles cause oxidative damage and histological changes in medaka (*Oryzias latipes*) after 14 days of exposure. *Environ Toxicol Chem* 2012 October 24.

169. Bar-Ilan O, Albrecht RM, Fako VE, Furgeson DY. Toxicity assessments of multisized gold and silver nanoparticles in zebrafish embryos. *Small* 2009 August 17;5(16):1897–1910.
170. Lee KJ, Browning LM, Nallathamby PD, Desai T, Cherukuri PK, Xu XH. In vivo quantitative study of sized-dependent transport and toxicity of single silver nanoparticles using zebrafish embryos. *Chem Res Toxicol* 2012 May 21;25(5):1029–1046.
171. Wu Y, Zhou Q. Dose- and time-related changes in aerobic metabolism, chorionic disruption, and oxidative stress in embryonic medaka (*Oryzias latipes*): Underlying mechanisms for silver nanoparticle developmental toxicity. *Aquat Toxicol* 2012 November 15;124–125:238–246.
172. Kashiwada S, Ariza ME, Kawaguchi T, Nakagame Y, Jayasinghe BS, Gartner K et al. Silver nanocolloids disrupt medaka embryogenesis through vital gene expressions. *Environ Sci Technol* 2012 June 5;46(11):6278–6287.
173. Chen X, Schluesener HJ. Nanosilver: A nanoproduct in medical application. *Toxicol Lett* 2008 January 4;176(1):1–12.

2

Polymer Nanocomposite Coatings: Effect of Crosslinkers and Fillers on the Microstructure and Gas Permeation

Vikas Mittal

CONTENTS

2.1 Introduction

Epoxide resins are used in a large number of fields, including in surface coatings, in adhesives, in potting and encapsulation of electronic components, in tooling, for laminates in flooring, and to a small extent in molding powders and in road surfacing [1–4]. The chemical resistance as well as other properties is as much dependent on the hardener as on the resin since these two determine the nature of linkages formed. A special mention of epoxy applications is their use in packaging laminates as adhesives. The crosslinked epoxy structure is transparent and also contributes to the gas barrier performance of the laminates (depending on the hardener and extent of crosslinking), which are important criteria for the packaging materials. Some approaches have also been reported to further enhance the permeation behavior of the epoxy polymers. One such approach relating to the molecular design is the insertion of rigid groups in the polymer backbone, which leads to enhanced resistance to permeant molecules diffusing through the polymer matrix along with improved thermal stability [5,6]. Another more simplified approach is the incorporation of inorganic fillers with high-aspect ratio, which leads to an increase in the flow path of the permeant molecules in the polymer matrix [7–9]. Layered silicate montmorillonite platelets with 1 nm thickness are one such example of high-aspect ratio fillers, which when dispersed in nanoscale have been reported to enhance the mechanical, thermal, and barrier properties of various polymer matrices [9–13]. To enhance the compatibility of such inorganic fillers with the low-surface energy polymer matrices, their high-energy hydrophilic surface is rendered organophilic by the exchange of their surface cations (like sodium) with long-chain alkyl ammonium ions, which also increases the interlayer spacing between the platelets [14,15].

A number of studies have reported the synthesis of epoxy nanocomposites after the incorporation of layered silicates, and enhanced properties have been ascribed to higher-aspect ratio filler platelets leading to much higher polymer filler interfacial contacts [16–20]. However, only mechanical properties have been given predominance, and less attention has been paid to barrier properties that do not automatically improve along with the improvement in mechanical properties. The goal of the current study was to thus study the synthesis and oxygen permeation of epoxy nanocomposite films (drawn on substrate foils) using different hardener amines as well as layered silicates. The fillers were modified with different surface modifications to study the effect of resulting interactions of the polymer and filler on the microstructure development and consequently the oxygen permeation performance of nanocomposites. Minerals with two different cation exchange capacities (CECs) were also used in order to analyze the effect of CEC of the filler on the morphology and properties of composites. The curing parameters such as time, temperature, and mole ratio of the curing agent to epoxy were also optimized using differential scanning calorimetry (DSC) in order to have the curing process that provides time for the intercalation of epoxy and amine in the clay interlayers, but is also quick enough to avoid excessive extra-gallery polymerization. Care was also taken not to hinder the transparency or to increase the bulkiness of the nanocomposite films owing to their potential application in packaging laminates.

2.2 Experimental

2.2.1 Materials

Organically modified montmorillonites with trade names Nanofil 804, Nanofil 32, Nanofil 15, Cloisite 30B, Cloisite 20A, and Cloisite 93A were supplied by Southern Clay Products (Gonzales, TX). The CEC of

Nanofil and Cloisite substrates was earlier measured to be 680 and 880 ueq/g respectively [21]. The epoxy resin, bisphenol A diglycidyl ether (4,4'-isopropylidenediphenol diglycidyl ether) with an epoxide equivalent weight 172–176, was supplied by Sigma (Buchs, Switzerland). Tetraethylenepentamine (TEPA), isophorone diamine (ID), ethylene dioxydiethyl amine (EDDA), tetrahydrofuran (THF), and dimethylformamide (DMF) were procured from Fluka (Buchs, Switzerland). Polypropylene (PP; 100 µm thick) foils to support the nanocomposite films were supplied by Alcan Packaging (Neuhausen, Switzerland). A surfactant (trade name BYK-307) was used to achieve better wetting and adherence of the neat epoxy coating to the substrate foils and was obtained from Christ Chemie (Reinach, Switzerland).

2.2.2 Nanocomposite Synthesis and Substrate Coating

The details on the nanocomposite synthesis can be found elsewhere [22]. The amount of the organically modified filler corresponding to 6 wt% was swollen with 25 g of solvent (THF or DMF) for 2 h. Subsequently, the filler dispersion was sonicated (ultrasound horn) twice at 70% amplitude for 5 min each time with 5 min pause in between. The epoxy resin solution (5.3 g in 5 g of solvent) was then added to the filler dispersion followed by similar swelling and sonication cycles. The curing agent (TEPA, ID, or EDDA) in an amount corresponding to the amine to epoxy mole ratio of 0.3:1 for TEPA and 0.75:1 for IsoDia and EDDA was then added to the prepolymer filler suspension. It was also required to degas the suspension by sonication for 10 s. The pure epoxy as well as epoxy nanocomposite films were drawn on the PP foils with the help of a bar coater (90 µm gap). The films were dried first at room temperature and atmospheric pressure for 15 min followed by drying under reduced pressure for another 15 min. The films were then transferred in a vacuum oven where they were cured at 70°C overnight, and postcured at 90°C for 4 h. The thickness of the dry films was approximately 10 µm; the method to measure is as reported earlier [22].

2.2.3 Nanocomposite Characterization

DSC runs for the epoxy crosslinker samples were obtained on DSC Q 1000 TA instrument. The cells were equilibrated at 30°C followed by heating to the chosen curing temperature. The cells were kept under isothermal conditions for 120 min at the curing temperature. Universal analysis program V3.8B TA was employed for the analysis.

The oxygen transmission rate through pure epoxy and nanocomposite coatings was measured using an OX-TRAN 2/20 ML (Mocon, Minneapolis, MN) at 23°C and 0% RH. The transmission rate of oxygen was normalized with respect to the total thickness of the epoxy or nanocomposite films and the PP substrate foil, and the transmission rate through the epoxy or nano-composite films was subsequently calculated as reported earlier [22]. An average of four measurements for each coating was reported.

Wide-angle x-ray diffraction was performed on a Scintag XDS 2000 diffrac-tometer (Scintag Inc., Cupertino, CA) using Cu Kα radiation ($\lambda = 0.15406$ nm) in reflection mode. The nanocomposite films were step-scanned (step width $0.02°\,2\theta$, scanning rate $0.06°$/min) at room temperature from $1.5°$ to $10°\,2\theta$. To determine the peak positions, the diffractograms were fitted with a split Pearson VII function (diffraction management system software 1.36b).

Bright-field TEM Zeiss EM 912 Omega microscope was used to study the microstructure of nanocomposites. Nanocomposite films were embedded in an epoxy matrix after etching with oxygen plasma for 3 min. Sections of 70 nm thickness were microtomed with a diamond knife (Reichert Jung Ultracut E) and were supported on 100 mesh grids sputter-coated with a 3 nm thick carbon layer.

2.3 Results and Discussion

As nanocomposites were developed with epoxy-based adhesives, therefore, no free standing films were generated, and the nanocomposite formulations were coated on PP substrates. To enhance the wetting and adhesion between the PP foils and epoxy formulations, apart from adding the surfactant in the formulation, the PP surface was corona-treated. No peeling of the coatings from the substrate was observed, which indicated that the epoxy adhesive did not lose its characteristic after the addition of inorganic fillers. Apart from that, no adhesion problem of films with filler on the substrates was observed, further confirming that the epoxy adhesive formulation was still suitable for the laminate production. The PP substrate was selected on the basis of its consumption in the packing laminates as well as owing to its high oxygen transmission rate so that it does not hinder the measurement of the perme-ation through the nanocomposite film. Figure 2.1 demonstrates the modi-fied fillers (with two different CECs) along with their surface modifications. Nanofil 804 and Cloisite 30B were modified with bis(2-hydroxyethyl)methyl hydrogenated tallow ammonium, whereas Nanofil 15 and Cloisite 20A had surface modification of dihydrogenated tallow dimethyl ammonium. Nanofil 32 was surface modified with benzyl dodecyl dimethyl ammonium; on the other hand, dihydrogenated tallow methyl ammonium was the modi-fication of Cloisite 93A. The modifications were differed in order to study the effect of different functional groups in the modifications on the morphology and properties of epoxy nanocomposites. For example, benzyl group was expected to generate stronger van der Waals attraction forces between the filler and the epoxy polymer, and the hydroxyethyl groups might react with

Nanofil 804/Cloisite 30B: $(C_{18}H_{37})CH_3\overset{+}{N}(CH_2CH_2OH)_2$

Nanofil 32: $(CH_3)_2\overset{+}{N}(CH_2Ph)C_{12}H_{25}$

Nanofil 15/Cloisite 20A: $(CH_3)_2\overset{+}{N}(C_{18}H_{37})_2$

Cloisite 93A: $(C_{18}H_{37})_2\,CH_3\overset{+}{N}H$

(a)

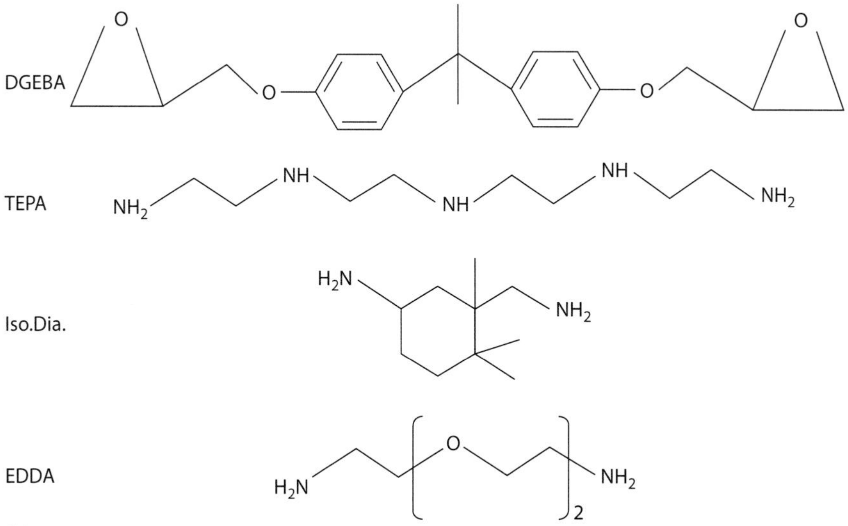

DGEBA

TEPA

Iso.Dia.

EDDA

(b)

FIGURE 2.1
(a) Chemical architectures of the surface modifications present on the filler surface (b) Chemical structures of the epoxy resin and amines used.

the epoxy groups thus tethering polymer chains to the silicate layers. The long alkyl chains, on the other hand, allow to generate higher basal plane spacing between the individual filler platelets increasing the chances of polymer intercalation. Different amines (as shown in Figure 2.1) to crosslink the epoxy structure to study the effect of the resulting network and interactions with the surface modifications were also used.

Optimization of the curing parameters such as curing time, temperature, and mole ratio of the ingredients is important owing to the fact that curing of the epoxy takes place inside as well as outside the filler interlayers, and the curing process should be designed to provide time for the intercalation of epoxy and amine in the filler interlayers, but it should also be fast so as to avoid excessive extra-gallery polymerization. Too high temperature should also be avoided as it may degrade the PP substrate foil. Figure 2.2 shows the DSC plots of epoxy curing runs at different temperatures of 40°C, 55°C, 60°C, 70°C, and 100°C, while keeping the TEPA amine to epoxy mole ratio constant at 1:1. Table 2.1 also shows the time to cure at each temperature, apart from the other details

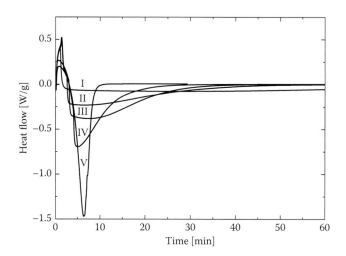

FIGURE 2.2

DSC thermograms showing the curing of epoxy at different temperatures at a fixed amine (TEPA) to epoxy mole ratio of 1:1; I: 40°C, II: 55°C, III: 60°C, IV: 70°C, and V: 100°C.

TABLE 2.1

DSC Analysis for the Optimization of Epoxy Curing Conditions

Amine	Amine:Epoxy Mole Ratio	Temperature (°C)	Filler	Time to Cure (min)
TEPA	1:1	40	No	>60
TEPA	1:1	55	No	55
TEPA	1:1	60	No	55
TEPA	1:1	70	No	35
TEPA	1:1	100	No	11
TEPA	0.6:1	70	No	41
TEPA	0.4:1	70	No	40
TEPA	0.3:1	70	No	45
TEPA	0.3:1	70	4 wt% Nanofil 15	30
TEPA	0.2:1	70	4 wt% Nanofil 15	80
ID	0.75:1	70	No	60
ID	0.75:1	70	4 wt% Nanofil 804	40
ID	0.5:1	70	4 wt% Nanofil 804	80

of DSC analysis. At 40°C, the time to cure was more than 60 min, whereas at 100°C, only 11 min were required to cure the epoxy resin. Apart from that, 100°C curing also led to wrinkles in the substrate foils. The trial runs at different amines (TEPA) to epoxy mole ratios, while keeping the temperature constant at 70°C, were also performed and shown in Figure 2.3. The curing time increased on decreasing the amine to epoxy mole ratio. Forty-five minutes were required for the complete curing in the case of amine to epoxy mole ratio

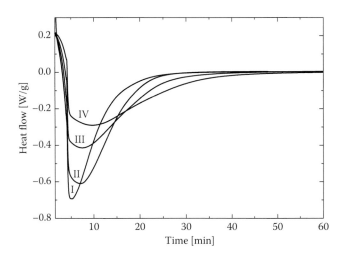

FIGURE 2.3
Curing of epoxy at 70°C using different amines (TEPA) to epoxy mole ratios; I: 1:1, II: 0.6:1, III: 0.4:1, and IV: 0.3:1.

of 0.3:1 at 70°C. From these results, curing temperature of 70°C and epoxy to amine ratio of 0.3:1 were chosen as optimal values as it provided enough curing time necessary for exfoliation (keeping in view the catalytic effect of filler on the curing behavior and thus further advancing the curing rate [2,23–25]), along with maintaining the thermal stability of substrate foils at the curing temperature. Trials were also made to cure epoxy with and without surfactant added to improve the adhesion of the epoxy and epoxy filler formulation with the substrate foils. As shown in Figure 2.4, the DSC thermograms of the epoxy curing in the presence or absence of surfactant were similar, and the time to cure was also near to 45 min in both the cases (TEPA to epoxy mole ratio of 0.3:1 at 70°C), indicating that the surfactant did not affect the curing behavior of epoxy. Similar to TEPA amine hardener, ID was also used to cure the epoxy resin. Owing to the presence of two crosslinking moieties in its structure as compared to the five sites in the case of TEPA, the amine to epoxy mole ratio of 0.75:1 was used in this case. Time to cure of 60 min was observed in this case.

As mentioned earlier, the presence of filler could also have catalytic effect on the curing rate [2,23–25]. The effect of the presence of filler on the overall curing process was also studied and demonstrated in Figures 2.5 and 2.6. The presence of dihydrogenated tallow dimethyl ammonium-modified clay led to the advancement of the curing significantly as shown by curve II as compared to curve I without the presence of filler in Figure 2.5 for the TEPA amine system (amine to epoxy mole ratio of 0.3:1 at 70°C). It was because of the acidic nature of the ammonium ions that have been reported to catalyze the epoxy amine polymerization reaction [2,23–25]. Although the extent of inter-gallery catalysis with quaternary ammonium ions was reported to be lower as compared to primary, secondary, and tertiary counterparts, it

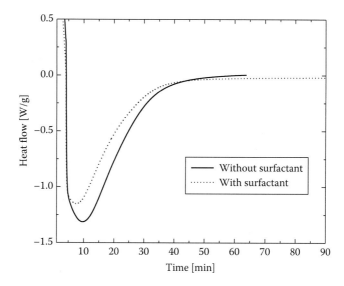

FIGURE 2.4
DSC thermograms showing the effect of surfactant on the curing behavior of epoxy; amine (TEPA) to epoxy mole ratio of 0.3:1, 70°C.

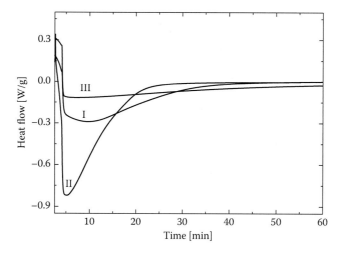

FIGURE 2.5
DSC thermograms indicating the effect of filler on curing at 70°C using TEPA amine; I: no filler, amine to epoxy mole ratio 0.3:1; II: 4 wt% Nanofil 15, amine to epoxy mole ratio 0.3:1; and III: 4 wt% Nanofil 15, amine to epoxy mole ratio 0.2:1.

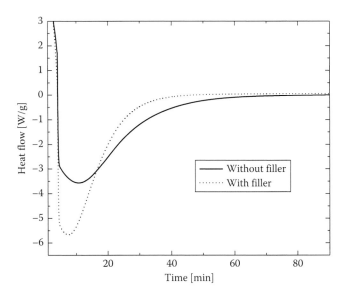

FIGURE 2.6
DSC thermograms indicating the effect of filler on curing at 70°C using isophorone diamine as crosslinker; I: no filler, amine to epoxy mole ratio 0.75:1; and II: 4 wt% Nanofil 804, amine to epoxy mole ratio 0.75:1.

could still be significant. Curve III indicated the curing process carried out at 70°C at a reduced amine to epoxy mole ratio of 0.2:1. The slowing down of the curing process was clearly visible, and it took more than 80 min to cure the resin. Similarly, as shown in Figure 2.6, bis(2-hydroxyethyl) methyl hydrogenated tallow ammonium-modified filler also enhanced the curing of epoxy with ID amine using an amine to epoxy mole ratio of 0.75:1 at 70°C. Complete curing could be achieved in 40 min. On the other hand, using the same amount of filler at lower amine to epoxy mole ratio of 0.5:1 led to sluggish curing and much longer curing time of 80 min. Thus, it was optimum to use the TEPA to epoxy mole ratio of 0.75:1 and ID to epoxy mole ratio of 0.75:1 as reasonable curing times of 30–40 min were achieved at a temperature of 70°C. The influence of curing temperature on the filler exfoliation was also analyzed through x-ray as shown in Figure 2.7. Using an amine (TEPA) to epoxy mole ratio of 0.3:1, the curing at 70°C led to a decrease in the peak intensity (related to the ordering in the filler platelets) in the 6 wt% Nanofil 15 nanocomposite as compared to the same material cured at 60°C, indicating that optimal curing temperature and time were the most important criteria for the microstructure evolution.

Table 2.2 shows the details of the formulations used to generate epoxy nanocomposites. The weight fraction of various fillers was fixed at 6 wt% in order to compare the microstructure as well as oxygen permeation behavior with each other.

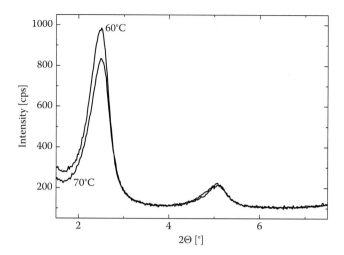

FIGURE 2.7
X-ray diffractograms of epoxy composites (6 wt% Nanofil 15, amine to epoxy mole ratio of 0.3:1) cured at 60°C and 70°C in order to study the effect of temperature on intercalation.

TABLE 2.2

Nanocomposite Formulations Containing 6 wt% of the Organically Modified Filler

OMMT	Epoxy	Amine	Amine:Epoxy Mole Ratio
Nanofil 15	DGEBA	TEPA	0.3:1
Nanofil 32	DGEBA	TEPA	0.3:1
Cloisite 20A	DGEBA	TEPA	0.3:1
Nanofil 804	DGEBA	ID	0.75:1
Cloisite 30B	DGEBA	ID	0.75:1
Cloisite 93A	DGEBA	ID	0.75:1
Nanofil 32	DGEBA	EDDA	0.75:1

It was also observed that TEPA amine agglomerated the Nanofil 804 and Cloisite 30B filler suspensions in solvent (with or without epoxy prepolymer). Other fillers with less polar surface modifications remained stable on the addition of TEPA amine to the solvent suspension. Such a phenomenon of "degellation" or "deexfoliation" of the organo-montmorillonite, containing bis(2-hydroxyethyl)methyl tallow ammonium as surface modification on the addition of polar additives such as amines as crosslinking agents, was also reported earlier [26]. It was suggested that bi- or multifunctional amine molecules may be able to bridge the silicate layers or the N–H groups in the primary and secondary amines may be polar enough to cause reaggregation of silicate layers. However, the use of isophorone diamine as crosslinker did not destabilize the Nanofil 804 and Cloisite 30B

TABLE 2.3

Filler Basal Plane Spacing and Oxygen Permeation through the
Nanocomposites Containing 6 wt% Organically Modified Filler

Composite	Amine	*d*-Spacing of Filler Powder (nm)	*d*-Spacing of Filler in Composite (nm)	Oxygen Transmission Rate (cm³ μm/m² d mmHg)
Epoxy	ID	—	—	4.81
Epoxy	TEPA	—	—	2.03
Epoxy	EDDA	—	—	3.02
Nanofil 804	ID	1.85	4.60 (light), 2.30 (d_{002})	3.82
Cloisite 30B	ID	1.82	3.54	3.55
Cloisite 93A	ID	2.47	4.44 (light), 2.22 (d_{002})	2.77
Cloisite 20A	TEPA	2.74	3.54	2.97
Nanofil 15	TEPA	2.74	3.53	3.12
Nanofil 32	TEPA	1.95	3.08	1.87
Nanofil 32	EDDA	1.95	3.04	2.67

filler suspensions indicating that the amine chemical structure as well as the number of functional groups may affect the formulation stability. Even bifunctional *m*-phenylene diamine did not destabilize the Nanofil 804 and Cloisite 30B filler suspensions, confirming that the amount of functional groups in the amine structure also need to be optimized when using polar surface modifications.

Table 2.3 shows the details of the basal plane spacing of filler powders and fillers in nanocomposites along with the oxygen transmission rates through the pure epoxy and epoxy nanocomposite films. Figures 2.8 and 2.9 also show the x-ray diffractograms of the pure fillers compared with nanocomposites. Nanofil 804 filler was observed to have a basal plane spacing of 1.85 nm, whereas the high CEC filler with same modification as Nanofil 804 was observed to have a basal plane spacing of 1.82 nm indicating no impact on the difference in the CEC. Similarly, Nanofil 15 and Cloisite 20A were also observed to have the same basal plane spacing values of 2.74 nm. Cloisite 93A with a surface modification of dihydrogenated tallow methyl ammonium demonstrated a basal plane spacing of 2.47 nm, which was significantly lower than Cloisite 20A with modification dihydrogenated tallow dimethyl ammonium, indicating that a slight change in the chemical architecture of the surface modification impacted the immobilization of the modification molecules on the filler surface. The basal plane spacing values increased in the nanocomposites indicating the polymer intercalation in the filler interlayers; however, the presence of basal plane spacing also signified that the fillers could not be completely exfoliated. The absence of signal corresponding to the filler powder in x-ray diffractograms also confirmed the absence of un-intercalated filler platelets. The basal plane spacing increased to maximum extent in the case of

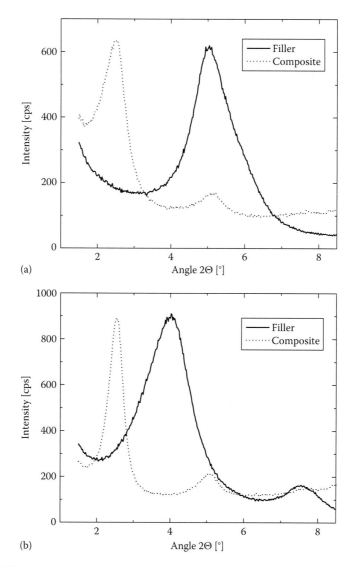

(a)

(b)

FIGURE 2.8
X-ray diffractograms of (a) Cloisite 30B and (b) Cloisite 20A fillers along with their 6 wt% epoxy composites.

Nanofil 804 filler, which also resulted from the better compatibility of the filler modification with the polar epoxy chains. Cloisite 30B composite, on the other hand, had a smaller enhancement in the basal plane spacing. The spacing in the case of Cloisite 93A was also observed to be 4.44 nm even though the surface modification was nonpolar and was not expected to have positive interactions with the polymer. However, owing to the higher acidic nature of the ammonium ion exchanged on the surface in this case,

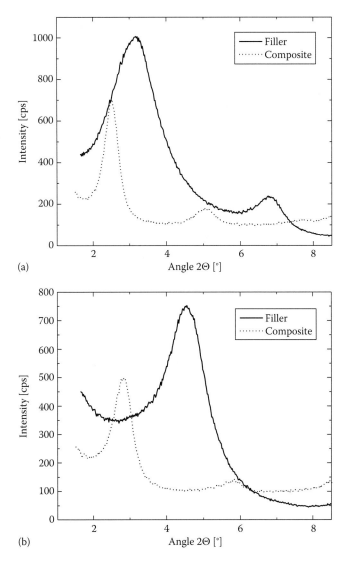

FIGURE 2.9
X-ray diffractograms of (a) Nanofil 15 and (b) Nanofil 32 fillers along with their 6 wt% epoxy composites. In the case of Nanofil 32 filler, composite made using TEPA as amine has been demonstrated.

its interaction with the epoxy chains can be expected to be much better. A basal plane spacing of only 3.54 and 3.53 was observed in the case of Cloisite 20A and Nanofil 15 composites even though the initial basal plane spacing of the filler was the highest, which has resulted from the lower extent of polymer intercalation in the filler interlayers owing to the polarity mismatch between the modification and polymer chains. Nanofil 32

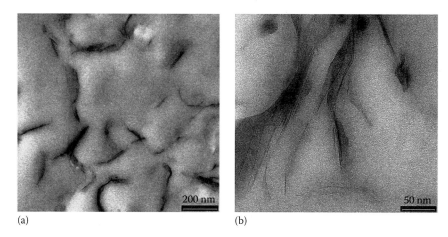

(a)   (b)

FIGURE 2.10
TEM micrographs of 6 wt% Nanofil 804 epoxy nanocomposite at different magnifications. The black lines represent the cross section of clay layers.

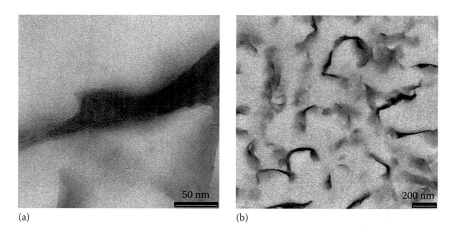

(a)   (b)

FIGURE 2.11
TEM micrographs of (a) 6 wt% Cloisite 30B and (b) Nanofil 32 (composite with TEPA as amine) epoxy nanocomposite at different magnifications. The black lines represent the cross section of clay layers.

composites were observed to have basal plane spacing of 3 nm irrespective of the amine used to cure the system.

To further supplement the findings of x-ray diffraction, Figures 2.10 and 2.11 also show the TEM micrographs of the nanocomposites with various fillers. Nanofil 804 composites were observed to contain large numbers of single exfoliated filler layers. Apart from that, filler tactoids of varying thicknesses were also present. The platelets were observed to be bent, folded, and misaligned. Similar to the case of Nanofil 804, Cloisite 93A, and Cloisite 30B nanocomposites also had extensive filler exfoliation. Thus, it indicated that even though x-ray diffraction indicated only filler intercalation, the microstructure contained also significantly exfoliated filler platelets. Nanofil 32 composites on the other hand were more intercalated than exfoliated, whereas Nanofil 15 and Cloisite 20A nanocomposites were extensively intercalated. The d-spacing of the intercalated tactoids coincided with the spacing values observed in the x-ray diffraction. These results thus coincided with the notion that more polar surface modifications led to higher extent of filler exfoliation, whereas decreased modification polarity led the microstructure to be more intercalated. Such findings were also reported earlier for epoxy and polyurethane nanocomposites [22,27].

The oxygen permeation values through the pure epoxy matrices cured by TEPA, ID, and EDDA amines were observed to be 4.81, 2.03, and 3.32 $cm^3 \mu m/m^2$ d mmHg respectively. It showed that the number of functional sites in the amine molecules affected the epoxy network structure accordingly. The use of TEPA with the highest number of crosslinking sites led to higher crosslinking density, which also resulted in the least oxygen permeation through the polymer films. On the other hand, the use of bifunctional EDDA and ID amines led to lower extent of crosslink density, and the oxygen permeation was higher than that of TEPA system. Even EDDA and ID amine systems led to different oxygen transmissions through the polymer films owing to their chemical architecture. A decrease of 21% in the oxygen permeation as compared to the pure polymer was observed in Nanofil 804, whereas the higher CEC Cloisite 30B filler with same modification as Nanofil 804 resulted in a decrease of 26% in the oxygen permeation. A much higher decrease of 43% in the oxygen permeation was observed in Cloisite 93A nanocomposite. Thus, the catalytic nature of the surface modification and the polar surface modification were observed to enhance the filler exfoliation, which subsequently resulted in significant oxygen permeation reduction. Nanofil 32 also led to a reduction of 8% and 12% in the TEPA and EDDA cured nanocomposite systems. These findings also coincided with the TEM findings, where Nanofil 32 filler was observed to be less intercalated than the Cloisite 93A, Nanofil 804, and Cloisite 30B nanocomposites. Cloisite 20A and Nanofil 15 led to an increase in the oxygen permeation as compared to the pure epoxy. Such a behavior was expected owing to the mismatch between the polarities of the surface modification and the filler [22,27], and it also confirmed the TEM findings. It was also worth noting that the oxygen permeation through the pure epoxy matrices generated by curing with different amines was already very low as compared to other conventional materials used in packaging laminates, for example, PP. Therefore, further reduction in oxygen permeation by the incorporation of layered silicate fillers was challenging. By designing the interface between the polymer and the filler, varying extents of reduction in the oxygen permeation could be achieved. The films also retained their optical properties as the pure epoxy and nanocomposite films were transparent and indistinguishable. The films also did not have increase in the density owing to filler addition. The pure polymer density was measured to be 1.18 g/cm^3, which increased to 1.20 g/cm^3 in nanocomposites.

2.4 Conclusions

Optimum curing time, temperature, and amine to epoxy mole ratio are required for the filler delamination in polymer matrix. The presence of filler further enhances the rate of curing, thus this effect should also be taken into consideration while optimizing the curing conditions. The curing time should not be very short as it may hinder the polymerization taking place in the interlayers; similarly, the curing temperature should not be too high for the same reasons. The different amines led to the generation of epoxy networks with different crosslink densities owing to their structure and the number of functional groups. Polar and acidic nature surface modifications led to higher interlayer spacing values as well as exfoliation. On the other hand, modifications with completely nonpolar nature were not compatible with the polymer chains thus leading to intercalation of only limited amount. The oxygen permeation through the epoxy networks was also affected by the network density as the highest crosslink density network had the lowest oxygen permeation. Coinciding with the TEM and x-ray findings, the oxygen permeation reduction was much better when bis(2-hydroxyethyl)methyl hydrogenated tallow ammonium and dihydrogenated tallow methyl ammonium were used as surface modifications. Owing to relatively weaker van der Waals forces with the epoxy matrix, the performance of the benzyl dodecyl dimethyl ammonium modification was moderate. On the other hand, owing to the mismatch between the polymer and apolar surface modification, dihydrogenated tallow methyl ammonium-modified filler led to an increase in the oxygen permeation. The CEC of the filler substrates had a minor effect on the oxygen permeation performance. The nanocomposite films retained their optical clarity as well as density even after the addition of filler thus confirming their potential use in packaging applications.

Acknowledgment

This work has been published in the *Journal of Reinforced Plastics and Composites*, 2012, Volume 31, 739–747, Sage Publishers.

References

1. Brydson, J. A. *Plastic Materials*, 3rd edn., Newnes-Butterworths, London, U.K., 1975.
2. May, C. A. *Epoxy Resins Chemistry and Technology*, 2nd edn., Dekker, New York, 1988.
3. Lee, H. and Neville, K. *Handbook of Epoxy Resins*, McGraw-Hill, New York, 1967.
4. Ellis, B. *Chemistry and Technology of Epoxy Resins*, Blackie Academic & Professional, London, U.K., 1993.
5. Sivis, H. C. *Trends Polym. Sci.* 1997, *5*, 75.
6. Brennan, D. J., Haag, A. P., White, J. E., and Brown, C. N. *Macromolecules* 1998, *31*, 2622.
7. Eitzman, D. M., Melkote, R. R., and Cussler, E. L. *AIChE J.* 1996, *42*, 2.
8. Fredrickson, G. H. and Bicerano, J. *J. Chem. Phys.* 1999, *110*, 2181.
9. Gusev, A. A. and Lusti, H. R. *Adv. Mater.* 2001, *13*, 1641.
10. Yano, K., Usuki, A., Okada, A., Kurauchi, T., and Kamigito, O. *J. Polym. Sci., Part A: Polym. Chem.* 1993, *31*, 2493.
11. Alexandre, M. and Dubois, Ph. *Mater. Sci. Eng. R* 2000, *28*, 1.
12. Giannelis, E. P. *Adv. Mater.* 1996, *8*, 29.
13. LeBaron, P. C., Wang, Z., and Pinnavaia, T. *J. Appl. Clay Sci.* 1999, *15*, 11.
14. Jasmund, K. and Lagaly, G., Eds. *Tonminerale und Tone*, Steinkopff-Verlag, Darmstadt, Germany, 1993.
15. Osman, M. A., Ploetze, M., and Suter, U. W. *J. Mater. Chem.* 2003, *13*, 2359.
16. Brown, J. M., Curliss, D., and Vaia, R. A. *Chem. Mater.* 2000, *12*, 3376.
17. Zilg, C., Mulhaupt, R., and Finter, J. *Macromol. Chem. Phys.* 1999, *200*, 661.
18. Zerda, A. S. and Lesser, A. J. *J. Polym. Sci., Part B: Polym. Phys.* 2001, *39*, 1137.
19. Kornmann, X., Lindberg, H., and Berglund, L. A. *Polymer* 2001, *42*, 1303.
20. Kong, D. and Park, C. E. *Chem. Mater.* 2003, *15*, 419.
21. Osman, M. A., Mittal, V., and Suter U. W. *Macromol. Chem. Phys.* 2007, *207*, 68.
22. Osman, M. A., Mittal, V., Morbidelli, M., and Suter U. W. *Macromolecules* 2004, *37*, 7250.
23. Lan, T., Kaviratna, P. D., and Pinnavaia, T. J. *Chem. Mater.* 1995, *7*, 2144.
24. Kamon, T. and Furakaw, H. *Adv. Polym. Sci.* 1986, *80*, 177.
25. Barton, J. M. *Adv. Polym. Sci.* 1985, *72*, 120.
26. Messersmith, P. B. and Giannelis, E. P. *Chem. Mater.* 1994, *6*, 1719.
27. Osman, M. A., Mittal, V., Morbidelli, M., and Suter U. W. *Macromolecules* 2003, *36*, 9851.

3

Carbon Nanotubes–Polytetrafluoroethylene Nanocomposite Coatings

Natthakan Rungraeng and Soojin Jun

CONTENTS

3.1 Background

3.1.1 Structures, Properties, and Characteristics of Carbon Nanotubes

Carbon nanotubes (CNTs) were discovered by a Japanese physicist, Sumio Iijima, in 1991. Since then, a number of studies investigating the properties and applications of CNTs have been reported. CNTs are apparently elongated fullerenes having either single or multiple hexagonal rolled-up sheets of carbon atoms (graphene) making up the walls with capped ends (Xie et al., 2005). Single-walled carbon nanotubes (SWCNTs) and multiwalled carbon nanotubes (MWCNTs) are commonly known as the common types of CNTs. Specifically, SWCNTs have only a single layer of graphene-sheet wall located around a central hollow core (tube). On the other hand, MWCNTs are generally composed of more than one graphene shell coaxially located along the tube. The diameters of the nanotubes are in the range of 1–10 s nm, depending on the number of walls, with high length-to-diameter ratio (generally more than 1000). CNTs are traditionally produced by thermal degradation of a carbon source in the presence of a metal catalyst using laser ablation or an arc discharge approach. The limitations of these methods are widely reported; specifically, laser ablation is not applicable for large-scale production while CNTs produced by the arc discharge process contain high-level contaminations of carbonaceous impurities as well as metal catalysts (Meyyappan et al., 2003).

Kumar and Ando (2010) reported that the synthesis of CNTs by the chemical vapor deposition (CVD) method has been widely employed. The benefits of the CVD technique are low capital cost, high production yield, easy to scale up, and high versatility of production. The CVD system can be designed to fabricate CNTs in various forms such as powder, film, and aligned and entangled tubes, depending on the configurations of the metal catalyst and the substrate. The process parameters of CNT growth in CVD are type of hydrocarbon precursors and catalysts, temperature, pressure, flow rate, deposition time, and reactor's dimension. The simplified configurations of the CVD system are shown in Figure 3.1. The growth of CNTs by the CVD system on a continuous basis is achieved by flowing a vapor of a select hydrocarbon precursor (C_xH_y), that is, methane, ethylene, acetylene,

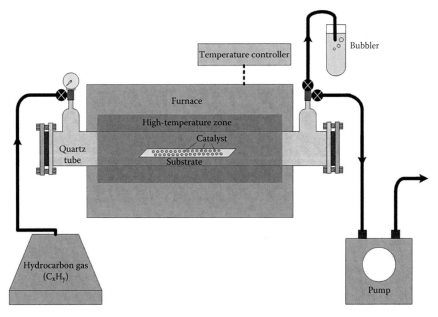

FIGURE 3.1
(See color insert.) Schematic diagram of simplified CVD system for growing CNTs.

benzene, xylene, or carbon monoxide, over a substrate containing a metal catalyst on its surface in a cylindrical reaction chamber. The temperature within the reactor is typically controlled to be 600°C–1200°C using an electronic temperature source with a controller.

At this step, the decomposition of hydrocarbon vapor occurs when it reacts with the metal catalyst at high temperature producing carbon and hydrogen species within the chamber. The hydrogen is not involved in passing through the outlet of the reactor while the carbon is readily dissolved into the metal catalyst. The formation of nanotubes occurs when the concentration of dissolved carbon species within the catalyst exceeds the maximum solubility limit. As a result, there is a formation of a cylindrical network of carbon atoms that grow out from the surface of the catalyst. The growth models of CNTs can be categorized into two main classes, namely "tip-growth model" and "base-growth model." As shown in Figure 3.2a, the first case happens when the interfacial interaction between the catalyst and the substrate is weak. Therefore, the CNT grows from the catalyst–substrate interface, pushing the catalyst particle away from the substrate. The elongation of the nanotube continues from its tip where the metal catalyst is located. On the other hand, the latter model is used to describe the CNT growth that takes place when the interaction of the catalyst–surface is stronger than the force produced by CNT precipitation. Therefore, the CNT precipitates out from the top of the catalyst and extends its length by forming a tube from the bottom (Figure 3.2b). The size of the metal catalyst for growing SWCNTs is typically in the range of a few nanometers while the size of a few tens of nanometers is needed for growing MWCNTs.

Both SWCNTs and MWCNTs have exceptional mechanical, thermal, chemical, magnetic, optical, and electrical properties. Some of these CNT properties are shown in Table 3.1. According to the data shown in Table 3.1, the tensile strength of CNTs is found to be 100 times that of steel with approximately one order of magnitude of steel density. Therefore, CNTs are well known as lightweight yet high-performance materials suitable for many applications ranging from biosensors to space exploration. Recently, fabrications of CNT–polymer composites have been intensively studied in order to expand the availability of CNTs. However, aggregation of CNT particles is a challenging task facing the fabrication of CNT–polymer composites. The formations of bundles and ropes occur during the production of CNTs due to van der Waals attraction among nanoparticles. This problem results in the limited exhibition of CNT potential. Therefore, proper pretreatment is needed to appropriately disintegrate and disperse individual nanotubes before mixing in order to enhance the uniformity of CNTs in the surrounding polymer matrix.

3.1.2 Dispersion of CNTs in the Matrix of Polymer

The most commonly used techniques to disperse CNTs in polymer matrices are described in the following subsections (Xie et al., 2005).

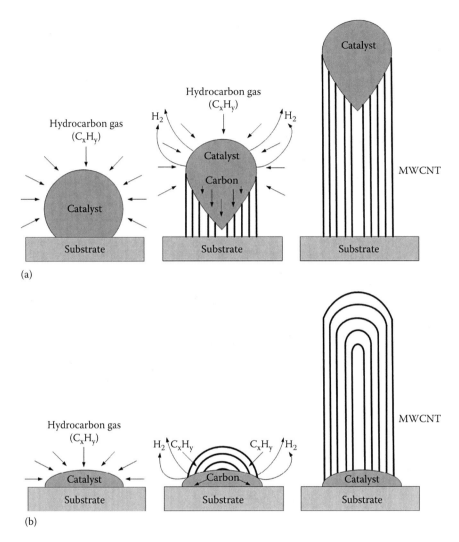

FIGURE 3.2
(See color insert.) Growth mechanisms of CNTs in CVD system: (a) tip-growth model and (b) base-growth model.

3.1.2.1 Physical Blending

High-power disaggregation methods are required to properly disperse nano-fillers such as CNTs in a polymer matrix. An ultrasonicator and high shear mixer are commonly employed to achieve this requirement. The ultrasonic waves generate cavitation in CNT suspension and subsequently disaggregate CNT bundles to form individual nanotubes. However, excessive use of ultrasonication might shorten both SWCNTs and MWCNTs. Although the

TABLE 3.1

Select Properties of Carbon Nanotubes

Property	CNTs
Specific gravity	0.8 g/cm^3 for SWCNT
	1.8 g/cm^3 for MWCNT (theoretical)
Elastic modulus	Approximately 1 TPa for SWCNT; 0.3–1 TPa for MWCNT
Strength	50–500 GPa for SWCNT; 10–60 GPa for MWCNT
Resistivity	5–50 μΩ cm
Current density	>10^8 A/cm^2 (Zhong et al., 2012)
Thermal conductivity	3000 W/m/K (theoretical)
Thermal expansion	Negligible (theoretical)
Thermal stability	>700°C (in air); 2800°C (in vacuum)
Specific surface area	10–20 m^2/g
Aspect ratio	Up to 132,000,000:1

shortening mechanism of CNTs is still unclear, it was found that shortening is more likely to occur near the center of mass (CoM) of CNTs, which generally happens in the shortening of polymers during ultrasonication (Pagani et al., 2012). In addition, reaggregation of both SWCNTs and MWCNTs will occur after ultrasonication is removed (Lahelin et al., 2011). Therefore, surfactants or heat may be applied to enhance the dispersion stability of CNTs in a polymer matrix.

3.1.2.2 Chemical Functionalization

Chemical functionalization is typically recognized as a promising method to magnify the mechanical properties of CNT-based composites. The method is also called the "grafting to" approach. Generally, a polymer matrix is allowed to react with oxidized or pre-functionalized CNTs. CNTs grown by this method may contain other undesirable impurities, which impact the quality of CNTs, such as fullerenes, nanocrystalline graphite, as well as metallic catalysts (Feng et al., 2003). Therefore, these impurities should be removed before the functionalization process. Linking of a carboxyl group (–COOH) or hydroxyl group (–OH) on the surface of CNTs is the primary step of CNT functionalization. These groups can be further replaced by various other functional groups that can provide homogeneous dispersion for high-performance CNT–polymer composites. Furthermore, it was found that the dispersibility of COOH-functionalized CNTs in water and other organic solvents is better than pristine CNTs. In addition, acids used in this process also increase the purity of CNTs by removing any remaining metal catalysts from the nanotubes. To obtain COOH-functionalized CNTs, nanotubes are typically treated with concentrated organic acids including nitric acid, sulfuric acid, and a mixture of nitric and sulfuric acids. Sonication is then applied to the mixture for a few hours before being agitated at high temperature for several hours.

3.1.2.3 In Situ Polymerization

In situ polymerization, which is also called the "grafting from" approach, is an effective way to uniformly disperse nanocomposites throughout a polymer matrix. The method is faster and also less expensive than chemical functionalization of nanomaterials including CNTs. Therefore, it is the most practical way to process CNT–polymer composites on a large production scale, in particular, the emulsion/suspension polymerization method (Lahelin et al., 2010). This polymerization technique consists of several steps including immobilizing the sidewalls and the edges of CNT particles with a monomer in the presence of a surfactant and an initiator. Subsequently, the CNT–polymer mixture is polymerized to create uniformly distributed CNT–polymer composites at a high level of CNT loadings.

3.1.3 PTFE Polymer

Polytetrafluoroethylene (PTFE or Teflon®) is a thermoplastic polymer (F_2C–$CF_2)_n$ which is synthesized by the polymerization of tetrafluoroethylene monomers, $n \cdot F_2 = CF_2$. The C–F bond that presents in PTFE has exceptionally high bonding energy, typically >100 kg/mol, resulting in a very low surface free energy of PTFE (18 mN/m, Lau et al., 2003). Because the melting point of PTFE is approximately 327°C, it is commonly employed as an inert coating on the surface of various types of materials including plastics and metals. PTFE is commercially introduced into food-processing equipments such as nonstick cookware coating, which enables minimum adhesion of a biological fouling layer on the PTFE-coated surface. In addition, PTFE coating also provides an excellent easy-to-clean attribute to the surface of a frying pan. However, the applications of PTFE coating are restricted because of several limitations, such as poor thermal conductivity as well as high wear rate. Therefore, PTFE composites are developed by the conjugation of PTFE with other materials, that is, either a filler or a matrix, in order to improve the wear resistance of PTFE. Chen et al. (2003) reported that the wear rate and friction coefficient of MWCNT–PTFE composites dramatically decreased when CNT content was increased. The lowest wear rate of the developed composites was only 1/290 that of pristine PTFE when 20 vol.% of CNT was introduced into the PTFE matrix. Other than the improvement in wear rate, the conjugation of CNT nanocomposites within the PTFE matrix was also utilized to develop a self-cleaning superhydrophobic surface in many applications. The term hydrophobic surface is typically used to describe any surface having a water contact angle (WCA) greater than 90°. The WCA of a superhydrophobic surface is generally more than 140° and the water sliding angle (WSA) is less than 10°. The unique attribute of a superhydrophobic surface is self-cleaning, which is also called the "lotus effect," and is caused by the formation of microscale roughness of the material, that is, a CNT that allows air to be trapped within its rough structure, thus preventing the direct contact between water on the superhydrophobic surface and the substrate underneath. Therefore, a water stream flowing on top of the superhydrophobic surface would quickly pick up dust and other solid impurities from the superhydrophobic surface. The fabrications and applications demonstrating the uses of superhydrophobic CNT–PTFE composites are further described in the following sections.

3.2 Recent Applications of CNT–PTFE Nanocomposite Coating

3.2.1 Reduction of Milk Fouling in Plate Heat Exchanger

Milk fouling is an undesirable adsorption of heat-denatured milk protein onto the surfaces involved in all indirect heating processes of milk at the temperature of 70°C–74°C or above (Bansal and Chen, 2006). The surface deposition of milk is mainly caused by natural milk whey protein, beta-lactoglobulin (β-Lg). In general, the concentration of β-Lg in cow milk is 0.32%–0.37% (de Jong et al., 2002; Bansal and Chen, 2006). During milk pasteurization using a test plate heat exchanger (PHE) unit, milk foulant on the surface is a soft-white solid that tends to accumulate at low-flow areas such as corners and gasket over the heating time. However, surface fouling can be eliminated by a routine clean-in-place (CIP) program in which considerable amounts of chemicals, water, and electrical energy are required. It is known that hydrophobic coating is one of the effective surface treatments to reduce the biological fouling on a heat exchanger surface (Santos et al., 2004; Rosmaninho et al., 2007; Balasubramanian and Puri, 2009; Ozden and Puri, 2010).

The utilization of hydrophobic and superhydrophobic surface coatings on PHE for liquid foods including milk and other dairy products has been recently elucidated by a number of researchers. For instance, Beuf et al. (2003) investigated that the fouling rate on uncoated stainless steel 316 during milk product pasteurization was not significantly different from that on surfaces that were modified by eight different coatings including diamond-like carbon (DLC), silica, SiO_2, Ni–P–PTFE, Excalibur, Xylan, and ion implantations (SiF^+, MoS_2). However, they found that surface energy and surface roughness of coatings played important roles in the cleaning efficiency of fouled surfaces. Rosmaninho et al. (2007) also confirmed that fouled PHE surfaces caused by calcium phosphate, β-Lg, and milk-based fluids could be easily cleaned if they had hydrophobic Ni–P–PTFE coating. Balasubramanian and Puri (2009) investigated that food-grade hydrophobic polymer-based coatings including Ni–P–PTFE, Lectroflour™-64, and AMC148-18 could significantly diminish surface deposition and thermal energy consumption rates of the PHE system for milk pasteurization. Ozden and Puri (2010) found that the mass of foulants on test PHE surfaces of superhydrophobic CNT-coated stainless steel was much lower than the control and other hydrophobic-coated surfaces, that is, Ni–P–PTFE, TM117P, and AMC148.

Many other researchers (Lau et al., 2003; Santos et al., 2004; Huang et al., 2005; Li et al., 2010; Luo et al., 2010; Men et al., 2010) have also reported that superhydrophobicity could be obtained by applying aligned carbon nanotubes (ACNTs) to the metal substrate. However, the usage of CNTs as coating material on the heat transfer surfaces of a PHE is challenging because of the high wear rate of CNTs. Due to its chemical inertness and high melting temperature, PTFE has been commonly used for nonstick cookware and often as a filler and a matrix. As a matrix, it has been successfully filled with nanoparticles such as alumina and CNTs and, interestingly, the wear resistance of Al_2O_3–PTFE nanocomposites as surface coating was reported to be 600 times higher than unfilled PTFE (Sawyer et al., 2003). Similar improvements of the wear resistance for CNT–PTFE nanocomposite coatings have been observed as well (Chen et al., 2003; Burris et al., 2007). The development of uniformly distributed CNT–PTFE nanocomposite films on metal substrates can be fabricated by ultrasonic wave and high temperature annealing applications (Show and Takahashi, 2009).

The PHE system requires the surface to be of high durability because of its harsh operating environment associated with high temperature and pressure. A long operation period with minimal downtime of CIP processes is preferred from the industrial standpoint. The anti-milk-fouling performance of CNT–PTFE composite coatings for milk-fouling reduction in PHE was recently reported by Rungraeng et al. (2012). According to this study, the composite coating was fabricated using a water-based PTFE emulsion containing 60% w/w of 0.2–0.5 μm PTFE particles. CVD-grown MWCNT (>95% nanotubes, 1–2 μm long, 10–30 nm outer diameter) was well distributed in the PTFE solution by ultrasonication using a 400 W sonicator for 30 min. The final concentrations of CNT in the PTFE solution were controlled to range from 0% to 5% (w/w). The prepared CNT–PTFE nanocomposite solution was spin-coated on the stainless-steel surface. Uniform CNT–PTFE film on the stainless-steel surface was obtained by two-step processes: the plate was constantly held at 100°C for 10 min for water evaporation and was then heated at 360°C for an hour to melt down all PTFE contents onto the surface as the melting point of PTFE particles is approximately 327°C.

A field emission scanning electron microscopy (FESEM) image of a PTFE coating ensured that the melting condition was sufficient because little unmelted PTFE appeared. The ultrasonicating and annealing processes played a crucial role in the uniform distribution of CNT nanoparticles in the PTFE matrix. These preparation steps successfully built a new hybrid nanocomposite surface, which possesses high wear resistance as well as

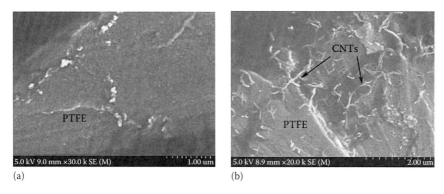

FIGURE 3.3
FESEM images of (a) PTFE and (b) 5 wt.% CNT–PTFE nanocomposite films on stainless steel 316 surface.

non-wettability. The morphologies of PTFE and CNT–PTFE coatings determined by FESEM are shown in Figure 3.3a and b, respectively. The results show that well-distributed MWCNT nanoparticles were partially anchored with PTFE leaving some parts uncovered by the PTFE matrix. The PTFE-merging parts of CNTs significantly strengthened the overall durability of the CNT nanocomposite coating while the unconcealed parts increased the micro/nanoscale surface roughness to characterize superhydrophobicity. WCAs were measured by sessile droplets of deionized (DI) water of uncoated, PTFE-coated, and 5% w/w CNT–PTFE-coated surfaces and were 71.2°, 119.6°, and 141.1°, respectively. The CNTs played an important role in increasing micro/nanoscale roughness of the base PTFE surface by creating numerous voids which acted as a boundary layer between water and PTFE surfaces. These voids limited the chance of water surface coming into contact with both base PTFE and stainless-steel substrates, resulting in magnification of the water repellent and self-cleaning surface properties (Xie et al., 2005; Nosonovsky and Bhushan, 2009). Nanocomposite coatings can prevent water droplets from being adsorbed onto the surface. The water repellent phenomenon observed in nature, which is also called the "lotus effect," led to the development of a self-cleaning surface on which the surface impurities will be swept away by means of flowing water. Self-cleanability of the targeted surface is achievable when the WCA value approaches 140° or more. The water droplets falling on the CNT–PTFE nanocomposite-coated surface actually had sufficient momentum to leave the surface, making several more bounces, eventually bouncing off without ever coming to rest on the surface. On the other hand, no bouncing of water droplets was found on the control surface. The photographs of water drops on the control and superhydrophobic surfaces are shown in Figure 3.4.

Foulant mass per unit area on the control surface after the first hour run was 23.41 mg/cm². Foulant masses per unit area on the uncoated plate linearly

(a)

(b)

FIGURE 3.4
Snapshots of a water droplet falling on (a) CNT–PTFE nanocomposite and (b) the control surfaces.

increased to 43.71 and 62.00 mg/cm^2 after 3 and 5 h of pasteurization. As the developed test unit had small dimensions, occupying only a single-channel heating section, the pressure drop across the heat transfer surface between the inlet and outlet was considerably negligible. Therefore, this test unit was not a pressurized system that could be normally found in the industrial-scale PHEs. Velocity and Reynolds number of milk flow found during the test trials were 0.0054 and 170.84 m/s, respectively. As the flow behavior of milk was laminar-type, the mass of fouling would be higher than industrial PHEs operating at turbulent flow profiles. However, the fouling deposition rates were reduced when using the coated surfaces, depending on their hydrophobicities. For PTFE coating, the masses of milk deposits collected after a continuous flow of milk for 1, 3, and 5 h were 18.29, 27.44, and 35.34 mg/cm^2, respectively. For the CNT-PTFE nanocomposite surface, the attachment of protein nucleation on the CNT-PTFE matrix was weaker than those onto uncoated and PTFE coated surfaces. Therefore, the shear force of the milk stream flowing over the surface could easily sweep away protein molecules deposited on the surface, resulting in a low nucleation rate and eventually the inhibition of the development of milk foulants at the same operation hours. Weighed foulants on CNT–PTFE nanocomposite surfaces were 4.96, 16.74, and 18.45 mg/cm², respectively. The heat transfer rates of the test PHE unit were enhanced by the CNT–PTFE nanocomposite layer; thus, the total energy consumption decreased from 3168 to 2844 kJ. Milk foulant development during pasteurization retards heat transfers on PHE surfaces, which causes an increase in the energy requirement of dairy industries by around 8% (Balasubramanian and Puri, 2009).

The amount of energy required for pasteurization significantly depends on the milk deposits accumulated on heat exchanger surfaces. There were decreases in the total heat energy consumption required to accomplish milk pasteurization for a period of 5 h by 3.4%–10.2% when PTFE and CNT–PTFE nanocomposite-coated surfaces were applied. The use of a superhydrophobic surface could save the energy consumption for thermal processes by lowering the nucleation and accumulation rates of milk deposits on PHE surfaces. In addition, MWCNT also has high Young's modulus (1.8 TPa, Chen et al., 2003) and thermal conductivity (600 W/m/K, Moisala et al., 2006). Therefore, the addition of CNT into a PTFE matrix enhanced the mechanical properties of pristine PTFE. The estimation of thermal conductivity of dispersed MWCNTs in a polymer matrix can be estimated using the following correlation provided by Moisala et al. (2006):

$$\frac{K_e}{K_m} = 1 + \frac{f K_{CNT}}{3 K_m} \tag{3.1}$$

where
K_e is the effective thermal conductivity of the composite (W/m/K)
K_m is the thermal conductivity of the PTFE matrix (0.25 W/m/K, Xie et al., 2005)
K_{CNT} is the thermal conductivity of the nanotube (W/m/K)
f is the mass fraction of the nanotube filler

Based on the proposed estimation, the thermal conductivity of CNT–PTFE composites with 5% CNTs would be 10.25 W/m/K. It may be noted that the thermal conductivity of CNT–PTFE nanocomposites is closer to that of food-grade stainless steel (approximately 16 W/m/K), compared to unmodified PTFE. The developed nanocomposite coating technique has a great potential for dairy applications. Fouling always creates the need for frequent CIP processes to keep the heating surfaces clean. Therefore, the developed coating technique is expected to reduce the frequency of routine CIP processes to insure a chemical-exempt food product, minimization of water usage, and reduction in the use of cleaning agents. However, for industrial implementation, the completeness of this study can be further fulfilled by investigating the wear resistance of developed surface coatings.

3.2.2 Comparison of Antibacterial Adhesion Performances on Superhydrophobic CNT–PTFE Composite and Superhydrophilic Coatings

Bacterial cell adhesion onto a metal substrate is highly sensitive to the surface nature such as wettability and surface energy. Biofilm or biofouling formation on the surfaces of food-processing equipments has been recognized as a widespread problem. The development of a surface consisting of antimicrobial agents with significant lethal effects has been widely studied. However, the major drawback of these antimicrobial surfaces is the formation of a surface debris layer of dead cells resulting in much lower lethal efficiency of the antimicrobial surface after a certain period of time (Crick et al., 2011). The prevention of bacterial adhesion would also significantly inhibit its subsequent biofouling formation (Zhao et al., 2005). Fabrication of self-cleaning surfaces is one example of most anticipated alternative ways to effectively retard the adhesion rate of bacteria on the surface. A superhydrophobic surface equipped with both low surface energy and micro/nanoscale roughness has a unique self-cleaning attribute, the so-called lotus effect. Micro/nanoscale surface roughness is one of the facile techniques that have been used to fabricate such superhydrophobic surfaces unveiling an outstanding water repellency attribute. A cluster of CNTs with substantial surface roughness is well regarded as a superhydrophobic material which has been proven to provide excellent antifouling properties as well as other desirable attributes, such as strength, flexibility, and thermal/electrical conductivity. However, the antifouling effect of a superhydrophobic surface requires a flow stream of liquid (Patel et al., 2010). On the other hand, an extremely water-attracting nanocomposite surface coated with TiO_2 nanoparticles, for example, could also perform another antifouling activity via its superhydrophilicity such that tight water-stretching reactions are able to form a stable bacteria-free hydration layer on the top of metal substrates.

Bacterial adhesion tests on both superhydrophobic and superhydrophilic surfaces were performed in a custom-designed parallel plate flow channel. The deposition of TiO_2 nanocomposites onto the stainless-steel surface was carried out using a sol–gel method which was slightly modified from Shen (2005). In brief, 20 mL ethanol was mixed with 1 mL ethyl acetoacetate (EAcAc) at room temperature; 4 mL of tetra-n-butyl titanate was added to the mixture with constant stirring for an hour. A quantity of 0.2 mL of DI water was gently added to the solution for hydrolysis. After aging for 24 h, the prepared TiO_2 solution was spin-coated on a steel plate at a rotational speed of 200 rpm. The wetted stainless-steel surface was heated at 150°C for 15 min to evaporate water from the surface. The surface coating and primary drying steps were repeated three more times in order to create the desired thickness of TiO_2 layers on the surface. The dried steel plate was further heated to a final drying step, annealing, in an electronic furnace at 450°C for 30 min where the superhydrophilic feature of the stainless-steel plate was achieved by the oxidation process of TiO_2 particles. In addition, the TiO_2-coated plate was further hydrothermally treated by submerging into boiling water for 10 min followed by reheating at 450°C for 10 min in order to prevent TiO_2 carriers from cracking on the surface. The plate was cooled down in the air until it reached room temperature before being rinsed with DI water to wash away excessive TiO_2 particles from the surface. The WCA of the stainless steel decreased from 71.2° to 0° (superhydrophilic). To fabricate a superhydrophobic surface, 20% w/w of CNT–PTFE solution was prepared by ultrasonicating COOH-functionalized multiwalled CNTs (>95% purity, 5–20 μm long, 30 ± 10 nm in an outer diameter) for 3 h with a water-based PTFE suspension (Teflon® PTFE30). The mixture was then spin-coated at 500 rpm on a pre-cleaned steel plate. CNT–PTFE nanocomposites were also kept inside the furnace at 360°C for 1 h for completion of the annealing process. The WCA value of CNT–PTFE-coated stainless steel was 154.6° (superhydrophobic). FESEM was used to visualize the micro/nanoscale surface morphologies of the developed superhydrophobic and superhydrophilic surfaces. In Figure 3.5a and b, it may be noticed that only air could penetrate into the entangled nanocomposite structure (approximate average size between 0.1 and 0.2 μ) between the CNT–PTFE during the adhesion test, while larger particles including bacteria merely rolled over the structure. For TiO_2-coated surfaces, it can be seen that the nanocomposite coating was tightly aligned to obtain a highly uniform nanoparticulate matrix. Average particle sizes of TiO_2 composites were seen to be approximately less than 0.1 μ, resulting in the occurrence of many tiny voids around groups of agglomerated TiO_2, which are the so-called micropapillae.

To perform the adhesion tests, 250 mL of *Escherichia coli* solution at room temperature was pumped through the test unit at the flow rates of 0 (stagnant) and 200 mL/min (continuous) separately. Adhesion times for both runs

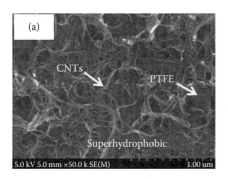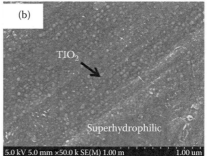

FIGURE 3.5
SEM images of (a) superhydrophobic and (b) superhydrophilic plates.

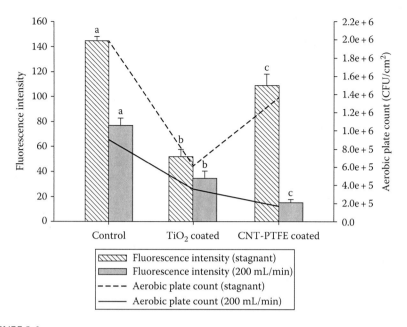

FIGURE 3.6
Fluorescence intensities reflecting a number of adhered bacteria on test plates after DAPI staining compared to total number of microorganisms by aerobic plate count method (significant differences ($P < 0.05$) within data at the same flow condition indicated by dissimilar letters).

were controlled to be an hour. Subsequently, six plates with bacterial attachments were gently rinsed with 1 mL of sterile DI water in order to remove loosely attached *E. coli* cells from the surfaces. After the plates were dried in air, one drop of fluorescence stain, namely 4',6-diamidino-2-phenylindole (DAPI), was added. It may be noted that an increase in the flow rate significantly decreased the number of bacteria adhering on all surfaces. This was due to the changes in wall shear rates in the flow channel. The calculated shear rates in stagnant and flow environments were 0 and approximately 16,000 s^{-1}, respectively. Figure 3.6 shows the fluorescence intensity values of all adhered surfaces which seemed in good agreement with the aerobic plate count. It illustrates that the anti-adhesion mechanisms on superhydrophobic and superhydrophilic surfaces were significantly different from each other. According to Peng et al. (2006), the existence of a hydration layer/shield around the superhydrophilic surface in an aqueous solution would limit the surface from interacting with other materials. Therefore, the TiO$_2$ plate had the lowest FI value among the three surfaces in the stagnant or dry environment (approximately 65% lower than the control). On the other hand, the FI value for the CNT–PTFE surface was approximately 80% lower than the control in a continuous mode due to little surface energy and air voids in the coating matrix, promoting water droplets to roll over without being adsorbed onto the substrate. It may be clearly observed that microbial adhesion rates on metal substrates were simultaneously affected by both wall shear rates and surface chemistry. The superhydrophobic CNT–PTFE coating had the lowest amounts of adhered bacteria in the highly sheared wall due to its lotus effect attribute. On the other hand, maximum anti-adhesion effect in stagnant environment was found on the superhydrophilic TiO_2 coating by means of a water boundary layer tightly located around TiO_2 surface.

3.3 Conclusion

SWCNT and MWCNT are cylindrical structures of carbon atoms having single and multiple graphene walls, respectively. The aspect ratio (length (L)/diameter (D)) of these nanotubes are typically higher than 1000. Therefore, they possess a number of exceptional attributes making them suitable for various applications. The synthesis of CNTs done by the CVD process involves the thermal decomposition of hydrocarbon vapor at high temperature in a quartz tube producing carbon and hydrogen species. The isolated carbon atom subsequently reacts with the metal catalyzer on the substrate resulting in initiation and growth of CNTs on the surface of the substrate. PTFE is a low surface-free energy thermoplastic commonly used as a coating material on metal surfaces creating hydrophobic nonstick surfaces, in particular for food-processing equipment. However, the major drawbacks that limit the use of PTFE are low thermal conductivity as well as wear resistance. This reflects the short lifetime of the PTFE coating layer onto the substrate. To overcome this problem, dispersion of CNTs in a PTFE matrix is found to significantly lessen the wear rate of PTFE by several hundred times. Physical blending, *in situ* polymerization, and chemical functionalization are the main dispersion methods that have been used to efficiently fabricate uniform CNT–PTFE composites. Other than mechanical improvement, properly aligned CNT particles in a matrix also greatly improve the hydrophobic property of CNT–PTFE composites. The presence of CNT particles in a PTFE matrix sufficiently produces a rough surface that accommodates a number of air gaps. This promotes a self-cleaning attribute, the so-called lotus effect, which significantly reduces the amounts of biological foulants on the CNT–PTFE-coated surface. Anti-biofouling performances of CNT–PTFE composite coatings were demonstrated in two distinct situations that are generally associated with food processing, namely (i) milk fouling on the surface of plate heat during pasteurization caused by heat denaturation of natural whey protein and (ii) reduction of bacterial adhesion in liquid flow channel. According to the results obtained from both experiments, CNT–PTFE nanocomposite coatings seem to be a promising anti-biofouling technique that substantially reduces the need of tedious process downtime for CIP protocols in the dairy industry during milk pasteurization as well as preventing the risk of cross-contamination of pathogenic bacteria from biofilms formed on the surfaces of food-processing equipments, thus improving food safety in ready-to-eat products.

References

Balasubramanian, S. and Puri, V. M. (2009). Thermal energy savings in pilot-scale plate heat exchanger system during product processing using modified surfaces. *Journal of Food Engineering, 91*, 608–611.

Bansal, B. and Chen, X. D. (2006). A critical review of milk fouling in heat exchangers. *Comprehensive Reviews in Food Science and Food Safety, 5*, 27–33.

Beuf, M., Rizzo, G., Leuliet, J.C., Müller-Steinhagen, H., Yiantsios, S., Karabelas, A., and Benezech, T. (2003). Fouling and cleaning of modified stainless steel plate heat exchangers processing milk products. In *Proceedings of Heat Exchanger Fouling and Cleaning: Fundamentals and Applications*, Santa Fe, NM.

Burris, D. L., Bosel, B., Bourne, G. R., and Sawyer, W.G. (2007). Polymeric nanocomposites for tribological applications. *Macromolecular Materials and Engineering, 292*, 387–402.

Chen, W. X., Li, F., Han, G., Xia, J. B., Wang, L. Y., Tu, J. P., and Xu, Z. D. (2003). Tribological behavior of carbon-nanotube-filled PTFE composites. *Tribology Letters, 15*, 275–278.

Crick, C.R., Ismail, S., Pratten, J., and Parkin, I. (2011). An investigation into bacterial attachment to an elastomeric superhydrophobic surface prepared via aerosol assisted deposition. *Thin Solid Films, 519*, 3722–3727.

Feng, Y., Zhou, G., Wang, G., Qu, M., and Yu, Z. (2003). Removal of some impurities from carbon nanotubes. *Chemical Physics Letters, 375*, 645–648.

Huang, L., Lau, S. P., Yang, H. Y., Leong, E. S. P., Yu, S. F., and Prawer, S. (2005). Stable superhydrophobic surface via carbon nanotubes coated with a ZnO thin film. *Journal of Physical Chemistry B, 109*, 7746–7748.

de Jong, P., te Giffel, M. C., Straatsma, H., and Vissers, M. M. (2002). Reduction of fouling and contamination by predictive kinetic models. *International Dairy Journal, 12*, 285–293.

Kumar, M. and Ando, Y. (2010). Chemical vapor deposition of carbon nanotubes: A review on growth mechanism and mass production. *Journal of Nanoscience and Nanotechnology, 10*, 3739–3758.

Lahelin, M., Annala, M., Nykänen, A., Ruokolainen, J., and Seppälä, J. (2011). In situ polymerized nanocomposites: Polystyrene/CNT and poly(methyl methacrylate)/CNT composites. *Composites Science and Technology, 71*, 900–907.

Lau, K. K. S., Bico, J., Teo, K. B. K., Chhowalla, M., Amaratunga, G. A. J., Milne, W. I., McKinley, G. H., and Gleason, K. K. (2003). Superhydrophobic carbon nanotube forests. *Nano Letters, 3*, 1701–1705.

Li, G., Wang, H., Zheng, H., and Bai, R. (2010). A facile approach for the fabrication of highly stable superhydrophobic cotton fabric with multi-walled carbon nanotubes-azide polymer composites. *Langmuir, 26*, 7529–7534.

Luo, Z., Zhanga, Z., Wanga, W., Liua, W., and Xuea, Q. (2010). Various curing conditions for controlling PTFE micro/nano-fiber texture of a bionic superhydrophobic coating surface. *Materials Chemistry and Physics*, *119*, 40–47.

Men, X., Zhang, Z., Yang, J., Wang, K., and Jiang, W. (2010). Superhydrophobic/superhydrophilic surfaces from a carbon nanotube based composite coating. *Applied Physics A*, *98*, 275–280.

Meyyappan, M., Delzeit, L., Cassell, A., and Hash, D. (2003). Carbon nanotube growth by PECVD: A review. *Plasma Sources Science and Technology*, *12*, 205–216.

Moisala, A., Li, Q., Kinloch, I.A., and Windle, A.H. (2006). Thermal and electrical conductivity of single- and multi-walled carbon nanotube-epoxy composites. *Composites Science and Technology*, *66*, 1285–1288.

Nosonovsky, M. and Bhushan, B. (2009). Superhydrophobic surfaces and emerging applications: Non-adhesion, energy, green engineering. *Current Opinion in Colloid & Interface Science*, *14*, 270–280.

Ozden, H.O. and Puri, V.M. (2010). Computational analysis of fouling by low energy surfaces. *Journal of Food Engineering*, *99*, 250–256.

Pagani, G., Green, M.J., Poulin, P., and Pasquali, M. (2012). Competing mechanisms and scaling laws for carbon nanotube scission by ultrasonication. *Proceedings of the National Academy of Sciences*, *109*, 11599–11604.

Patel, P., Choi, C.K., and Meng, D.D. (2010). Superhydrophilic surfaces for antifogging and antifouling microfluidic devices. *Journal of the Association for Laboratory Automation*, *15*, 114–119.

Peng, C., Song, S., and Fort, T. (2006). Study of hydration layers near a hydrophilic surface in water through AFM imaging. *Surface and Interface Analysis*, *38*, 975–980.

Rosmaninho, R., Santos, O., Nylander, T., Paulsson, M., Beuf, M., Benezech, T., Yiantsios, S., Andritsos, N., Karabelas, A., Rizzo, G., Müller-Steinhagen, H., and Melo, L.F. (2007). Modified stainless steel surfaces targeted to reduce fouling–evaluation of fouling by milk components. *Journal of Food Engineering*, *80*, 1176–1187.

Rungraeng, N., Cho, Y.C., Yoon, S. H., and Jun, S. (2012). Carbon nanotube-polytetrafluoroethylene nanocomposite coating for milk fouling reduction in plate heat exchanger. *Journal of Food Engineering*, *111*, 218–224.

Santos, O., Nylander, T., Rosmaninho, R., Rizzo, G., Yiantsios, S., Andritsos, N., Karabelas, A., Müller-Steinhagen, H., Melo, L., Boulangé-Petermann, L., Gabet, C., Braem, A., Trägårdh, C., and Paulsson, M. (2004). Modified stainless steel surfaces targeted to reduce fouling–surface characterization. *Journal of Food Engineering*, *64*, 63–79.

Sawyer, W. G., Freudenberg, K. D., Bhimaraj, P., and Schadler, L. S. (2003). A study on the friction and wear behavior of PTFE filled with alumina nanoparticles. *Wear*, *254*, 573–580.

Shen, G.X., Chen, Y.C., and Lin, C.J. (2005). Corrosion protection of 316 L stainless steel by a TiO_2 nanoparticle coating prepared by sol-gel method. *Thin Solid Films*, *489*, 130–136.

Show, Y. and Takahashi, K. (2009). Stainless steel bipolar plate coated with carbon nanotube (CNT)/polytetrafluoroethylene (PTFE) composite film for proton exchange membrane fuel cell (PEMFC). *Journal of Power Sources*, *190*, 322–325.

Xie, X. L., Mai, Y. W., and Zhou, X. P. (2005). Dispersion and alignment of carbon nanotubes in polymer matrix: A review. *Material Science and Engineering, 49,* 89–112.

Zhao, Q., Liu, Y., Wang, C., Wang, S., and Müller-Steinhagen, H. (2005). Effect of surface free energy on the adhesion of biofouling and crystalline fouling, *Chemical Engineering Science, 60,* 4858–4865.

Zhong, G., Warner, J.H, Fouquet, M., Robertson, A.W., Chen, B., and Robertson, J. (2012). Growth of ultrahigh density single-walled carbon nanotube forests by improved catalyst design. *ACS Nano, 6,* 2893–2903.

4

Organic–Inorganic Hybrid Coatings with Enhanced Scratch Resistance Properties Obtained by the Sol–Gel Process

Massimo Messori and Marco Sangermano

CONTENTS

4.1 Organic–Inorganic Hybrid Materials

Organic–inorganic hybrid materials can be defined as nanocomposites comprising organic and inorganic components intimately mixed over length scales ranging from a few Angstroms to a few tens of nanometers. The properties of these materials are not only the sum of the individual contributions of organic and inorganic phases, but the role of the interfaces could be predominant. The nature of the interface is usually used for a generic classification into two distinct classes. In class I materials, organic and inorganic components are simply embedded in each other, and only weak bonds (hydrogen, van der Waals, etc.) are responsible for the cohesion to the whole structure. In class II materials, the two phases are linked together through strong chemical bonds (covalent or ionic–covalent bonds). Several synthetic strategies can be adopted for the preparation of organic–inorganic hybrid materials according to the so-called bottom-up or top-down approaches, respectively:

- The top-down approach corresponds to the assembling or the dispersion of well-defined nano-building blocks that consist of perfectly calibrated preformed objects that keep their integrity in the final material
- The bottom-up approach generally corresponds to very convenient soft chemistry-based routes including conventional sol–gel chemistry, the use of specific bridged and polyfunctional precursors, and hydrothermal synthesis

A third possibility is the self-assembling procedures, which correspond to the organization or the texturation of growing inorganic or hybrid networks, templated by organic surfactants.

The different procedures permit to obtain both massive bulk materials and thin films and coatings.

Inorganic oxides with particle size under 100 nm represent a promising group of fillers for the top-down approach. Nanoparticles have been shown to improve the mechanical properties even at low loadings, and due to their small particle size, they do not affect the transparency of clear coats. These nanoparticles are usually modified with organosilanes to render them hydrophobic and thereby improve their distribution and dispersive mixing into the polymer matrix [1,2]. Inorganic oxide nanoparticles such as SiO_2, ZrO_2, and TiO_2 have been embedded in curable resins, resulting in improved scratch and abrasion resistance of the obtained coatings [2,3]. One of the main limits of this approach is that commercially available powders often contain aggregates that cannot be destroyed during processing and therefore deteriorate properties like transparency.

The extensive surface area that characterizes the nanometric size inorganic particles is the reason for their strong tendency of aggregation. The aggregation, dispersion, and the homogeneous distribution within the polymeric matrix are a key challenge in the preparation of nanostructured coatings with enhanced surface properties. Furthermore, the high volume fraction content will adversely modify the rheology of the formulation with an important deterioration in processability. In this respect, the *in situ* generation of the inorganic domains through sol–gel process could represent an appealing strategy to avoid nanoparticles aggregation and the increase in the formulation viscosity.

The sol–gel process is a chemical method to prepare inorganic materials, initially employed to synthesize high-purity inorganic networks as glasses and ceramics [4,5]. Mild conditions characterize this method, which becomes strategic when organic materials are involved in the process permitting to avoid their thermal degradation. Typical precursors are metal alkoxides, which when reacting with water in the presence of an adequate catalyst lead to nanoparticles having narrow grain size distribution with dimensions ranging between 5 and 100 nm.

The most used method is based on the so-called aqueous sol–gel reaction, which can be schematically divided into two distinct steps: the first one named hydrolysis, which produces hydroxyl groups, and the second one named condensation, which involves the polycondensation of hydroxyl groups and residual alkoxyl groups to form a three-dimensional network as follows:

Hydrolysis: $M(OR)_x + nH_2O \rightarrow M(OR)_{x-n}(OH)_n + nROH$

Condensation:

$\equiv M-OH + HO-M \equiv \rightarrow \equiv M-O-M \equiv + H_2O$ (water condensation)

and/or

$\equiv M-OR + HO-M \equiv \rightarrow \equiv M-O-M \equiv + ROH$ (alcohol condensation)

The presence of an organic oligomer or polymer (bearing or not bearing suitable groups reactive toward the sol–gel process) in the reactive system leads to the formation of organic–inorganic hybrid structures composed of metal oxide and organic phases intimately mixed with each other. The optical, physical, and mechanical properties of these nanocomposites are strongly dependent not only on the individual properties of each component, but also on important aspects of the involved chemistry such as uniformity, phase continuity, domain size, and the molecular mixing at the phase boundaries.

The morphologies of the hybrid materials are strictly dependent on the characteristics of the organic polymer (or oligomer or monomer) such as its molecular weight, the presence of and the number of reactive functionalities, as well as its solubility in the sol–gel system.

In spite of these advantages, it is necessary to underline that the crystallinity of the inorganic phase is generally very low. For this reason, during the last years, a lot of researches have been done on the so-called non-hydrolytic sol–gel (NHSG) reaction, to obtain very pure and crystalline silica [6] and other metal oxides [7].

The NHSG reaction can be divided into two steps, as well as the aqueous route. The first reaction step involves either a metal halide or a metal alkoxide with an organic oxygen donor (such as alcohols and ethers). The second step, called condensation, can follow different pathways depending on the alkoxide employed. The most important and the most used are the condensations through alkyl halide elimination and/or ether elimination [8]:

Metal halide – alcohol reaction: $MX_m + nROH \rightarrow M(X)_{m-n}(OR)n + nHX$

Condensation:

$\equiv M-OR + RO-M \equiv \rightarrow \equiv M-O-M \equiv + R-O-R$ (by ether elimination)

and/or

$\equiv M-OR + X-M \equiv \rightarrow \equiv M-O-M \equiv + R-X$ (by alkyl halide elimination)

In the present chapter, an exhaustive literature review on organic–inorganic hybrid coatings with enhanced scratch resistance properties, obtained through a bottom-up approach by sol–gel process, is reported.

4.2 Scratch-Resistant Organic–Inorganic Hybrid Coatings Obtained by Sol–Gel Process (Bottom-Up Approach)

The possibility to use organic–inorganic hybrid materials prepared by sol–gel process as functional and/or protective coatings was reported more than 10 years ago by the researchers of Fraunhofer-Institut für Silicatforschung (Germany) [9–11]. In this respect, ORMOCER® is a trademark of Fraunhofer-Gesellschaft zur Förderung der angewandten Forschung e.V. and designates organic–inorganic polymers synthesized by a two-step sol–gel processing method that can be applied as thin transparent coatings on various substrates (polymers, metals, and ceramics). The curing of ORMOCER® can be carried out by thermal heating (at temperatures below 200°C) or by UV-radiation. Several applications are claimed in these papers, also concerning anti-abrasion and anti-scratch coatings. Unfortunately, notwithstanding these claims, no quantitative data concerning scratch resistance were reported and in order to avoid the mismatch between claimed properties and experimental data, in the following paragraphs, only papers reporting quantitative evaluation of scratch resistance will be reported and illustrated.

Taking into account that the mechanical properties of the coatings can be strongly affected by their adhesion to the substrate, a reasonable classification can be made on the basis of the type of substrate (metal, plastic, or glass) considered in the study [12]. Generally, coatings for metals are proposed for corrosion protection, and in this view, the scratch resistance of the coating is a very important parameter for the maintenance of its anticorrosion properties. In the case of plastic substrates, which are intrinsically soft materials, the organic–inorganic hybrid coatings are typically proposed and characterized for the specific enhancement of the scratch resistance of the substrate. Finally, when glass is proposed as the substrate, intrinsic physical–mechanical properties of the organic–inorganic hybrids are usually studied without the specific aim to protect the glass.

4.2.1 Coatings for Metallic Substrates

Knowles and coworkers [13] reported a paper concerning the investigation on the effect of four different organic–inorganic hybrid material compositions prepared according to the Vitresyn® method [14]. Authors tried to determine how the different proportions of organic and inorganic materials in the starting composition affect the coating thickness, microstructure, and

TABLE 4.1

Composition of Starting Materials

Sample	TEOS (wt%)	TMSPM (wt%)	Resin (wt%)
1	75	23	2
2	45	52	3
3	55	16	29
4	29	34	37

scratch behavior of the hybrid coatings when deposited on pure aluminum substrates. Tetraethoxysilane (TEOS) and an aliphatic urethane acrylate resin were used as the primary inorganic precursor and the source of the organic component respectively, and 3–trimethoxysilylpropyl methacrylate (TMSPM, also referred to in the literature as MPTMA, γ-MPS, MSMA, TPM, MEMO, MPTS) was used both as a secondary inorganic source and silane coupling agent (see Table 4.1 for composition details). Wet coatings were applied by means of angled flow coating method, in which the liquid mixtures were slowly poured carefully and systematically over one surface of the aluminum panels, with the substrates at an angle of 45° to the vertical and with the liquid allowed to flow from one edge to the opposite edge. After drying and UV-curing, optically transparent, homogeneous, amorphous materials with no evidence of any silica nanodomains were obtained.

The critical loads at which the coatings shown in Figure 4.1 exhibited failure were 0.67, 1, 1.93, and 2.33 N for Samples 1–4, respectively. Notwithstanding the significantly thicker coating of Sample 4 (24 μm) in comparison with Samples 1–3 (3.5, 6, and 8 μm, respectively), the trends in these values are exactly what might be expected from the consideration of the proportions of TEOS, TMSPM, and urethane acrylate in the starting sol compositions. High levels of TEOS in the starting formulations embrittle the coatings, although for a given weight fraction of TEOS, this can be offset by increasing the level of urethane acrylate relative to TMSPM.

In a previous work [15], the same authors also investigated the effect of UV-curing time (from 2 to 40 min) on Sample 1 composition. The applied normal load was 10 N for each experiment. Coatings UV-cured for 20 min or longer on the aluminum delaminated totally, exposing the underlying substrate. Coatings cured for less than 10 min partially peeled off, whereas the coating cured for 10 min remained mostly intact. Analysis of the scratch tracks and the scratch debris showed that, at the applied load and scratch speed used, the longer the UV-curing time, the more brittle the coating behavior. This is consistent with an increasing amount of inorganic network formation (as confirmed by NMR analysis) as a function of UV-induced condensation reactions. Further testing on plastic substrates at a higher testing load of 40 N confirmed the trends seen in the coatings applied on aluminum.

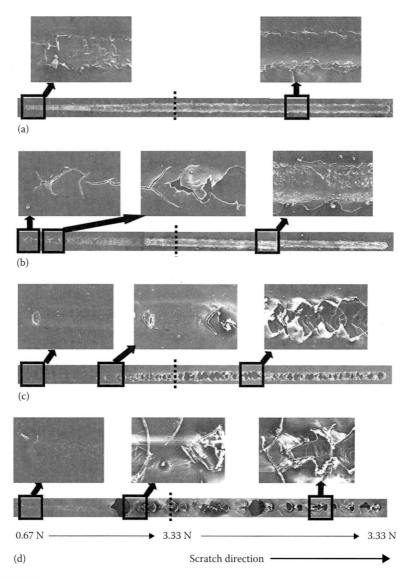

0.67 N ──────────────► 3.33 N ──────────────────► 3.33 N

(d)   Scratch direction ────────────►

FIGURE 4.1
SEM micrographs of scratch tracks on Al substrates coated with (a) Sample 1, (b) Sample 2, (c) Sample 3, and (d) Sample 4. The normal load increased linearly up to the dotted line (0.67 N–3.33 N), 4 mm away from the start of the scratch track on the far left of each micrograph. (Reprinted from Han, Y.H. et al., *Surf. Coat. Technol.*, 202, 1859, 2008. With permission.)

TABLE 4.2

Critical Load Values Obtained
for Sample 1 Coated on
Different Substrates

Substrate	Critical Load (N)
Aluminum	0.3
Brass	1.1
Stainless steel	1.8
Nickel	1.8

Knowles's research group also evaluated the scratch resistance and the delamination behavior of Sample 1 coating applied on different metallic and nonmetallic substrates [16]. Concerning the adhesion to metals, they found critical load values significantly dependent on the type of substrate, as reported in Table 4.2, confirming that adhesion between coating and substrate is a key parameter for the scratch resistance of the whole system. Authors also reported that scratch resistance of this organic–inorganic coating was much higher in the case of plastic substrates such as polycarbonate (PC), polyethylene terephthalate, and polymethyl methacrylate (PMMA) (see dedicated section).

Ahmad et al. reported the preparation and characterization of corrosion-protective polyurethane fatty amide/silica hybrid coatings [17]. The hybrid coatings were prepared at room temperature starting from Linseed oil fatty amide and TEOS (20–30 phr), followed by the addition of toluene-2,4-diisocyanate and subsequently applied onto steel substrates and thermally cured. The coating performance was studied in terms of inorganic precursor content. The system with the highest inorganic content (TEOS 30 phr) revealed excellent physical–mechanical performance among all the studied compositions. These coatings revealed excellent corrosion resistance properties than whole organic polyurethane fatty amide coating. The inorganic domains strictly interconnected to the organic polymer network provided improved mechanical properties to coatings such as scratch hardness (here measured according BS 3900 technical standard) and also impart excellent barrier action against different corrosive media.

The same authors also reported the preparation and characterization of organic–inorganic hybrid coatings based on castor oil/diglycidyl ether of bisphenol A (DGEBA) epoxy blend (as organic precursor) with TEOS as inorganic precursor [18]. After application on mild steel strips and thermal curing at 120°C, transparent and glossy coatings with good scratch hardness were produced.

Subasri et al. prepared organic–inorganic hybrid sol–gel-based silica coatings derived from hydrolysis and condensation of organically modified silane precursors like phenyltrimethoxysilane (PhTMOS) and methyltriethoxysilane

(MTES) along with TEOS [19]. The coatings were deposited on SS 316 grade stainless steel substrate, using dip coating technique and heat treated at 150°C for 2 h in air (final coating thickness of 2–3 μm). Authors investigated the effect of different organic functional groups like methyl (aliphatic) and phenyl (aromatic) on the properties of coatings derived from hydrolysis and condensation reaction of PhTMOS and MTES, along with TEOS. They observed that more water condensation reaction occurred in sol synthesized using MTES when compared to that using PhTMOS, which means that the three-dimensional network was more complete for the former. Accordingly, the coating from MTES sol showed higher pencil scratch hardness when compared to B from PhTMOS sol. This has been attributed to a complete network formation when methyl-modified silane was used, due to the higher hydrolysis and condensation rates of MTES when compared to PhTMOS.

The preparation of hybrid materials based on hydroxyl-terminated aliphatic polyester (AP) and TEOS as silica precursor and their use as protective coatings for steel and aluminum were reported by van der Linde and coworkers [20]. The coatings were applied with a doctor blade and thermally cured at 200°C. The dry coating thickness was found dependent on the TEOS content ranging from 20 to 40 μm. The scratch resistance of the systems was evaluated (on glass substrates) by using the König pendulum apparatus according to DIN 53 157 technical standard. The determined hardness as a function of the measured silica content is shown in Figure 4.2 from which a clear increase in hardness with increasing silica content, which seems to reach a maximum around 10–12 wt%, can be observed. The increase in hardness and T_g for the polyester/TEOS system has been attributed to an increasing network density, caused by silica cluster formation around the end-groups of the polyesters.

Soucek and coworkers reported an accurate study on organic–inorganic hybrid coatings formulated using multifunctional vinyltrimethoxysilane (VTMS) oligomers and acrylated AP. A radical photo-initiator was added to the VTMS/AP formulation, and the films were cross-linked via UV-radiation after deposition onto aluminum substrates [21]. Aluminum panels coated with hybrids having different contents of VTMS (in the range of 0–20 wt%) were characterized in terms of surface rubbing abrasion resistance (Taber Abraser Test) and pencil hardness according to ASTM D 3363-74 standard. Results indicated an increase in pencil hardness (from 1H to 4H) and a decrease in abrasion weight loss by increasing the VTMS content. This behavior has been attributed to the high hardness of VTMS colloids and the strong interaction and interpenetrated nature between chains of organic phase and inorganic particles. However, the high content of inorganic phase is limited by the impact resistance of the films.

The use of organic–inorganic hybrids for bioinert coating of implantable electronic devices has been recently proposed by using an NHSG approach for the preparation of hybrids [22]. The hybrids were prepared from diphenylsilanediol (DPSD) mixed with different ratios of TMSPM

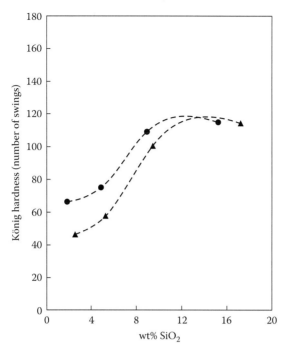

FIGURE 4.2
König hardness as a function of the silica content for the polyester–TEOS coatings, with a dry coating thickness of 20 μm (squares) and 40 μm (triangles). (Reprinted from Frings, S. et al., *Prog. Org. Coat.*, 33, 126, 1998. With permission.)

and (3-glycidoxypropyl)trimethoxysilane (GPTS), in the presence of barium hydroxide as catalyst for their condensation. The hybrids were then applied onto silicon wafer and photo-cured, exploiting the presence of methacryloyl groups. The adhesion strength of the coatings was evaluated by the maximum load to resist in scratch test. Authors reported that the adhesion properties increased with the addition of GPTS, and the best behavior was obtained from the hybrids containing 5–10 mol% of GPTS.

4.2.2 Coatings for Plastic Substrates

The relatively poor wear and scratch resistances typical of transparent plastics such as polystyrene, PC, and PMMA often represents their main limiting factor in all those applications in which the contact with dust or abrasive agents cannot be avoided. In order to enhance the scratch resistance of these materials without affecting the optical properties, transparent organic–inorganic hybrid coatings can be applied to their surface. This approach tries to combine the compatibility of the organic component of the coating with respect to the substrate (key parameter for a good adhesion) and the intrinsically high hardness of the inorganic component of the coating (in order to impart superior anti-scratch properties).

Fabbri and coworkers reported the preparation of hybrid coatings for PC based on TEOS and polyethylene oxide functionalized with alkoxysilane end-groups in order to enhance its compatibility/reactivity toward the silica precursor during the sol–gel reaction [23]. The curing reaction was carried out with a thermal treatment or with microwave (MW) irradiation, with variable coating thickness values in the range of 0.1–0.3 μm. NMR analysis showed that both treatment options were effective in promoting the cross-linking of the hybrid network, even if the same degree of curing was reached in 24 h at 140°C and only 35 s when MW irradiated. The scratch resistance was evaluated with a scratch tester on the basis of both the L_{C1} critical load (at which a scratch track appears) and the L_{C2} critical load (at which coating detachment occurs). All the coated samples exhibited a large increase in the L_{C1} values, relative to uncoated PC, which is indicative of the suitability of these materials as protective coatings. The coatings show the highest scratch resistance for thicknesses in the range of 0.10–0.25 μm for both thermal- and MW-cured systems.

The same authors also investigated similar systems based on TEOS and alkoxysilane-terminated polyethylene oxide or polycaprolactone [24]. Both oven heating and MW irradiation were effective curing methods, and the scratch resistance (here evaluated in terms of penetration depth) was found significantly higher with respect to uncoated PC and almost independent on the type and duration of the treatment. Interestingly, MW heating was emphasized due to its high rate and heating selectivity, because of the mismatch of dielectric properties between the coating and the candidate substrates to be used, thus offering interesting possibilities of scale-up of the process to many low-loss dielectrics.

Wu et al. reported the development of scratch-resistant coatings for PC substrates based on GPTS and TEOS with a cross-linking agent (ethylenediamine) and different amounts of colloidal silica applied by dip coating and thermally cured [25]. Scratch resistance was evaluated in terms of the pencil hardness grade, which increased by four grades on the same layer thickness of 5 μm. Authors concluded that the major factors influencing pencil scratch resistance are substrate hardness/strength, coating thickness, elasticity modulus, fracture toughness, and intrinsic hardness of the coating material. In particular, pencil hardness was increased from grade 2B to 5H by adjusting these parameters.

A very similar approach was also proposed by Etienne et al., who reported the preparation of protective coatings for PC based on GPTS and TEOS or preformed silica nanoparticles [26]. Young's modulus and hardness of the coating were measured and compared to literature models. Authors reported that the agreement between models and experimental values depends on the method of preparation of the nanocomposite coating (bottom-up or top-down approach) as a consequence of the different structures and morphologies obtained.

Concerning UV-cured hybrid coatings, hybrid coating materials were synthesized by Kim and coworkers [27] by using acrylate end-capped polyester, 1,6-hexanediolacrylate, TEOS, and TMSPM. The hybrid materials were cast onto PC substrates and cured by UV-irradiation. The pencil hardness of all of the samples examined was higher than 1H, whereas that of uncoated PC substrate was 6B.

Gururaj et al. investigated the effect of plasma pretreatment on adhesion and mechanical properties of UV-curable coatings on PC and PMMA [28]. The sol was synthesized by the hydrolysis and condensation of an UV-curable silane (TMSPM) in combination with Zr-*n*-propoxide. Coatings deposited by dip coating were cured using UV-radiation followed by thermal curing between 80°C and 130°C. Authors reported that pencil scratch hardness of the coating was 2H for PC and 1H in the case of PMMA and was unaffected by the surface activation, suggesting that the pencil scratch hardness is more dependent on the extent of densification or network formation in the coating (hardness) rather than bonding or adhesion with the substrate, and since coating on PC was densified at higher temperature, scratch hardness of coated PC is higher than that of PMMA.

Protective and functional coatings for PMMA were also proposed by Messori et al., who reported the preparation of hybrid coatings based on triethoxysilane-terminated polycaprolactone and TEOS [29,30]. The coatings were applied by dipping and were thermally cured. Compact and well-adherent coatings of thickness varying from 1 to 15 μm were obtained (see Figure 4.3). Interestingly, a preferential segregation of silica onto the outer surface was found while the coating–PMMA interface was found to be richer in PCL segments. Significant improvements of the anti-scratch properties were noted only in the case of PMMA coated with silica-rich ceramers (more than 50% by weight).

Chen et al. reported the preparation of protective and functional coatings for PVC [31]. Coatings were obtained after the incorporation of α-FeOOH

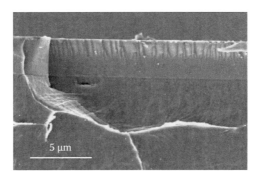

FIGURE 4.3
SEM micrograph of the edge view of the fracture surface of PMMA slabs coated with polycaprolactone/silica hybrid coating. (Reprinted from Messori, M. et al., *Polymer*, 44, 4463, 2003. With permission.)

nanoneedles within an organic–inorganic hybrid matrix based on TEOS and GPTS. Strong UV-absorption, high transparency, and inherent hardness of α-FeOOH nanoneedles improved the properties of the coatings. Correspondingly, scratch and aging resistance of the coated PVC substrate were significantly enhanced.

Toselli et al. reported a detailed investigation on the possibility to use hybrid coatings for the improvement of scratch resistance of polyethylene (PE) without any surface pretreatment to increase its surface polarity [32]. As organic component, several polymers were tested after functionalization with trialkoxysilane groups in order to improve the formation of covalent bonds between organic and inorganic phases. Only polyethylene-*b*-poly ethylene glycol (PE-*b*-PEG) block copolymers led to good-quality coatings. Various hybrid coatings with different organic–inorganic ratios and nano-sized domains were prepared by reacting α-triethoxysilane-terminated PE-*b*-PEG copolymers with TEOS. It was also observed that these hybrid coatings were able to significantly improve the scratch resistance of PE, while preserving its transparency.

4.2.3 Coatings for Glass Substrates

Concerning glass substrates, taking into account that their scratch resistance is intrinsically high and that anti-corrosion properties are usually not required, the application of organic–inorganic coatings onto glass is usually carried out only for the purpose of characterization of the coatings themselves. As already mentioned, mechanical properties of coated systems are strongly dependent on the adhesion between coating and substrate, and, as a consequence, data reported for coatings applied onto glass should be transferred to other substrates (metals or plastics) with caution because of the completely different adhesive behavior.

A partial exception to this preliminary consideration is represented by the work of Duran and coworkers, which reported an investigation on the mechanical behavior of glass reinforced with silica hybrid sol–gel coatings [33]. Authors studied hybrid sols prepared from TEOS, MTES, TMSPM, and 2-hydroxyethyl methacrylate and free-radically cured.

Amerio et al. presented an interesting investigation by comparing photo-cured acrylic coatings containing silica inorganic domains introduced either by dispersing preformed silica nanoparticles or by *in situ* generation via sol–gel process [34]. Excellent scratch-resistant coatings characterized by high critical load, small cracks, and high elastic recovery were obtained by UV and sol–gel dual-curing process. On the contrary, coatings with very poor scratch resistance characterized by large plastic deformation, severe cracking, and weak elastic recovery were obtained by dispersing preformed silica nanoparticles into the acrylic resin by sonication and mechanical mixing.

Sangermano and coworkers deeply investigated the preparation and characterization of organic–inorganic hybrid coatings obtained by a dual-curing process combining the sol–gel reaction with the epoxy UV-induced polymerization technique. Photoluminescent polymeric films with improved scratch resistance were prepared by introducing photoluminescent tetracopper iodide clusters into an acrylic photo-curable formulation based on bisphenol A ethoxylate diacrylate, TEOS, and TMSPM [35].

The same authors also reported the preparation of similar methacrylated UV-cured coatings with enhanced scratch resistance properties [36]. In this paper, methacrylic-based organic–inorganic hybrid coatings containing silica, titania, or alumina domains were *in situ* generated via a sol–gel process starting from the corresponding metal alkoxide precursors (TEOS, titanium tetrapropoxide, and aluminum tripropoxide, respectively). Scratch test showed a general increase in first and second critical load values for hybrid coatings in good agreement with the obtained Persoz hardness values. The scratch resistance enhancement was found not linear with the inorganic content and almost independent of the intrinsic hardness of the different inorganic domains, but overall, the hybrid coatings always showed a better anti-scratch behavior with respect to the completely organic reference coating.

Also advanced multifunctional coatings were prepared by UV-curing of epoxy-based formulations containing hyperbranched polymers (HBPs) and an epoxy functionalized alkoxysilane additive [37]. The addition of HBP to the UV-curable epoxy resin induced a significant flexibilization of the glassy epoxy network with an increase in toughness of the cured polymeric coatings. After the addition of the functionalized alkoxysilane into the UV-curable formulations, an increment of surface hardness was obtained without strongly affecting the flexibilization and the toughness achieved by the addition of the HBP additive. The increase in surface hardness was accompanied with an increase in scratch resistance and modulus. During scratch test, the pristine epoxy coating was damaged at very low normal load (0.17 N) while the dual-cured hybrid films appeared damaged at a higher load ranging from 0.72 up to 1.09 N as a function of the initial content of the inorganic precursor. These results indicated a significant improvement on scratch resistance properties, with a good correlation between the scratch resistance and the silica content achieved by the condensation of the siloxane groups present in the inorganic precursor. Advanced scratch-resistant and tough nanocomposite epoxy coatings were obtained by properly selecting the components of the formulation.

Dual cross-linking of the inorganic and organic sites through a sol–gel and radical photo-polymerization process was also investigated by the research group coordinated by Croutxe-Barghorn [38]. They prepared hybrid coatings starting from ethoxylated bisphenol A dimethacrylate and TMSPM in the presence of two distinct photo-initiators: a diaryl iodonium salt and a hydroxyphenylketone, which ensured, respectively, the catalysis of the sol–gel reaction and the buildup of the organic network through a radical process. The simultaneous photolysis of both photosensitive molecules upon UV irradiation

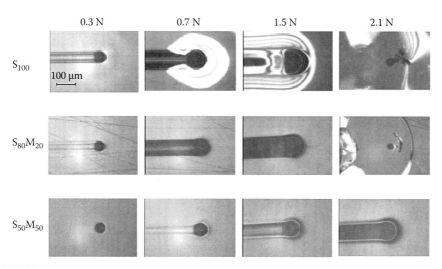

FIGURE 4.4
In situ photographs of the moving tip for specific normal loads during the scratch experiments. (Reprinted from Belon, C. et al., *Macromol. Mater. Eng.* 296, 506, 2011. With permission.)

allows the generation of a solid polymethacrylate/polysiloxane hybrid film in a single step. The hybrid precursor TMSPM acted as reinforcing agent by increasing the cross-linking density and by inducing a great improvement of the scratch resistance of the organic film without changing its stiffness. In particular, the occurrence of delamination and cracks were delayed in the presence of TMSPM with an optimal behavior found for the film prepared with a content of TMSPM of 50 wt% (S50M50 sample in Figure 4.4).

The same approach was used for the dual reactions of an acrylate and a methacrylate silane precursor 3-(trimethoxysilyl) propyl acrylate (TMSPA) and TMSPM [39]. When exposed to an increasing normal load scratch, TMSPA-based coating was found to be more resistant. Its high organic polymerization rate combined with a reasonable inorganic cross-linking density may account for this result.

Concerning the UV-cured hybrid coatings, Bautista et al. investigated the correlation between the scratch resistance and the chemical structure of organic–inorganic hybrid coatings [40]. TEOS, TMSPM, and 1,6-hexanedioldiacrylate (HDDA) were used as starting materials for obtaining UV-cured coatings. The best scratch resistance was obtained by the coating prepared with TMSPM. The inorganic domains formed by opened polyhedral structures resulted connected by the reacted organic chains of the TMSPM. This opened inorganic polyhedral decreases stiffness, without loss in mechanical properties. Partial substitution of the trialkoxysilane by HDDA decreases the mechanical properties of the coating also decreasing the scratch resistance. Partial substitution of the trialkoxysilane by TEOS increases the mechanical properties of the coating, nevertheless the brittleness is increased and the scratch resistance of the coating is decreased. This has been attributed to the higher condensation degree obtained with the presence of TEOS, which results in a more rigid chemical structure.

Moriya reported the preparation of organic–inorganic hybrid resins obtained by the copolymerization of common organic monomers such as styrene, acrylonitrile, and methyl methacrylate with a polysiloxane derivative with methacryloyloxy groups that was obtained from the reaction of polymeric tributylstannyl ester of silicic acid and (3-methacryloyloxypropyl) dimethylchlorosilane [41]. Scratch resistance of hybrids based on methyl methacrylate was improved from 3H to 7H with increasing silica content.

The scratch resistance of organic–inorganic coatings obtained from fluorinated oligomers as organic counterpart and/or from different metal alkoxides as inorganic precursors was investigated by Messori and coworkers. α,ω-Triethoxysilane terminated perfluoropolyether (PFPE) was used in combination with TEOS to prepare thermally cured coatings having the peculiar properties of PFPE (hydro- and oleo-phobicity and low friction) and of silica (hardness) [42,43]. Hybrid films were characterized by an enhanced scratch resistance that was attributed to the strong decrease in friction coefficient due to a preferential surface segregation of PFPE segments at the air–coating interface.

Organic–inorganic hybrids were prepared with silica, zirconia, or titania *in situ* generated within epoxy resins based on DGEBA and Jeffamine® by means of the aqueous sol–gel process [44]. The highest scratch resistance was shown by silica-based hybrids for which L_{C1} values were higher than 10 N, approximately one order of magnitude higher than that of unfilled epoxy. A significant decrease was observed in the case of the highly filled sample (L_{C1} 3.9 N for a silica content of 50 phr). Also titania- and zirconia-based hybrids showed a significantly increased scratch resistance with respect to pristine epoxy with highest values of L_{C1} depending on inorganic content.

The effect of coating thickness and preparation conditions on the failure of organic–inorganic hybrid coatings on glass was analyzed by Malzbender and de With [45,46]. The experiments were carried out using glass coated with thermally cured organic–inorganic hybrid materials prepared from methyltrimethoxysilane, silica nanoparticles, and a little amount of TEOS. Authors reported a detailed discussion on the failure mode of this type of coatings.

4.3 Quantitative Evaluation of Scratch Resistance

In general, the coating damage mechanisms involve elastic and plastic deformations, cracking, chipping, and delamination. Many industrial test methods and technical standards exist to evaluate coating resistance against scratch damage as described in the following text.

ASTM D3363-05(2011)e2 and ISO 15184:1998 technical standards describe the procedures for the determination of film hardness of an organic coating on a substrate in terms of drawing leads or pencil leads of known hardness. Pencil hardness measurements have been used by the coatings industry for many years to determine the hardness of clear and pigmented organic coating films. This test method has also been used to determine the cure of these coatings. This test method is especially useful in developmental work and in production control testing in a single laboratory, taking into account that the results obtained may vary between different laboratories when different pencils as well as panels are used. Even if these standards are specifically devoted to the characterization of surface hardness, sometimes they are also used to evaluate the scratch resistance of polymeric coatings.

There are still other technical standards [47–49] describing different procedures for the characterization of the hardness of polymeric coatings that can be used also for a rough evaluation of their scratch resistance. All these methods present evident deficiencies due to their dependence on the competency of the operator and oversimplification of the complex material response in polymeric coating systems.

In recent years, the use of instrumented scratch testers has become a more popular and meaningful way to address coating damage. In this concern, ASTM D7027-05 technical standards describe the use of an instrumented scratch machine to produce and quantify surface damage under controlled conditions for the evaluation of scratch resistance of polymeric coatings and plastics. Similarly, ISO 1518-1:2011 and ISO 1518-2:2011 technical standards describe laboratory procedures (constant- and variable-loading methods) using instrumented scratch machines for the determination of scratch resistance of paints and coatings.

In all cases, the scratch-inducing and data acquisition process is automated to avoid user-influenced effects that may affect the results and thus overcoming the deficiencies found in other more subjective test methods.

In general terms, a scratch tester is based on the generation of a controlled scratch with a sharp tip on a selected area. The tip (commonly diamond or hard metal) is drawn across the surface under constant, incremental, or progressive normal load, as schematically depicted in Figure 4.5.

ASTM D7027-05 presents two basic test modes. According to Test Mode A, a scratch is applied onto the specimen surface under an increasing load from 2 to 50 N, over a distance of 100 mm and at a constant rate of $0.1~\mathrm{ms^{-1}}$. This test mode is intended to determine the critical normal load at which whitening will occur for a material system. Whitening phenomenon is defined as the visible damage along the scratch groove of the surface caused by microcracking, voiding, crazing, and debonding. For materials that do not show whitening phenomena, the normal load at which a predetermined scratch width exists will be used as a basis for comparison.

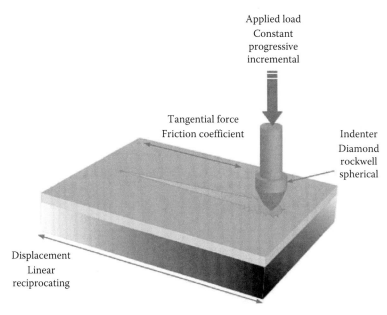

FIGURE 4.5
Schematic representation of an instrumented scratch tester. (Reprinted from Sangermano, M. and Messori, M., *Macromol. Mater. Eng.*, 295, 603, 2010. With permission.)

To compare and rank different materials, the normal load shall be plotted as a function of the scratch width.

According to Test Mode B, a scratch is applied onto the specimen surface under a constant normal load of 30 N, over a distance of 100 mm and at a constant rate of 0.1 ms^{-1}. This modality is intended to evaluate the homogeneous response of the material and establish the scratching coefficient of friction, defined as the ratio of the measured friction (tangential) force to the normal load.

Scratched surface can be visually inspected or by using evaluation tools to study the surface damage. For Test Mode A, the critical normal load is determined by the onset of the whitening of the material due to scratch. Measurement of the scratch widths, or depths, or both shall also be taken to aid the quantification of scratch resistance.

Notwithstanding the presence of these specific technical standards and their detailed experimental conditions, papers published in scientific journals are usually characterized by significant adaptations of the experimental conditions particularly concerning a shorter scratch length due to specimen size restrictions.

In general terms, it is important to underline that a comparison among published results obtained with different materials, characterization techniques, and experimental conditions is often difficult if not impossible, taking into account that experimental results from scratch testing on coatings are strongly influenced by several parameters such as coating thickness, mechanical properties of the underlying substrate, the interface bonding strength, and test conditions (i.e., the scratch speed, the applied normal load, and the type of indenter used).

4.4 Conclusions

In this chapter, the state of the art reported in literature on the scratch resistance improvements of hybrid coatings is reviewed. The main focus is on the effect of nanometric size fillers on scratch resistance improvement, describing the achievement of nanostructured organic–inorganic hybrid coatings via a bottom-up approach.

The *in situ* formation of the inorganic domain by using sol–gel process has been deeply reported. It has been shown that this way is appealing in order to obtain a good dispersion and distribution of the inorganic phase within the polymer matrix with a strong interaction between the organic and inorganic domains: this requirement is generally needed in order to limit the mechanical damage, and in particular scratching damage in such systems, and to avoid high viscous formulations to be coated.

The main parameters influencing the anti-scratch properties of organic–inorganic hybrid coatings obtained with sol–gel process are the chemical composition and in particular the ratio between organic and inorganic components, the coating thickness, the adhesion to the substrate, and also the curing conditions (thermal, UV, or MW).

List of Abbreviations

DGEBA   diglycidyl ether of bisphenol A
DPSD   diphenylsilanediol
GPTS   3-glycidoxypropyltrimethoxysilane (also referred to in the literature as GPTMS, GLYMO)
HDDA   1,6-hexanedioldiacrylate
MTES   methyltriethoxysilane
PhTMOS phenyltrimethoxysilane
TEOS   tetraethoxysilane
TMSPM   3-(trimethoxysilyl)propyl methacrylate (also referred in the literature as MPTMA, γ-MPS, MSMA, TPM, MEMO, MPTS)
VTMS   vinyltrimethoxysilane

References

1. Bauer, F., Ernst, H., Decker, U., Findeisen, M., Glasel, H.J., Langguth, H., Hartmann, E., Mehnert, R., and Peuker, C. 2000. Preparation of scratch and abrasion resistant polymeric nanocomposites by monomer grafting onto nanoparticles, 1–FTIR and multi-nuclear NMR spectroscopy to the characterization of methacryl grafting. *Macromolecular Chemistry and Physics* 201:2654–2659.
2. Glasel, H.J., Bauer, F., Ernst, H., Findeisen, M., Hartmann, E., Langguth, H., Mehnert, R., and Schubert, R. 2000. Preparation of scratch and abrasion resistant polymeric nanocomposites by monomer grafting onto nanoparticles, 2–Characterization of radiation-cured polymeric nanocomposites. *Macromolecular Chemistry and Physics* 201:2765–2770.
3. Sepeur, S., Kunze, N., Werner, B., and Schmidt, H. 1999. UV curable hard coatings on plastics. *Thin Solid Films* 351:216–219.
4. Brinker, C.J. and Shrerer, G.W. 1990. *Sol-Gel Science: The Physics and Chemistry of Sol-Gel Processing*, Boston, MA: Academic Press.
5. Wen, J.Y. and Wilkes, G.L. 1996. Organic/inorganic hybrid network materials by the sol-gel approach. *Chemistry of Materials* 8:1667–1681.
6. Bourget, L., Corriu, R.J.P., Leclercq, D., Mutin, P.H., and Vioux, A. 1998. Non-hydrolytic sol-gel routes to silica. *Journal of Non-Crystalline Solids* 242:81–91.
7. Bilecka, I. and Niederberger, M. 2010. New developments in the nonaqueous and/or non-hydrolytic sol-gel synthesis of inorganic nanoparticles. *Electrochimica Acta* 55:7717–7725.
8. Mutin, P.H. and Vioux, A. 2009. Nonhydrolytic processing of oxide-based materials: Simple routes to control homogeneity, morphology, and nanostructure. *Chemistry of Materials* 21:582–596.
9. Haas, K.H., Amberg-Schwab, S., and Rose, K. 1999. Functionalized coating materials based on inorganic-organic polymers. *Thin Solid Films* 351:198–203.
10. Haas, K.H., Amberg-Schwab, S., Rose, K., and Schottner, G. 1999. Functionalized coatings based on inorganic-organic polymers (ORMOCER (R) s) and their combination with vapor deposited inorganic thin films. *Surface and Coatings Technology* 111:72–79.
11. Amberg-Schwab, S., Katschorek, H., Weber, U., Burger, A., Hansel, R., Steinbrecher, B., and Harzer, D. 2003. Inorganic-organic polymers as migration barriers against liquid and volatile compounds. *Journal of Sol-Gel Science and Technology* 26:699–703.
12. Sangermano, M. and Messori, M. 2010. Scratch resistance enhancement of polymer coatings. *Macromolecular Materials and Engineering* 295:603–612.
13. Han, Y.H., Taylor, A., and Knowles, K.M. 2008. Characterisation of organic-inorganic hybrid coatings deposited on aluminium substrates. *Surface and Coatings Technology* 202:1859–1868.
14. Taylor, A. 2001 *Coating materials*. Patent WO & 01/25343 A1.
15. Han, Y.H., Taylor, A., Mantle, M.D., and Knowles, K.M. 2007. UV curing of organic-inorganic hybrid coating materials. *Journal of Sol-Gel Science and Technology* 43:111–123.
16. Han, Y.H., Taylor, A., and Knowles, K.M. 2009. Scratch resistance and adherence of novel organic-inorganic hybrid coatings on metallic and non-metallic substrates. *Surface and Coatings Technology* 203:2871–2877.

17. Ahmad, S., Zafar, F., Sharmin, E., Garg, N., and Kashif, M. 2012. Synthesis and characterization of corrosion protective polyurethanefattyamide/silica hybrid coating material. *Progress in Organic Coatings* 73:112–117.

18. Sharmin, E., Akram, D., Ghosal, A., Rahman, O.U., Zafar, F., and Ahmad, S. 2011. Preparation and characterization of nanostructured biohybrid. *Progress in Organic Coatings* 72:469–472.

19. Kumar, N., Jyothirmayi, A., Soma Raju, K.R.C., and Subasri, R. 2012. Effect of functional groups (methyl, phenyl) on organic-inorganic hybrid sol-gel silica coatings on surface modified SS 316. *Ceramics International* 38:6565–6572.

20. Frings, S., Meinema, H., van Nostrum, C., and van der Linde, R. 1998. Organic-inorganic hybrid coatings for coil coating application based on polyesters and tetraethoxysilane. *Progress in Organic Coatings* 33:126–130.

21. He, J.Y., Zhou, L., Soucek, M.D., Wollyung, K.M., and Wesdekmiotis, C. 2007. UV-curable hybrid coatings based on vinylfunctionalized siloxane oligomer and acrylated polyester. *Journal of Applied Polymer Science* 105:2376–2386.

22. Kim, I.Y., Nomura, K., Kikuta, K., Ohta, J., Tokuda, T., and Ohtsuki, C. 2012. Organic-inorganic hybrids for bioinert coating on implantable electronic devices. *Key Engineering Materials* 493–494:508–512.

23. Fabbri, P., Messori, M., Toselli, M., Veronesi, P., Rocha, J., and Pilati, F. 2008. Enhancing the scratch resistance of polycarbonate with poly(ethylene oxide)-silica hybrid coatings. *Advances in Polymer Technology* 27:117–126.

24. Fabbri, P., Leonelli, C., Messori, M., Pilati, F., Toselli, M., Veronesi, P., Morlat-Therias, S., Rivaton, A., and Gardette, J.L. 2008. Improvement of the surface properties of polycarbonate by organic-inorganic hybrid coatings. *Journal of Applied Polymer Science* 108:1426–1436.

25. Wu, L.Y.L., Chwa, E., Chen, Z., and Zeng, X.T. 2008. A study towards improving mechanical properties of sol-gel coatings for polycarbonate. *Thin Solid Films* 516:1056–1062.

26. Etienne, P., Phalippou, J., and Sempere, R. 1998. Mechanical properties of nanocomposite organosilicate films. *Journal of Materials Science* 33:3999–4005.

27. Lee, S., Oh, K.K., Park, S., Kim, J.S., and Kim, H. 2009. Scratch resistance and oxygen barrier properties of acrylate-based hybrid coatings on polycarbonate substrate. *Korean Journal of Chemical Engineering* 26:1550–1555.

28. Gururaj, T., Subasri, R., Raju, K.R.C.S., and Padmanabham, G. 2011. Effect of plasma pretreatment on adhesion and mechanical properties of UV-curable coatings on plastics. *Applied Surface Science* 257:4360–4364.

29. Messori, M., Toselli, M., Pilati, F., Fabbri, E., Fabbri, P., and Busoli, S. 2003. Poly(caprolactone)/silica organic-inorganic hybrids as protective coatings for poly(methyl methacrylate) substrates. *Surface Coatings International Part B-Coatings Transactions* 86:181–186.

30. Messori, M., Toselli, M., Pilati, F., Fabbri, E., Fabbri, P., Busoli, S., Pasquali, L., and Nannarone, S. 2003. Flame retarding poly(methyl methacrylate) with nanostructured organic-inorganic hybrids coatings. *Polymer* 44:4463–4470.

31. Wang, Q., Zhang, D., Wu, Z., Tian, Y., and Chen, Y. 2009. Preparation of scratch and aging resistant nanocomposite coating on PVC substrate. *Journal of Nanoscience and Nanotechnology* 9:1250–1253.

32. Toselli, M., Marini, M., Fabbri, P., Messori, M., and Pilati, F. 2007. Sol-gel derived hybrid coatings for the improvement of scratch resistance of polyethylene. *Journal of Sol-Gel Science and Technology* 43:73–83.

33. Pellice, S., Gilabert, U., Solier, C., Castro, Y., and Duran, A. 2004. Mechanical behavior of glass reinforced with SiO₂ hybrid sol-gel coatings. *Journal of Non-Crystalline Solids* 348:172–179.
34. Amerio, E., Fabbri, P., Malucelli, G., Messori, M., Sangermano, M., and Taurino, R. 2008. Scratch resistance of nano-silica reinforced acrylic coatings. *Progress in Organic Coatings* 62:129–133.
35. Roppolo, I., Messori, M., Perruchas, S., Gacoin, T., Boilot, J.P., and Sangermano, M. 2012. Multifunctional luminescent organic/inorganic hybrid films. *Macromolecular Materials and Engineering* 297:680–688.
36. Sangermano, M., Gaspari, E., Vescovo, L., and Messori, M. 2011. Enhancement of scratch-resistance properties of methacrylated UV-cured coatings. *Progress in Organic Coatings* 72:287–291.
37. Sangermano, M., Messori, M., Martin-Gallego, M., Rizza, G., and Voit, B. 2009. Scratch resistant tough nanocomposite epoxy coatings based on hyperbranched polyesters. *Polymer* 50:5647–5652.
38. Belon, C., Chemtob, A., Croutxe-Barghorn, C., Rigolet, S., Le Houerou, V., and Gauthier, C. 2011. A simple method for the reinforcement of UV-cured coatings via sol-gel photopolymerization. *Macromolecular Materials and Engineering* 296:506–516.
39. Belon, C., Chemtob, A., Croutxe-Barghorn, C., Rigolet, S., Le Houerou, V., and Gauthier, C. 2010. Combination of radical and cationic photoprocesses for the single-step synthesis of organic-inorganic hybrid films. *Journal of Polymer Science Part A-Polymer Chemistry* 48:4150–4158.
40. Bautista, Y., Gomez, M.P., Ribes, C., and Sanz, V. 2011. Relation between the scratch resistance and the chemical structure of organic-inorganic hybrid coatings. *Progress in Organic Coatings* 70:358–364.
41. Moriya, O., Sasaki, Y., Sugizaki, T., Nakamura, Y., and Endo, T. 2001. Synthesis of hybrid resins via polysiloxane with methacryloyloxy groups. *Journal of Polymer Science Part A-Polymer Chemistry* 39:1–7.
42. Fabbri, P., Messori, M., Montecchi, M., Nannarone, S., Pasquali, L., Pilati, F., Tonelli, C., and Toselli, M. 2006. Perfluoropolyether-based organic-inorganic hybrid coatings. *Polymer* 47:1055–1062.
43. Fabbri, P., Messori, M., Montecchi, M., Pilati, F., Taurino, R., Tonelli, C., and Toselli, M. 2006. Surface properties of fluorinated hybrid coatings. *Journal of Applied Polymer Science* 102:1483–1488.
44. Bondioli, F., Darecchio, M.E., Luyt, A.S., and Messori, M. 2011. Epoxy resin modified with *in situ* generated metal oxides by means of sol-gel process. *Journal of Applied Polymer Science* 122:1792–1799.
45. Malzbender, J. and de With, G. 2001. Analysis of scratch testing of organic-inorganic coatings on glass. *Thin Solid Films* 386:68–78.
46. Malzbender, J. and de With, G. 2001. Scratch testing of hybrid coatings on float glass. *Surface and Coatings Technology* 135:202–207.
47. ASTM, *ASTM D2134–93(2007) Standard Test Method for Determining the Hardness of Organic Coatings with a Sward-Type Hardness Rocker*, 2007.
48. ASTM, *ASTM D1474–98(2008) Standard Test Methods for Indentation Hardness of Organic Coatings*, 2008.
49. ISO, *ISO 1522:2006, Paints and Varnishes–Pendulum Damping Test*, 2006.

5

Chitosan–Magnesium Aluminum Silicate Nanocomposite Coatings

Thaned Pongjanyakul

CONTENTS

5.1 Introduction

Film coatings in pharmaceutical dosage forms mainly consist of polymers that are applied to the surface of cores, such as tablets and pellets, in the form of solutions or dispersions. Droplets of the polymer solutions spread, coalesce, and adhere on the core surface. After the solvent evaporation in drying process, a uniform thin film polymers is formed (Mehta, 1997). The purposes of film coating of the tablets are protection of active pharmaceutical ingredients from environmental conditions, avoidance of local side effects in gastrointestinal (GI) tract, and controlled or sustained drug release in GI tract. Moreover, the coated films can enhance the mechanical strength of tablets during production, packaging, and transportation (Bauer et al., 1998).

Polymers used for coating material are dissolved or dispersed in water-based and solvent-based systems. The organic solvents used have been considered undesirable for this purpose due to the risk of explosion during coating process, environmental pollutions, and residue in the coated films (Nagai et al., 1997). Nowadays, water-based polymers have been developed and they are widely used in tablet coating process. Synthetic cellulose derivatives, such as hydroxypropyl methylcellulose (Cao et al., 2004; Sangalli et al., 2004), and natural polymers, such as sodium alginate (Pongjanyakul et al., 2005) and chitosan (CS) (Nunthanid et al., 2002), have been previously used as coating materials for tablets. In the recent years, nanocomposite materials composed of polymer and clay have been developed and characterized for improving the physical properties of polymeric films, such as mechanical properties and water vapor permeability (WVP). CS is one of the polysaccharides that is prepared and characterized in the form of nanocomposites with clays for use as biosensors (Fan et al., 2007; Zhao et al., 2008), packaging materials (Rhim et al., 2006), superabsorbent materials (Ruiz-Hitzky et al., 2005), and drug carrier systems (Wang et al., 2007).

In this chapter, the basic knowledge about molecular interaction between CS and magnesium aluminum silicate (MAS), one type of clay, in the form of dispersions and films is summarized. The characteristics of the CS–MAS nanocomposite films and the application of these films as tablet coating material in the pharmaceutical field are also discussed.

5.2 Chitosan and Magnesium Aluminum Silicate

Chitosan (CS) is a polysaccharide that consists of N-acetyl-D glucosamine and D-glucosamine (Figure 5.1a). It is obtained from chitin, that is, reacted via deacetylation in alkaline condition. CS is insoluble at neutral and alkaline pHs since its pK_a is in the range of 6.2–7.0 (Hejazi and Amiji, 2003). It dissolves and swells in acidic media due to ionization of the amino groups of the CS molecules. CS has been extensively used in many fields, for example, agriculture, water and waste treatment, food and beverages, cosmetics, and pharmaceutics (Rinaudo, 2006), since it possesses biodegradability, biocompatibility, and nontoxicity (Illum, 1998). Due to the high solubility in acidic media, CS films can rapidly swell and dissolve in gastric condition. Thus, the tablets coated with CS films cannot sustain drug release in stomach. For this reason, CS is blended with other substances in order to enhance the acid stability of films. It is found that CS can form polyelectrolyte complexes with anionic polymers, such as pectin (Fernández-Hervás and Fell, 1998; Macleod et al., 1999) and polyalkylenoxide–maleic acid copolymer (Yoshizawa et al., 2005), via electrostatic interaction. This interaction leads to retardation of acid swelling and an improvement of film stability in gastric fluid.

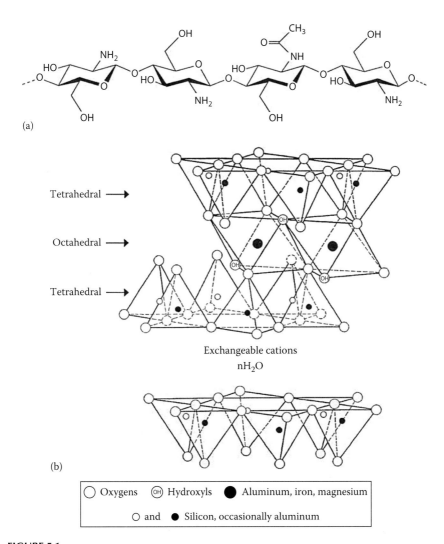

FIGURE 5.1
Molecular structure of (a) CS and (b) montmorillonite clays.

MAS is a mixture of montmorillonite and saponite clays. It consists of three-lattice layers, a central octahedral sheet of aluminum or magnesium, and two external silica tetrahedron layers (Alexandre and Dubois, 2000), as shown in Figure 5.1b. It has been washed with water to optimize purity and performance and is employed as a pharmaceutical excipient due to its non-toxicity and nonirritation at levels used in drug formulations (Kibbe, 2000). The silicate layer surface of clay has negatively charged faces, but weak positive charges are presented on the edges of silicate layers. The silicate layers of clay can be separated and can form 3-D structures when they are hydrated in

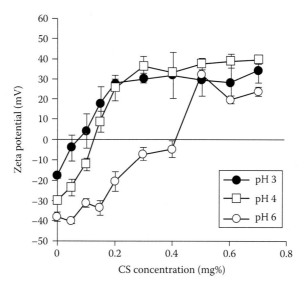

FIGURE 5.2
Effect of CS on the zeta potential of MAS in acetate buffer solution at different pHs.

water. Due to the negative charges of MAS, it can electrostatically interact with a positively charged amine drug, such as nicotine (Suksri and Pongjanyakul, 2008; Pongjanyakul et al., 2009) and propranolol hydrochloride (Rojtanatanya and Pongjanyakul, 2010). Furthermore, MAS also interacts with anionic polymers, such as sodium alginate (Pongjanyakul and Puttipipatkhachorn, 2007) and xanthan gum (Ciullo, 1981), via hydrogen bond formation.

Because of an opposite charge of MAS and CS, strong molecular interaction via electrostatic force occurs when mixing them in the dispersion form. This interaction is dependent upon the pH of the medium due to the different ionization levels of CS and MAS at different pHs (Figure 5.2). The lower the pH of the medium, the smaller the zeta potential of MAS because hydronium ions enriched in low-pH medium can adsorb on the surface of the silicate layers, resulting in a decrease in the zeta potential of MAS. In contrast, the lower pH medium causes higher ionization of the amino groups of CS, leading to a higher zeta potential of CS. Therefore, CS in the lower pH can rapidly neutralize the zeta potential of MAS to zero when compared to those at higher pHs (Khunawattanakul et al., 2008). Furthermore, when MAS and CS in the form of dispersions are mixed at acidic pH medium, CS that can reduce the zeta potential of MAS induces a flocculation of MAS particles by bridging the mechanism of long-chain molecules of CS (Roussy et al., 2005; Günister et al., 2007). The size–frequency distributions of the CS–MAS flocculates provide greater particle size and polydispersity index (Figure 5.3a). The amount of MAS added does not strongly affect the flocculate size. Moreover, a narrower size distribution and smaller size

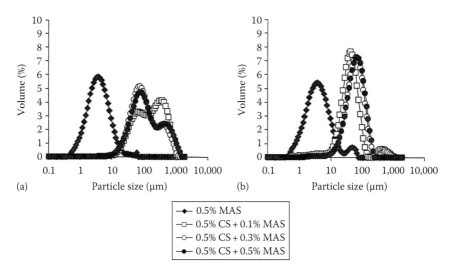

FIGURE 5.3
Size-frequency distribution of CS-MAS flocculates in composite dispersions (a) without and (b) with heat treatment at 60°C for 48 h.

of the CS–MAS flocculates are found after heating at 60°C (Figure 5.3b). Additionally, incorporation of MAS into CS dispersions causes an increase in viscosity and a shift of CS flow type from Newtonian to pseudoplastic flow with thixotropic properties. Heat treatment brings about a significant decrease in viscosity and thixotropic properties of the composite dispersions (Khunawattanakul et al., 2008).

The interaction of polymer and clay in the dispersion form can predict formation of composite types, such as micro- or nanocomposites. The flocculation of MAS particles after mixing CS is a unique characteristic to form the exfoliated or intercalated nanocomposite films (Khunawattanakul et al., 2010). On the other hand, sodium alginate can interact with MAS, but cannot induce a large size of MAS flocculates (Pongjanyakul and Puttipipatkhachorn, 2009), resulting in a phase-separated microcomposite film (Pongjanyakul et al., 2005; Pongjanyakul, 2009). Therefore, the characteristics of polymer-clay composite dispersion, especially zeta potential and particle size, are very important for the nanocomposite formation in films.

5.3 Nanocomposite Formation

CS–MAS nanocomposite films can be prepared using casting/solvent evaporation method. The solid-state characterization of the CS–MAS films is investigated using FTIR spectroscopy, ^{29}Si NMR spectroscopy, and powder

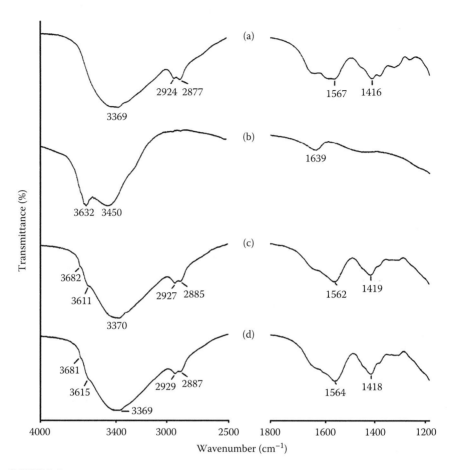

FIGURE 5.4
FTIR spectra of (a) CS films, (b) MAS powder and CS-MAS (1:1) films prepared using (c) non-heated and (d) heated dispersions.

x-ray diffractometry for describing molecular interaction and nanocomposite formation. FTIR spectra of the CS film shows important peaks at around 2877–2922 cm^{-1} for a CH stretching peak, 1567 cm^{-1} for a NH$_2$ bending (amide II) peak of primary amine, and 1416 cm^{-1} for a CH$_2$ bending peak (Figure 5.4a). A lower wavenumber shift of the NH$_2$ bending peak of the CS films is found when adding MAS (Figure 5.4c), indicating the electrostatic interaction between the negatively charged MAS and the protonated amine groups of CS. A shift of the CH stretching and the CH$_2$ bending peaks of CS, and an OH stretching peak of SiOH groups of MAS at 3632 cm^{-1} (Figure 5.4b) moved to 3611–3615 cm^{-1} in the CS–MAS films (Figure 5.4c and d), it suggests a rearrangement of hydrogen bonds of the primary hydroxyl groups of CS (Harish Prashanth et al., 2002; Kasaai, 2008). This change displays intermolecular hydrogen bonding between the silanol groups on the surface of the

MAS silicate layers and CS. Interestingly, the CS–MAS films prepared using both heated and nonheated dispersions presented a new peak at around 3682 cm^{-1} (Figure 5.4c and d). This absorption peak represents the stretching of free SiOH groups (Patel et al., 2007), when the thickness of the silicate layers increases and it cannot be observed in the spectra of MAS powder.

A change of negative charges of MAS when interacting with CS can be specifically characterized using the solid-state ^{29}Si NMR spectroscopy. The ^{29}Si NMR spectra gives evidence to electronic changes in the tetrahedral sheet of MAS. The slightly negative change in the ^{29}Si chemical shift indicates a decrease in the charge of the montmorillonite silicate layers (Gates et al., 2000). The ^{29}Si chemical shift of MAS is located at −94.07 ppm (Figure 5.5a). The signal of this chemical shift shows stronger intensities when increasing MAS content in the films, and a slightly negative change of this chemical shift in the CS–MAS films is found (Figure 5.5b through d). These changes suggest that the negative charge on the MAS silicate layers electrostatically interact with the protonated amine groups of CS. Moreover, an increase in the negative change of this chemical shift of the CS–MAS films prepared using a heated dispersion (Figure 5.5e) is found compared with those prepared using nonheated dispersions.

The nanocomposite formation between CS and MAS in the films is investigated using PXRD. MAS powder shows the basal spacing peak at 7.0° 2θ (Figure 5.6a), indicating a thickness of the MAS silicate layer of 1.26 nm. The CS–MAS (1:0.2) films prepared using nonheated and heated dispersions have no obvious basal spacing peak of MAS (Figure 5.6c, left and right panels), suggesting that the MAS silicate layers are completely separated in the CS film matrix, and an exfoliated nanocomposite film is formed (Alexandre and Dubois, 2000). This type of nanocomposite is formed because a low content of clay is dispersed in the films (Ray and Okamoto, 2003). The basal spacing peak of MAS clearly displays at 5.3° and 3.5° 2θ for the CS–MAS films at the ratios of 1:0.6 and 1:1 prepared by using non-heated dispersions (Figure 5.6d and e, left panel), respectively, resulting in an increased thickness of the MAS silicate layers of 1.67 and 2.52 nm, respectively. This indicates the formation of intercalated nanocomposites in which CS can intercalate into the MAS silicate layers. The CS–MAS (1:0.6) films prepared using a heated dispersion are intercalated nanocomposites as well (Figure 5.6d, right panel). Additionally, multiple layers of CS can intercalate into the MAS silicate layers as evidenced by the peaks at 2.4° 2θ of the films prepared at the ratio of 1:1 using heated dispersion (Figure 5.6e, right panel), leading to the highest thickness of the MAS silicate layers in this study (3.68 nm). For the nanocomposite formation between CS and MAS, the schematic presentation of the CS–MAS intercalated nanocomposites is presented in Figure 5.7. Furthermore, it is not necessary to use the heat treatment in composite dispersions before film casting, but the heat treatment can induce multilayers of CS intercalated into the MAS silicate layers when using the 1:1 ratio of CS and MAS.

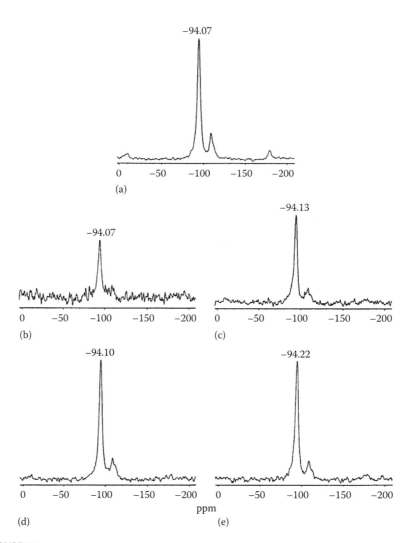

FIGURE 5.5
Solid-state ^{29}Si NMR spectra of (a) MAS powder and CS-MAS films in the ratios of (b) 1:0.2, (c) 1:0.6, and (d) 1:1 prepared using nonheated dispersions, and (e) CS-MAS (1:1) films prepared using a heated dispersion.

5.4 Characteristics of CS–MAS Nanocomposite Films

5.4.1 Appearance and Morphology

CS films obtained are transparent and yellowish, whereas the incorporation of MAS results in an opaque appearance of the CS–MAS films. The CS films

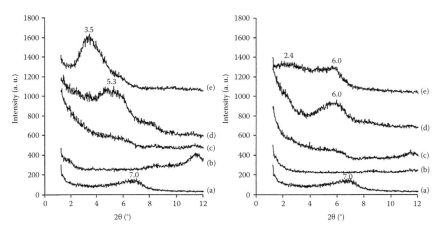

FIGURE 5.6
PXRD pattern of (a) MAS powder, (b) CS film and CS-MAS films in the ratios of (c) 1:0.2, (d) 1:0.6, and (e) 1:1. The films were prepared using nonheated dispersions (left panel) and heated dispersion (right panel).

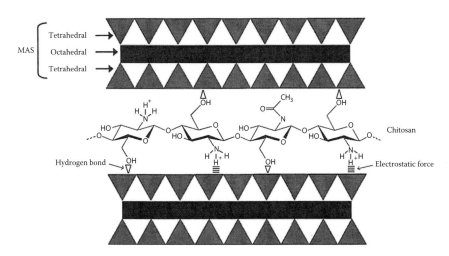

FIGURE 5.7
Schematic representation of CS-MAS intercalated nanocomposites.

show smooth surface and dense matrix morphology when they are viewed using SEM (Figure 5.8a). On the other hand, increasingly rougher surfaces are observed when the MAS content in the films increases. The matrix morphology of the CS–MAS films has a layer structure (Figure 5.8b). The CS–MAS (1:1) film prepared using a nonheated dispersion showed a similar surface morphology and matrix structure as that prepared using a heated dispersion (Khunawattanakul et al., 2010).

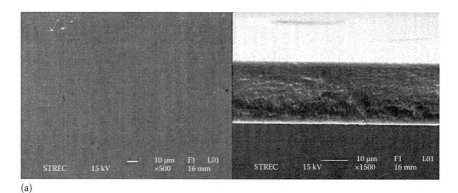

(a)

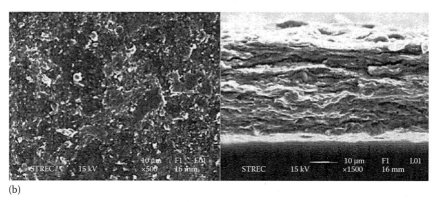

(b)

FIGURE 5.8
Surface and matrix morphology of (a) CS film and (b) CS-MAS film at the ratios of 1:1 prepared using nonheated dispersions.

5.4.2 Mechanical Properties

The tensile strength of the CS films prepared using nonheated dispersions obviously decreases by adding a small amount of MAS (in the ratio of 1:0.2), whereas the tensile strengths of the CS–MAS films in the higher ratios of MAS present a greater tensile strength than the CS films (Figure 5.9). The % elongation of all CS–MAS films is higher than that of the CS films (Khunawattanakul et al., 2010). The exfoliated nanocomposite structure of the CS–MAS (1:0.2) films decreases the tensile strength of the CS film because the MAS interacts with CS and disperses completely in the CS matrix, interrupting the intermolecular hydrogen bonding of CS. However, increasing MAS content in the CS–MAS films provides greater tensile strength than the CS films, indicating that the intercalation of CS into the MAS silicate layers can form a strong matrix structure in films. Moreover, the exfoliated and intercalated nanocomposites of the CS–MAS films in the ratios of 1:0.2 and 1:0.6 have obviously higher % elongation than the CS films (Figure 5.9). The increased flexibility of these films may be due to the formation of dangling

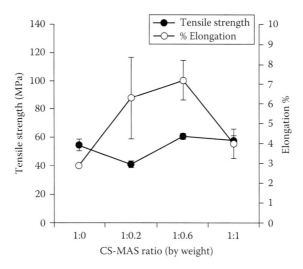

FIGURE 5.9
Tensile strength and percentage of elongation of CS and CS-MAS films at various ratios of CS and MAS prepared using nonheated dispersions.

chains and conformational effects at the clay–matrix interface (Alexandre and Dubois, 2000). However, the highest content of MAS in the films causes a reduction in the % elongation of the nanocomposite films. On the other hand, the CS–MAS (1:1) film prepared using heated dispersions provides the highest % elongation when comparing to those prepared using the same method. This suggests that the formation of CS multilayers intercalated into the MAS silicate layer can enhance the flexibility of the films. Furthermore, the layer matrix structure of the CS–MAS nanocomposite films observed using SEM can generally improve not only the flexibility, but also the strength of the films. The heat treatment can enhance the tensile strength, but it reduces the % elongation of the CS–MAS films. This is likely due to the depolymerization of CS chains during the heat treatment (Holme et al., 2008), while shorter CS chains can form a denser matrix film. This leads to high strength in the films but also causes a decrease in the flexibility of films.

5.4.3 Water Uptake Properties

Incorporation of MAS into the CS films decreases water uptake of the films in deionized water, except for the CS–MAS (1:0.2) films that show the highest water uptake. This suggests that the interaction of CS with MAS to create the intercalated nanocomposite possesses a denser matrix structure with smaller water-filled channels. However, the exfoliated nanocomposite of the CS–MAS films can highly absorb water when compared with the intercalated nanocomposites. The completely dispersed MAS in the CS matrix provides larger water-filled channels in the film, resulting in higher water uptake efficiency. Additionally, the use of pH 6.8 phosphate buffer gives an obviously lower water uptake of the films than the use of deionized water because CS can be cross-linked by phosphate anions (Nunthanid et al., 2001). This cross-linking process results in denser matrix structures and restricted water penetration into the films.

5.4.4 Water Vapor and Drug Permeabilities

The exfoliated CS–MAS nanocomposite films have slightly higher WVP than the CS films and the CS–MAS films with higher content of MAS (intercalated nanocomposite). This indicates that the exfoliated nanocomposite has a looser matrix structure, and water vapor can easily permeate through the films. The increase in MAS ratio in the CS–MAS films does not lead to different WVP coefficients when compared with the CS films.

The drug permeability study is very important for the tablet coating films that are used for modifying drug release from the coated tablets. The propranolol HCl (PPN) permeation profiles in pH 6.8 phosphate buffer are presented in Figure 5.10. Increasing MAS ratio in the nanocomposite films causes a lower PPN permeation flux, but a longer lag time is found (Khunawattanakul et al., 2010). This is due to the high affinity of PPN, a positively charged drug with MAS (Rojtanatanya and Pongjanyakul, 2010). Moreover, the dense matrix structure of the films occurs when CS is interacted with MAS silicate layers and is cross-linked with phosphate ions in the medium, resulting in lower PPN permeability.

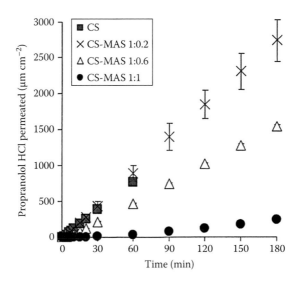

FIGURE 5.10
Permeation profiles of propranolol HCl across CS films and CS-MAS films at different ratios in pH 6.8 phosphate buffer.

The permeation mechanism across the CS–MAS films of the positively charged drug is different from nonelectrolyte and negatively charged drugs. For positively charged drugs, the adsorption (partition) of the drug onto the MAS particles in the CS–MAS films is the first process, followed by a drug diffusion in the CS matrix across the films (Pongjanyakul, 2009). In contrast, diffusion through pores is used to explain the permeation of the nonelectrolyte and the negatively charged drugs. The drug is presumed to diffuse through microchannels within the structure of the hydrated CS–MAS films (Thacharodi and Panduranga Rao, 1993).

5.5 Application of Tablet Coatings

5.5.1 Appearance of Coated Tablets

Tablets coated using CS and CS–MAS nanocomposite films at various ratios have different visual appearances as presented in Figure 5.11. The CS coated tablets show a yellow and shrunken film, whereas the CS–MAS coated tablets have a brown and opaque film. Moreover, the CS–MAS coated tablets have lesser film defects than the CS coated tablets. The microscopic surface and

(a) (b)

(c) (d)

FIGURE 5.11
(See color insert.) Appearance of PPN-coated tablets with (a) CS film and CS-MAS films in the ratios of (b) 1:0.2, (c) 1:0.6, and (d) 1:1 at 4.3 mg cm^{-2} coating level.

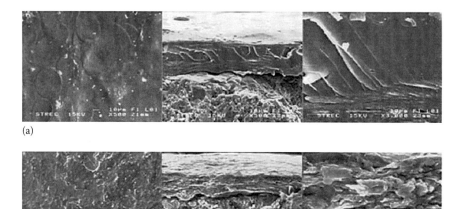

(a)

(b)

FIGURE 5.12
SEM photographs of surface and film matrix morphology of PPN-coated tablets with (a) CS film and (b) CS-MAS (1:1) films at 4.3 mg cm^{-2} coating level.

matrix morphology of the coated films investigated using SEM are shown in Figure 5.12. The CS coated tablets display a smooth surface and homogeneous film matrix, suggesting that the tablets can be coated with CS, and visually observed film defect is called picking (Figure 5.11a) (Rowe, 1997). Picking can occur when the film pulls away from the surface of core tablets in the initial period of the film coating process. In contrast, the CS–MAS coated films show rougher surfaces than the CS coated films, especially when using a higher ratio of MAS (Khunawattanakul et al., 2011). The CS–MAS coated tablets display fewer film defects than the CS coated tablets because the stickiness of the coated tablets is reduced during the coating process when incorporating MAS into the CS dispersions. The matrix morphology of the CS–MAS films show a layer structure that is similar to free CS–MAS films and it is prepared using the casting/solvent evaporation method. The rough surface and layer structure of the films occur because of the formation of CS–MAS flocculate particles in the dispersion (Khunawattanakul et al., 2010).

5.5.2 Effect of CS–MAS Ratio

The water uptake process of the coated tablets immediately occurs after the coated tablets are exposed to the medium. A surrounding medium penetrates across the coated film into the core tablets, and the drug particles inside the tablets are dissolved. The drug concentration in the coated tablets creates a concentration gradient for driving drug diffusion through the coated films, and after that, drug molecules are released out of the coated tablets.

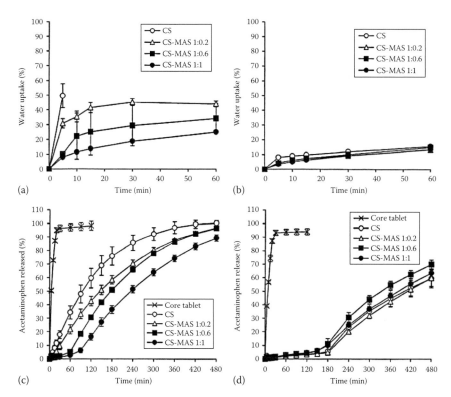

FIGURE 5.13
Effect of CS-MAS ratio on (a, b) water uptake and (c, d) drug release of ACT-coated tablets at 4.3 mg cm^{-2} coating level in (a, c) 0.1 M HCl and (b, d) pH 6.8 phosphate buffer.

Thus, water uptake determination of the coated tablets is essential for describing drug release from the coated tablets. The water uptake of acetaminophen (ACT) coated tablets in 0.1 M HCl and pH 6.8 phosphate buffer is shown in Figure 5.13. In 0.1 M HCl, water uptake of the CS coated tablets rapidly increases within the first 5 min of the test (Figure 5.13a). CS with a pK$_a$ of 6.2–7.0 can ionize and swell in acidic medium, leading to a loose CS film matrix and allowing water molecules to penetrate into the film. However, water uptake of the CS coated tablets cannot be determined after that because swelling and erosion of the CS films cause a rupture of swollen tablets upon handling. On the other hand, tablets coated with the CS–MAS films at all ratios can be handled during the test, suggesting that the CS–MAS nanocomposite formation can retard the swelling and erosion of CS in the composite films in acidic medium. The greater the MAS ratio added into the CS films, the lower the water uptake of the coated tablets (Khunawattanakul et al., 2011). Water uptake of the CS and CS–MAS coated tablets obviously decreases in pH 6.8 phosphate buffer (Figure 5.13b) when compared with acidic medium. This indicates that CS is cross-linked with phosphate anions bringing about an insoluble and stable film (Nunthanid et al., 2001) and lead-ing to a restriction of water penetration into the coated tablets.

Core tablets of ACT show an immediate release of the drugs in both 0.1 M HCl and pH 6.8 phosphate buffer (Figure 5.13c and d). In 0.1 M HCl, the CS coated tablets provide fast release of drug without lag time (Figure 5.13c) because of higher water uptake and swelling of the CS film in acidic condi-tions, leading to greater drug permeability of the films. On the other hand, increasing the MAS ratio causes longer lag times and lower drug release rate of the CS–MAS coated ACT tablets in 0.1 M HCl (Khunawattanakul et al., 2011). This suggests that the ACT tablets coated with CS–MAS (1:0.2) films contain enough water to dissolve the drug particles and build a high drug concentration gradient, resulting in a higher release rate and shorter lag time when compared with ACT tablets coated with CS–MAS films at higher MAS ratios. The coated films with higher MAS ratio can restrict water penetration into the coated tablets, leading to slower dissolution of ACT particles. The CS–MAS coated ACT tablets have an obviously longer lag time and lower drug release rate in pH 6.8 phosphate buffer when compared with the drug release of the same tablets in acidic medium. Increasing the MAS ratio in the films do not affect the release characteristics of ACT in pH 6.8 phosphate buffer. The release characteristics of the CS and CS–MAS coated tablets in pH 6.8 phosphate buffer is different in 0.1 M HCl because the cross-linking of CS in the films with phosphate anions bring about a stable film, restricting water uptake of the coated tablets. This leads to a similar drug release in the CS coated and CS–MAS coated tablets. For the surface and matrix morphol-ogy of the coated films after exposing with acidic medium, the appearance of MAS particles on the film's surface and the expansion of the layer structure of the film matrix are observed (Figure 5.14a). This suggests that erosion and swelling of the films clearly occur in acidic medium, which is different from using pH 6.8 phosphate buffer (Figure 5.14b)

In the case of PPN, a high water soluble drug, the water uptake of these coated tablets is greater than the ACT coated tablets and it decreases with increasing MAS ratio. However, the drug release of the PPN coated tablets is similar when varying CS–MAS ratios (Khunawattanakul et al., 2011). This similarity is due to the fast dissolution of PPN particles in the core tablets and a very high drug concentration gradient for drug diffusion.

5.5.3 Effect of Coating Level

In 0.1 M HCl, water uptake of the CS–MAS coated tablets increases rapidly within 5 min and increases gradually thereafter. The higher the coating level, the greater the water uptake of the coated tablets is found (Figure 5.15a). In contrast, the water uptake at different coating levels is similar in pH 6.8 phosphate buffer (Figure 5.15b). CS can swell in acidic medium although its chains form nanocomposites with MAS. Swelling of CS loos-ens the film matrix structure and increases the matrix volume for adsorbing

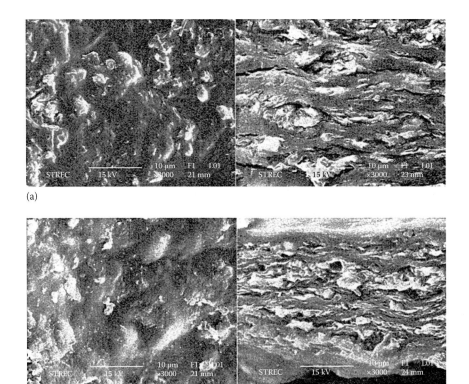

FIGURE 5.14
Surface and matrix morphology of the CS-MAS (1:1) films coated on the tablets after drug release testing for 1 h in (a) 0.1 M HCl and (b) pH 6.8 phosphate buffer.

water, leading to higher water uptake when increasing the coating level. On the other hand, CS does not swell in pH 6.8 phosphate buffer, resulting in no difference of water uptake of the coated films at all coating levels.

The drug release profiles of the PPN tablets coated with CS–MAS (1:1) films at different coating levels are shown in Figure 5.15c and d. An increase in coating level shows obviously longer lag time and lower drug release rate of the CS–MAS coated PPN tablets, especially in acidic medium (Khunawattanakul et al., 2011). This results from the lower water uptake into the tablet cores and longer diffusional path length for drug permeation.

5.5.4 Drug Solubility

The solubility of drug influences drug release characteristics from the CS–MAS coated tablets. The drug release profiles of the CS–MAS coated tablets containing PPN, ACT, and diclofenac sodium (DCF) in pH 6.8 phosphate buffer is presented in Figure 5.16. The PPN coated tablets show the fastest

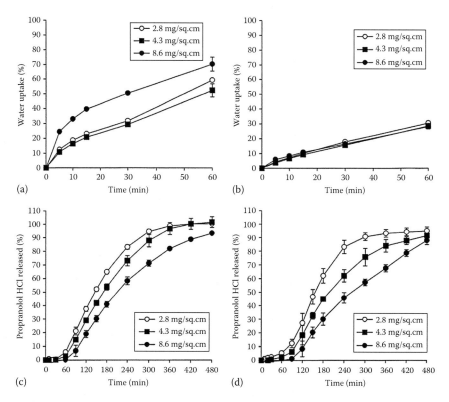

FIGURE 5.15
Effect of coating level on (a, b) water uptake and (c, d) drug release of PPN tablets coated with CS-MAS (1:1) films in (a, c) 0.1 M HCl and (b, d) pH 6.8 phosphate buffer.

drug release rate and the shortest lag time, whereas a very slow drug release and the longest lag time are obtained from the DCF-coated tablets. The reason for this is the difference of drug solubility in pH 6.8 phosphate buffer. The solubility of PPN, ACT, and DCF is 219.1, 20.4, and 2.8 mg/mL respectively (Khunawattanakul, 2011). The high solubility of PPN causes a fast dissolution of drug particles, and the higher concentration of PPN inside the tablet leads to increased water penetration into the tablet cores because of an osmotic pressure difference across the coated films. On the other hand, the lowsolubility drug cannot induce water penetration into the tablet cores, leading to slow drug dissolution and drug release.

5.5.5 Effect of Enzymes in GI Tract

The effects of pepsin and pancreatin on the release of PPN from the CS and CS–MAS (1:1) coated tablets are presented in Figure 5.17. PPN release from the CS and CS–MAS coated tablets is similar in simulated gastric fluid (SGF) with and without pepsin (Figure 5.17a). Pepsin has a high chitosanolytic activity at

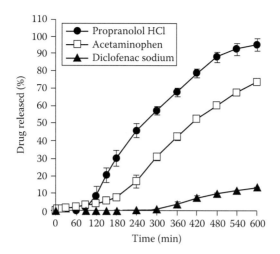

FIGURE 5.16
Drug release profiles of tablets coated with CS-MAS (1:1) films at 8.6 mg cm^{-2} coating level in pH 6.8 phosphate buffer.

pH 5, which is different from the conditions in stomach (pH of 1.2 and 37°C), leading to no effect of pepsin on the CS and CS–MAS film coated tablets. In contrast, the CS coated tablets in simulated intestinal fluid (SIF) with pancreatin provide obviously faster PPN release than those in the medium without pancreatin (Figure 5.17b), whereas the CS–MAS coated tablets show no difference of PPN release in SIF with and without pancreatin (Khunawattanakul et al., 2011). This result suggests that pancreatin in SIF affects the release of PPN from the CS film coated on the tablets because degradation of CS in SIF via depolymerization occurs from lipase in pancreatin (Zhang et al., 2002).

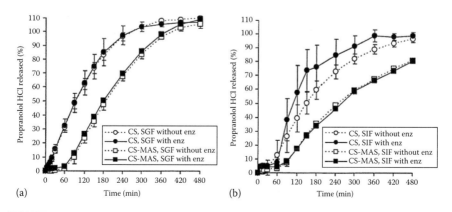

FIGURE 5.17
PPN release profile of CS and CS-MAS (1:1) coated tablets in (a) SGF and (b) SIF with or without enzyme.

Moreover, pancreatin that possesses a proteolytic activity can digest CS (McConnell et al., 2008). The drug release of the CS–MAS coated tablets is not affected by pancreatic lipase because the CS–MAS nanocomposite formation can reinforce the films' structure to form a denser matrix when compared with the CS films. Moreover, the penetration of pancreatic lipase into the CS–MAS nanocomposite film is limited because of a large molecule with low diffusivity of pancreatic lipase. Additionally, the molecule size of pancreatic lipase is too large for attacking the CS molecules intercalated in the silicate layer of MAS.

5.6 Conclusion

CS and MAS can electrostatically interact to form flocculation in the dispersion form, leading to exfoliated or intercalated nanocomposite formation in the films, which is dependent upon the MAS ratio added. This formation is not necessary to induce by using heat treatment. The CS–MAS nanocomposite films present better mechanical property, especially flexibility, and lower drug permeability than the CS films. For tablet film coating, the CS–MAS coated films have a rough surface and a layered matrix film, whereas a smooth surface and dense matrix film are found on CS coated films. However, the CS–MAS coated tablets provide fewer film defects, and lower swelling and erosion in acidic medium than the CS coated tablets. Moreover, the CS–MAS films on the tablets present good stability towards enzymatic degradation in SIF. The drug release pattern from the CS–MAS coated tablets can be modified by varying CS–MAS ratios and coating levels. Furthermore, drug solubility also influences drug release characteristics of the CS–MAS coated tablets. In conclusion, the CS–MAS nanocomposites demonstrate a strong potential for use in tablet film coating intended for modulating drug release from tablets.

References

Alexandre, M., Dubois, P., 2000. Polymer-layered silicate nanocomposites: Preparation, properties and uses of a new class of materials. *Mater. Sci. Eng.* 28, 1–63.

Bauer, K.H., Lehmann, K., Osterwald, H.P., Rothgang, G., 1998. *Coated Pharmaceutical Dosage Forms.* Medpharm GmbH Scientific Publishers, Stuttgart, Germany.

Cao, Q.R., Choi, H.G., Kim, D.C., Lee, B.J., 2004. Release behavior and photo-image of nifedipine tablet coated with high viscosity grade hydroxypropylmethylcellulose: Effect of coating conditions. *Int. J. Pharm.* 274, 107–117.

Ciullo, P.A., 1981. Rheological properties of magnesium aluminum silicate/xanthan gum dispersions. *J. Soc. Cosmet. Chem.* 32, 275–285.

Fan, Q., Shan, D., Xue, H., He, Y., Cosnier, S., 2007. Amperometric phenol biosensor based on laponite clay-chitosan nanocomposite matrix. *Biosens. Bioelectron.* 22, 816–821.

Fernández-Hervás, M.J., Fell, J.T., 1998. Pectin/chitosan mixtures as coatings for colon-specific drug delivery: An *in vitro* evaluation. *Int. J. Pharm.* 169, 115–119.

Gates, W.P., Komadel, P., Madejová, J., Bujdák, J., Stucki, J.W., Kirkpatrick, R.J., 2000. Electronic and structural properties of reduced-charge montmorillonite. *Appl. Clay Sci.* 16, 257–271.

Günister, E., Pestreli, D., Ünlü, C.H., Atıcı, O., Güngör, N., 2007. Synthesis and characterization of chitosan-MMT biocomposite systems. *Carbohydr. Polym.* 67, 358–365.

Harish Prashanth, K.V., Kittur, F.S., Tharanathan, R.N., 2002. Solid state structure of chitosan prepared under different N-deacetylating conditions. *Carbohydr. Polym.* 50, 27–33.

Hejazi, A., Amiji, M., 2003. Chitosan-based gastrointestinal delivery systems. *J. Control. Release* 89, 151–165.

Holme, H.K., Davidsen, L., Kristiansen, A., Smidsrød, O., 2008. Kinetics and mechanism of depolymerization of alginate and chitosan in aqueous solution. *Carbohydr. Polym.* 73, 656–664.

Illum, L., 1998. Chitosan and its use as a pharmaceutical excipient. *Pharm. Res.* 15, 1326–1331.

Kasaai, M.R., 2008. A review of several reported procedures to determine the degree of N-acetylation for chitin and chitosan using infrared spectroscopy. *Carbohydr. Polym.* 71, 497–508.

Khunawattanakul, W., 2011. Development and evaluations of chitosan-magnesium aluminum silicate nanocomposite films for pharmaceutical uses. PhD dissertation, Khon Kaen University, Khon Kaen, Thailand.

Khunawattanakul, W., Puttipipatkhachorn, S., Rades, T., Pongjanyakul, T., 2008. Chitosan-magnesium aluminum silicate composite dispersions: Characterization of rheology, flocculate size and zeta potential. *Int. J. Pharm.* 351, 227–235.

Khunawattanakul, W., Puttipipatkhachorn, S., Rades, T., Pongjanyakul, T., 2010. Chitosan-magnesium aluminum silicate nanocomposite films: Physicochemical characterization and drug permeability. *Int. J. Pharm.* 393, 220–230.

Khunawattanakul, W., Puttipipatkhachorn, S., Rades, T., Pongjanyakul, T., 2011. Novel chitosan-magnesium aluminum silicate nanocomposite film coatings for modified-release tablets. *Int. J. Pharm.* 407, 132–141.

Kibbe, H.A., 2000. *Handbook of Pharmaceutical Excipients*, 3rd edn. American Pharmaceutical Association, Washington, DC, pp. 295–298.

Macleod, G.S., Collett, J.H., Fell, J.T., 1999. The potential use of mixed films of pectin, chitosan and HPMC for bimodal drug release. *J. Control. Release* 58, 303–310.

McConnell, E.L., Murdan, S., Basit, A.W., 2008. An investigation into the digestion of chitosan (noncrosslinked and crosslinked) by human colonic bacteria. *J. Pharm. Sci.* 97, 3820–3829.

Mehta, A.M., 1997. Processing and equipment considerations for aqueous coatings. In: McGinity, J.W. (Ed), *Aqueous Polymeric Coatings for Pharmaceutical Dosage Forms.* Marcel Dekker, Inc., New York, pp. 287–326.

Nagai, T., Obara, S., Kokubo, H., Hoshi, N., 1997. Application of HPMC and HPMCAS to aqueous film coating of pharmaceutical dosage forms. In: McGinity, J.W. (Ed), *Aqueous Polymeric Coatings for Pharmaceutical Dosage Forms.* Marcel Dekker, Inc., New York, pp. 177–226.

Nunthanid, J., Puttipipatkhachorn, S., Yamamoto, K., Peck, G.E., 2001. Physical properties and molecular behavior of chitosan films. _Drug Dev. Ind. Pharm._ 27, 143–157.

Nunthanid, J., Wanchana, S., Sriamornsak, P., Limmatavapirat, S., Luangtana-anan, M., Puttipipatkhachorn, S., 2002. Effect of heat on characteristics of chitosan film coated on theophylline tablets. _Drug Dev. Ind. Pharm._ 28, 919–930.

Patel, H.A., Somani, R.S., Bajaj, H.C., Jasra, R.V., 2007. Preparation and characterization of phosphonium montmorillonite with enhanced thermal stability. _Appl. Clay Sci._ 35, 194–200.

Pongjanyakul, T., 2009. Alginate-magnesium aluminum silicate films: Importance of alginate block structures. _Int. J. Pharm._ 365, 100–108.

Pongjanyakul, T., Khunawattanakul, W., Puttipipatkhachorn, S., 2009. Physicochemical characterizations and release studies of nicotine-magnesium aluminum silicate complexes. _Appl. Clay Sci._ 44, 242–250.

Pongjanyakul, T., Priprem, A., Puttipipatkhachorn, S., 2005. Investigation of novel alginate-magnesium aluminum silicate microcomposite films for modified-release tablets. _J. Control. Release_ 107, 343–356.

Pongjanyakul, T., Puttipipatkhachorn, S., 2007. Sodium alginate-magnesium aluminum silicate composite gels: Characterization of flow behavior, microviscosity and drug diffusivity. _AAPS Pharm. Sci. Technol._ 8(3): E72.

Pongjanyakul, T., Puttipipatkhachorn, S., 2009. Polymer-magnesium aluminum silicate composite dispersions for improved physical stability of acetaminophen suspensions. _AAPS Pharm. Sci. Technol._ 10, 346–354.

Ray, S.S., Okamoto, M., 2003. Polymer/layer silicate nanocomposites: A review from preparation to processing. _Prog. Polym. Sci._ 28, 1539–1641.

Rhim, J.W., Hong, S.I., Park, H.M., Ng, P.K., 2006. Preparation and characterization of chitosan-based nanocomposite films with antimicrobial activity. _J. Agric. Food Chem._ 54, 5814–5822.

Rinaudo, M., 2006. Chitin and chitosan: Properties and applications. _Prog. Polym. Sci._ 31, 603–632.

Rojtanatanya, S., Pongjanyakul, T., 2010. Propranolol-magnesium aluminum silicate complex dispersions and particles: Characterization and factors influencing drug release. _Int. J. Pharm._ 383, 106–115.

Roussy, J., Vooren, M.V., Dempsey, B.A., Guibal, E., 2005. Influence of chitosan characteristics on the coagulation and the flocculation of bentonite suspensions. _Water Res._ 39, 3247–3258.

Rowe, R.C., 1997. Defects in aqueous film-coated tablets. In: McGinity, J.W. (Ed), _Aqueous Polymeric Coatings for Pharmaceutical Dosage Forms._ Marcel Dekker, Inc., New York, pp. 419–440.

Ruiz-Hitzky, E., Darder, M., Aranda, P., 2005. Functional biopolymer nanocomposites based on layered solids. _J. Mater. Chem._ 15, 3650–3662.

Sangalli, M.E., Maroni, A., Foppoli, A., Zema, L., Giordano, F., Gazzaniga, A., 2004. Different HPMC viscosity grades as coating agents for an oral time and/or site-controlled delivery system: A study on process parameters and _in vitro_ performances. _Eur. J. Pharm. Sci._ 22, 469–476.

Suksri, H., Pongjanyakul, T., 2008. Interaction of nicotine with magnesium aluminum silicate at different pHs: Characterization of flocculate size, zeta potential and nicotine adsorption behavior. _Colloid Surf. B_ 65, 54–60.

Thacharodi, D., Panduranga Rao, K., 1993. Release of nifedipine through crosslinked chitosan membranes. _Int. J. Pharm._ 96, 33–39.

Wang, X., Yumin, D., Luo, J., Lin, B., Kennedy, J.F., 2007. Chitosan/organic rectorite nanocomposite films: Structure, characteristic and drug delivery behaviour. *Carbohydr. Polym.* 69, 41–49.

Yoshizawa, T., Shin-ya, Y., Hong, K.J., Kajiuchi, T., 2005. pH-and temperature-sensitive release behaviors from polyelectrolyte complex films composed of chitosan and PAOMA copolymer. *Eur. J. Pharm. Biopharm.* 59, 307–313.

Zhang, H., Alsarra, I.A., Neau, S.H., 2002. An *in vitro* evaluation of a chitosan-containing multiparticulate system for macromolecule delivery to the colon. *Int. J. Pharm.* 239, 197–205.

Zhao, X., Mai, Z., Kang, X., Zou, X., 2008. Direct electrochemistry and electrocatalysis of horseradish peroxidase based on clay-chitosan-gold nanoparticle nanocomposite. *Biosens. Bioelectron.* 23, 1032–1038.

6

View from Inside to the Surface of Nanocomposite Coatings

Milena Špírková and Jiří Brus

CONTENTS

6.1 Introduction

6.1.1 Nanocomposite Definition

Composites are well-known common natural (wood, bond) and commercial (concrete) materials. Polymer composites are employed for long-term purposes and are used in profusion, for example, in construction, aircraft, and electric industries. However, the nature and composition of "classical" (i.e., micrometer-scale) composites rather result in property limitations. Consequently, nanocomposites started to be developed and intensely studied in the last three decades, particularly due to strict demands on current advanced materials. Novel tailor-made nanocomposite products can overcome the limiting property criteria specific for "classical" composite products. What is meant by the term nanocomposite? "Nanocomposites can be considered solid structures with nanometer-scale dimensional repeat distances between the different phases that constitute the structure. These materials typically consist of an inorganic (host) solid containing an organic component or vice versa. Or they can consist of two or more inorganic/organic phases in some combinatorial form with the constraint that at least one of the phases or features is in the nanosize. Nanocomposite materials can demonstrate different mechanical, electrical, optical, electrochemical, catalytic, and structural properties than those of each individual component. The multifunctional behavior for any specific property of the material is often more than the sum of the individual components" [1]. Different organic phases in polymer nanocomposites (PNCs) are well-known phenomena, for example, in segmented polyurethane elastomers or in polymeric blends, on the condition that they satisfy the aforementioned definition of nanocomposites. Nevertheless, PNCs mostly refer to systems containing polymeric organic components as the host solid, which in turn contain nanometer-scale (mostly) inorganic substance(s). Numerous inorganic nanoparticles used as nanofillers differing in nature, shape, size, and so on were prepared and characterized by both classic and trendy analytical methods. The possibility of a broad choice of starting materials (host solids and nanofillers) combined with the acquisition of products featuring the desired and tailored functional properties are two main reasons for the extensive long-term expansion of PNCs. Thin 2D PNC systems are frequently used in current coatings and film formulations due to their sometimes unique and targeted features, for example, electro-optical, barrier, or bactericidal properties, material reinforcement, flame resistance, and protection against corrosion, erosion, scratching, abrasion, and hydrolysis [2–23]. The introduction of inorganic nanofillers into an organic polymer matrix may further bring improved functional characteristics like self-cleaning and enhanced durability [21,22].

6.1.2 Basic Ways of PNC Preparation

The homogeneous distribution and building of nanofiller(s) in the organic polymeric matrix is crucial for efficient utilization of PNC systems and does not depend on the method of preparation.

Polymeric organic–inorganic nanocomposites can be formed by two basic methods: (i) by building up the inorganic nanoparticles or nanostructures *in situ* (using, for example, the sol–gel process [3–15,19,23–31]) within the organic phase, and (ii) by homogeneous dispersion of previously prepared (treated) nanoparticles in the organic matrix, distributed either in a ready-made polymer or in polymerizable monomer(s).

6.1.2.1 Organic Polymeric Matrix

An organic polymeric matrix, as a "host" solid of PNCs (named "film former" in coating terminology), can be of different origins—natural, modified natural, or synthetic. However, synthetic substances are utilized almost exclusively in current coating formulations. Generally, it is a macromolecular organic substance or substance(s) from macromolecules that are produced during film formation. An organic polymeric matrix is mostly of epoxy [14,15,17,25,32–41], polyacrylate [4,6,9,19,24,31,42–44], or polyurethane [45–52] origin, but matrices based on polyimide [16,23,53,54], polypropylene [55–57], polycaprolactone [2,58], polystyrene [59–61], polysiloxane [62,63], polylactide [64,65], or poly(vinylidene fluoride) [66,67] are used as well. The choice of material for the matrix depends on the function and use of the coating; the product can be either thermoplastic or thermoset.

6.1.2.2 Nanofillers

From a chemical composition aspect, nanofillers can be either of metal origin (gold, silver, platinum, iron, and titanium), semimetal origin (silica nanoparticles, layered silicates), or nonmetal (=organic) origin (carbon nanotubes [CNTs], nanofibres, and graphene). Concerning the choice of nanofiller right from the shape insight, all formations having at least one dimension ≤100 nm can be considered as nanoparticles. That means, 0D nanoparticles (=the size in all dimensions is below 100 nm), 1D fillers (= one particle dimension exceeds 100 nm), and 2D fillers (= two dimensions exceed 100 nm) can be used. Spherical silica or titanium dioxide nanoparticles [4,6,19,20,23,27,33, 36–38,41,43,45,48,51,68,69] and "cubic" polysilsesquioxane (POSS) cages [32,47] are typical representatives of 0D nanofillers. CNTs [44,69–72] are mostly used as 1D nanofillers. Clays [17,18,23,34,35,40,42,49,50,52,56,70,72–76] and currently also carbon grapheme sheets [77–79] are typical 2D nanoadditives. For applications in the coating industry, silicon- or titanium-based nanofillers are preferentially used.

As mentioned, the combinations of the functional (e.g., structural, optical, mechanical, electrical, electrochemical, catalytic) properties of PNCs is almost inexhaustible owing to the broad choice of starting materials for the host solid and nanometer-size filler. Analytical techniques enabling the detailed control of all the steps in PNC processing, the control of morphology of PNCs, and the control of the interface between the matrix and nanofiller are essential for PNC preparation featuring desired and tailor-made properties. These phenomena, that is, the view "inside," are necessary for both 2D and 3D systems. On the other hand, detailed surface characteristics are crucial especially for 2D products; hence the view "on the surface" is one from the key form of characterization of PNC coatings, films, membranes, and other thin material products. The sources dealing with PNC preparation and characterization on the research and application bases are enormous (e.g., papers [1–46,48–68,70,72,75,76,78,79], reviews [47,69,71,73,74,77,80–82], books [83–95], patents, conference contributions, etc.) and they multiply, exceed the extent of our contribution in the book. Hence, Section 6.2 will be a short overview of the basic analytical methods enabling the view "inside" and Section 6.3 the view "on the surface" of PNCs. Section 6.4 (a practical example of the view "inside" and "on the surface" of PNC coatings) will concentrate on an epoxide-based polymer organic matrix, and simultaneously on 0D and 2D nanofillers, that is, on PNC coatings and films with our own research experience.

6.2 View "Inside" Polymer Nanocomposite Coatings

Whether it is for the challenges faced by polymer scientists who wish to synthesize new polymers of predefined structure or for material scientists who are interested in understanding structure–property relationships, the possibility of polymers to exist in a wide range of solid forms including various polymorphic forms, amorphous phases, micro- or nanocomposites, as well as highly ordered complexes with unique properties has prompted interest in the physical characterization of these systems. Basic analytical techniques enabling a view inside polymer nanocomposites are solid-state nuclear magnetic resonance (ss-NMR) spectroscopy, small-angle x-ray scattering (SAXS) and wide-angle x-ray scattering (WAXS), and transmission electron microscopy (TEM). These techniques will be shortly described in the following paragraphs. ss-NMR spectroscopy will be described in more detail for two reasons: (i) it is a fundamental method for the description of PNC behavior, especially for the structure and dynamics starting from the segmental level; and (ii) ss-NMR spectroscopy is still not excessively used and fully exploited. In this way, the paragraphs devoted to current methods of ss-NMR spectroscopy should show further possibilities for PNC coating determination, especially on the sub-micrometer level.

6.2.1 Solid-State NMR Spectroscopy

6.2.1.1 Basic Principles

Traditional approaches (x-ray scattering, microscopy, spectroscopy, and thermal analysis) have all been used to examine different aspects of structure, dynamics, and energetics in solid polymers. Among these techniques, ss-NMR spectroscopy has emerged as a powerful technique for the characterization of both structure and dynamics [19,23,25,26,36,38,43,76]. This technique, which in many cases complements x-ray diffraction, has the unique ability to probe electronic environments of specific nuclei over a large timescale without the requirement of long-range periodic order and homogeneous samples. ss-NMR spectroscopy is thus not only suited for determining the chemical composition of prepared polymers, but this experimental approach also provides extremely detailed structural information that may be useful for rationalizing physical properties of complex multicomponent macromolecular (nano)composites.

The basic principles of NMR spectroscopy are the same for solution- and solid-state measurements; however, to obtain high-resolution solid-state spectra the standard solution-state methods must be modified to reach acceptable sensitivity of NMR measurements and to compensate missing molecular tumbling that is generally fast in solutions. Due to missing molecular motion, the anisotropic nuclear interactions such as heteronuclear and homonuclear dipolar couplings (e.g., ^1H–^{13}C, ^1H–^1H) and chemical shift anisotropy (CSA) come to dominate the evolution of nuclear spins, leading to severe broadening of NMR signals in rigid solids. The resulting NMR spectra are featureless, consisting of many broad and overlapping signals with a line width of several tens of kHz [96,97]. The lack of resolution, however, does not mean that there is an absence of structure information, but rather reflects its overcrowding to such an extent that we are not able to read or understand it. Due to this fact, various techniques have been proposed to increase the spectral resolution of NMR spectra in solid state. Among many of the recently developed experimental procedures, the most important techniques are (i) magic angle spinning (MAS), currently reaching up to 60–70 kHz in

1.2 mm probeheads [98,99]; (ii) heteronuclear dipolar decoupling, now implemented by a range of pulse schemes such as two-pulse phase modulated decoupling (TPPM) or SPINAL64 [100,101]; (iii) homonuclear decoupling, traditionally represented by frequency-switched Lee–Goldburg decoupling (FSLG) or phase-modulated Lee–Goldburg decoupling (PMLG) [102–108]; and finally (iv) cross-polarization (CP), allowing efficient signal enhancement of the nuclei with low isotopic abundance and low gyromagnetic ratio (e.g., ^{13}C, ^{15}N, ^{29}Si, etc.) [109,110]. Combining these techniques into a single experiment such as ^{13}C CP/MAS NMR has allowed the acquisition of high-quality 1D spectra of rare nuclei in a reasonable time. Similarly, typical 2D homonuclear or heteronuclear correlation experiments and/or spin-relaxation measurements can be currently performed with acceptable quality during overnight experimentations.

6.2.1.2 Kinetics and Mechanism of Polymerization

In the following few paragraphs, we would like to briefly demonstrate potentialities as well as limitations of the ss-NMR techniques that are currently utilized to describe the structure and dynamics of complex multicomponent macromolecular systems including polymer films. Among the many advantages of modern ss-NMR spectrometers, including the ingenius design of high-sensitivity probe heads, the ability to perform NMR experiments under moderately fast spinning frequency (5–7 kHz) not only for true solids but also for gel-like samples and even for liquids has opened new ways to probe many chemical and physical transformations. Consequently, one can easily follow the kinetics and mechanism of polymerization reactions continuously from the starting solution through the gel state up to an insoluble product if the functionality of monomers allows the formation of networks (Figure 6.1).

6.2.1.3 Primary Characterization of Polymeric Composites

In many cases, the prepared polymeric coatings are heterogeneous systems consisting of nanodomains in which segmental motions can be quite different from the motions expected in a polymer matrix. These differences significantly complicate quantitative and qualitative structural characterization. Therefore, the investigated nanocomposites should be characterized by three sets of NMR experiments: in the first step, the composites should be analyzed by a combination of MAS and CP/MAS NMR measurements of ^{13}C and ^{29}Si nuclei, making it possible to qualitatively and quantitatively determine the primary chemical structure and to recognize the "mobile" and "frozen" areas. Thereafter, the size of nanodomains, if present, can be determined in ^{1}H spin-diffusion experiments that currently include 1D carbon-detected Goldman–Shen experiment [112], 2D homonuclear ^{1}H–^{1}H NOESY or CRAMPS correlation techniques [113,114], and a range of 2D ^{1}H–^{13}C HETCOR

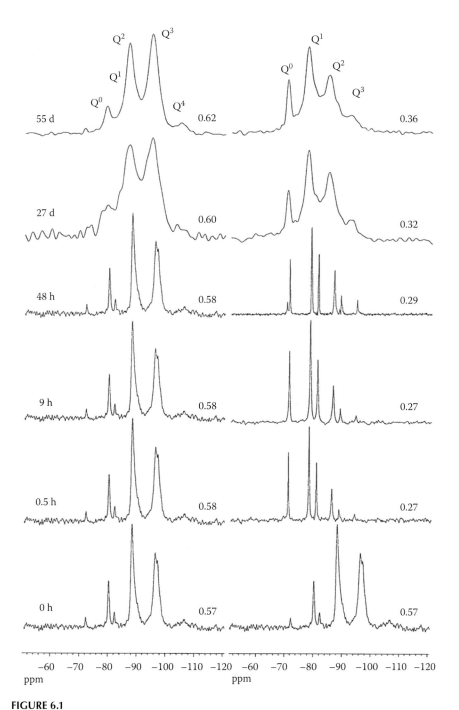

FIGURE 6.1

^{29}Si MAS NMR spectra of two differently prepared solutions of sodium silicate reordered during the 2-month polycondensation. (From Kobera, L. et al., *Ceramics-Silikáty*, 55, 343, 2011.)

experiments [115,116]. Finally, the molecular mobility should be investigated in detail, by studying segmental motion frequencies and amplitudes in the high- and low-frequency regions, thus giving a deep insight into how the molecular dynamics of the various nanocomposites' components and of their segments influence the material properties (modulus, glass transition). For such studies, various relaxation experiments combined with separation local field (SLF) techniques have been successfully applied [117–119].

A typical behavior of multicomponent polymeric nanocomposites exhibiting nonuniform segmental dynamics is demonstrated in Figure 6.2a, which shows the ss-NMR spectra of polypropyleneoxide network (PPO) reinforced by POSS cages [121]. In general, the highly mobile segments are reflected by narrow and intensive signals in ^{13}C MAS NMR spectra, whereas the same segments are barely detectable in ^{13}C CP/MAS NMR spectra. In contrast, highly rigid segments are easily detected in ^{13}C CP/MAS NMR spectra due to the high efficiency of the polarization transfer from protons to carbons. In this particular case, the segments of the PPO matrix are highly mobile as reflected by narrow and high intensive signals in the ^{13}C MAS NMR spectrum, while the phenyl substituents of $POSS_{Ph}$ cages must be substantially rigidified as demonstrated by relative signal enhancement of the corresponding signals in the ^{13}C CP/MAS NMR spectrum (Figure 6.2a). Consequently, one can conclude that the siloxane cages do not adopt the segmental dynamics of the PPO matrix and vice versa.

6.2.1.4 Segmental Dynamics—Relaxation Experiments and Frequency of Motion

It has long been recognized that besides the structure, molecular dynamics covering a wide range of timescales, from picoseconds to seconds, and amplitudes from small-angle fluctuations up to high-amplitude jumps determine the mechanical properties of polymers. That is why, after the qualitative characterization of molecular segment mobility, the motional dynamics of the investigated systems must be studied in detail.

Generally, two distinctive groups of segmental motions are easily recognized via spin relaxation mechanisms: high-frequency (ca. hundreds of MHz) motions (librations, rotations, and jumps of small groups), and low-frequency motions (ca. tens of kHz) involving movements of larger parts of polymer chains. The high-frequency motions yield valuable "diagnostic" information about the situation of the corresponding groups in a material. They can support some slower motions, and they can influence intermolecular or intersegmental interactions. The low-frequency motions are known to correlate strongly with the thermomechanical material properties like modulus or glass transition and they can also absorb considerable amounts of mechanical energy.

An insight into the correlation times of segmental motions is provided by NMR relaxation measurements. The high-frequency motions covering a

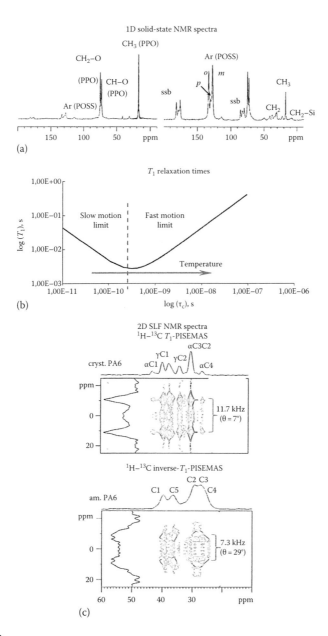

FIGURE 6.2
(a) 13C MAS and 13C CP/MAS NMR spectra of epoxy networks modified by monofunctional POSS units (POSSPh) containing phenyl substituents. (From Brus, J. et al., *Macromolecules*, 2, 372, 2008.) (b) Temperature dependence of T1 relaxation times. (c) 2D T1-filtered and inverse T1-filtered PISEMAS NMR spectra selectively recording ¹H–¹³C dipolar profiles for crystalline and amorphous domains of PA6/MMT nanocomposite. (From Brus, J. and Urbanová, M., *J. Phys. Chem. A*, 23, 5050, 2005; Brus, J. et al., *Macromolecules*, 16, 5400, 2006.)

(*continued*)

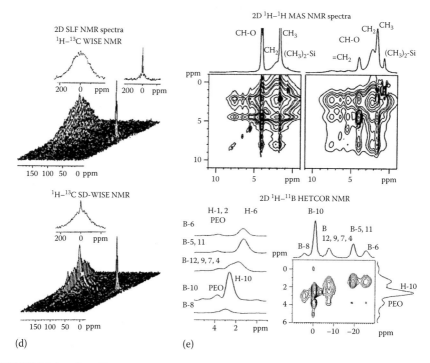

FIGURE 6.2 (continued)
(d) ^{1}H–^{13}C WISE NMR spectra of polyimide/polydimethylsiloxane composite recorded without and with spin-diffusion period (0.5 ms). (From Brus, J. et al., *Macromolecules*, 35, 1253, 2002.)
(e) ^{1}H–^{1}H MAS NMR spectra of POSS-reinforced epoxy networks and ^{1}H–^{11}B FSLG HETCOR NMR spectrum of sodium cobalt-bis-dicarbolide dispersed in PEO matrix. (From Brus, J. et al., *Macromolecules*, 2, 372, 2008; Matějíček, P. et al., *Macromolecules*, 10, 3847, 2011.)

frequency range from ca. 500 to 50 MHz at 11.7 T are probed by the measurements of spin-lattice T_1 relaxation times at laboratory reference frame of various nuclei starting from ^{1}H to ^{15}N. In contrast, the measurements of relaxation times at rotating frame $(T_{1\rho})$ applying a spin-locking B_1 field, whose intensity expressed in frequency units ranges from ca. 10 to 100 kHz, open the window into the analysis of mid-kilohertz motions. Considering the existence of motional heterogeneities, two types of relaxation experiments should be performed: the experiments with direct excitation of ^{13}C spins having the capability to detect highly mobile segments, and the techniques based on the CP probing rigid polymer units. The justifiability of this approach has been recently confirmed by the observed differences in $T_{1\rho}$ (^{13}C) relaxation times that revealed remarkable differences between mid-kilohertz frequency motions occurring in the "free" and in the "constrained" domains of the PPO matrix. Estimating correlation frequencies with higher precision, the dependence of relaxation times on temperature must be measured. For typical solids in slow motion regime, the T_1 relaxation times decrease with increasing temperature, whereas highly flexible segments in the fast motion regime exhibit an increase in T_1 values (Figure 6.2b).

6.2.1.5 Segmental Dynamics—SLF Experiments and Amplitudes of Segmental Motions

However, the analysis of standard spin-lattice relaxation experiments in terms of motional frequencies is practically impossible for a system in which multiple dipolar couplings involving many abundant and rare nuclei (e.g., ^1H, ^{10}B, ^{11}B, ^{23}Na, ^{13}C) act simultaneously or in which paramagnetic impurities such as Fe^{3+} ions are present. As an example of such polymeric systems, we can mention polymeric nanocomposites based on polyamide-6 (PA6) modified by natural montmorillonite (MMT) [122] or recently prepared highly ordered complexes of polyethylene oxide (PEO) and sodium 3-cobalt-bis-dicarbolides (NaCoD) [124]. In such cases, alternative site-specific measurements of motionally averaged ^1H–^{13}C one-bond dipolar couplings that carry information about motional amplitudes of molecular segments can be applied [125–127]. This measurement is allowed by the fact that any molecular motion with a correlation time shorter than ca. 40 µs causes averaging of one-bond ^1H–^{13}C dipolar interactions ($\omega_D/2\pi = 25$ kHz for a typical rigid C–H bond). Then the ratio of a motionally averaged dipolar coupling constant (D_{CH}) and the rigid-limit value ($D_{CH,rig}$) defines the order parameter (S^2). The order parameter can be converted to the amplitude of segmental motion (θ) using a range of relations that were derived for specific motional models. The simplest one assumes low-amplitude uniaxial rotational diffusion motion $S^2 = 1 - (3/2)\theta^2$ [128,129].

Applying the recently developed experimental techniques such as LGCP, PILGRIM, PISEMAS, inverse-T_1-filtered PISEMAS [121,125–127], or real-time T_{1C} PISEMA [130], the amplitudes of segmental reorientations were determined for a range of polymer systems including polyethylene (PE), PA6, PA6/MMT composites [121,122], polyelastine [128], POSS-reinforced polyepoxy networks [120], and PEO/NaCoD complexes [124]. The developed techniques, however, allow not only the site-specific measurement of dipolar profiles for each resolved segment but also domain-selective measurements for the segments in crystalline (rigid) domains and/or amorphous (flexible) matrices in semicrystalline, generally heterogeneous, polymeric composites (Figure 6.2c).

If only qualitative information is required, a ^1H wide-line separation experiment (WISE) allowing the extraction of a ^1H–^1H dipolar spectrum for each resolved carbon atom is an optimal choice [117,118]. In the resulting spectra, the rigid segments are characterized by a broad dipolar spectrum with the line width reaching 20–40 kHz, while flexible segments are reflected by very narrow signals (Figure 6.2d). Up to date, a wide range of complex polymeric systems have been characterized in this way [117,118,130–132]. The advantage of this technique follows from easy implementation because no specific optimization of experimental parameters is required, and from easy modification by inserting a spin-diffusion period [117,118]. This modification then allows monitoring the ¹H–¹H spin-diffusion process, which can be used to probe the miscibility of multicomponent systems and even precisely determine the size of domains in nanoheterogeneous systems for which this information can be hardly provided by other experimental approaches [133,134].

6.2.1.6 Local Order of Nanocomposite Networks via ¹H–¹H Spin Diffusion

Successful development of polymer nanocomposites requires a deep analysis of self-assembling processes and the understanding of their impact on material properties. In order to assess the size of inorganic domains formed in the investigated nanocomposites, ¹H spin diffusion is usually probed and a series of 2D experiments correlating spatially close to ¹H nuclei via through-space dipolar couplings are measured. In general, ¹H–¹H correlations allow tracing of interatomic contacts up to ca. 0.5 nm, and the analysis of ¹H spin diffusion makes it possible to probe the size of domains in heterogeneous systems up to a width of about 50 nm. The success of such experiments is, however, preconditioned by a sufficient spectral resolution and requires the removal of the dipolar broadening. Besides various homodecoupling techniques such as BR24 and w-PMLG, a high-speed ¹H–¹H MAS NOESY-type experiment is an elegant way to access the required cross-peaks separation (Figure 6.2e). Particularly, using the recently designed small-diameter probe heads that enable MAS frequencies up to 80–100 kHz, the obtained spectral resolution in ¹H dimension is very promising.

Usually the spin diffusion between dipolarly coupled ¹H spins is traced through the dependence of correlation signal intensity on the mixing time. Applying the initial-rate approximation, this evolution is described by the second Fick's law, and assuming a simple two-component system, the size of the dispersed domains of the component A can be calculated according to the following equation:

$$d_A = 2\frac{\varepsilon}{f_B}\left(\frac{1}{\pi}Dt_{eq}\right)^{1/2},$$

where
  f_B is the volume fraction of the continuous phase B
  t_{eq} is the time of magnetization equilibration
  D is diffusivity
  ε is the reflects dimensionality of the process.

The diffusivity, a parameter reflecting the average strength of dipolar couplings, must be determined for every component in each polymer network. In the literature, the procedures based on measurements of ¹H static line width or $T_2(^1H)$ relaxation times have been described and discussed in detail. However, it is noteworthy that dipolar couplings are effectively reduced by the sample spinning. Especially at very fast spinning frequencies above 20 kHz, the dipolar couplings are reduced in such a way that effective long-range ^{1}H–^{1}H polarization transfer requires either extremely long mixing time (>50 ms) and/or recoupling by using rotor-synchronized pulses.

A drawback of ^{1}H–^{1}H correlation experiments follows from the low dispersion of ^{1}H NMR chemical shifts which usually is not larger than 20 ppm. Consequently, in the case of disordered solids, ^{1}H NMR signals of individual molecular segments severely overlap, thus preventing detailed elucidation of the molecular structure and hierarchical architecture. The following technique nicely combines the advantage of ^{1}H nuclei (100% isotopic abundance and high magnetogyric constant allowing observation of spin diffusion and thus providing geometrical constraints) with the high resolution of ^{13}C NMR spectra. This 2D ^{1}H–^{13}C HETCOR experiment, in which ^{1}H–^{1}H spin exchange occurs during the mixing delay tm inserted after the first ^{1}H detection period t_1, provides a significantly better resolution afforded by ^{1}H–^{13}C chemical shift separation. In the spectra, we can observe in principle two types of cross-peaks. First, signals reflecting the shortest (usually one-bond) ^{1}H–^{13}C distance, and second, cross-peaks reflecting a long-range through-space polarization transfer, that is, long-range interatomic contacts. The latter signals can be used to estimate the geometry of the studied systems. Similar information can be derived also from heteronuclear experiments that involve other heteronuclei with wide dispersion of chemical shift such as ^{19}F, ^{11}B, ^{29}Si, and so on (Figure 6.2e).

6.2.2 Small- and Wide-Angle X-Ray Diffraction

From a historical point of view, x-rays were first discovered by Wilhelm Conrad Röntgen in 1895 and they were immediately applied to elucidate the structure of crystalline solids. It was Max von Laue who, around 1912, suggested that crystals behave like 3D gratings provoking the phenomenon of diffraction, proving that x-rays are electromagnetic waves of very short wavelength ($\approx$0.1 nm). X-ray diffraction was fully exploited in 1913 by William Henry Bragg and his son William Lawrence [136]. Currently, the advanced concepts of x-ray structure analysis involving the application of synchrotron radiation combined with the x-ray pair distribution function (PDF) method has become a formidable tool in the analysis of not only typical, highly crystalline organic and inorganic solids but also of complex or disordered and in some cases even nanocrystalline materials [135]. In this regard, x-ray diffraction methods occupy an outstanding position in the characterization of polymeric nanocomposites.

There is no doubt that the structure of polymeric nanocomposites at nanometer scale (1–100 nm) is a crucial factor for many technological properties, such as ductility, hardness, and scratch resistance. While classical WAXS or x-ray powder diffraction gives information about the molecular structure within crystalline domains, SAXS extends the scope to the characterization of the nano-domains. From this it follows that a combination of the two techniques represents a powerful tool for the characterization of multicomponent polymer systems such as nanocomposites [18,20,22,23,27,32,35,36,41,42,46–52,55,56,68,75–77,79].

SAXS is one of the most rapidly growing techniques in polymer characterization. Predominantly this is due to the fact that the structure of polymers at nanometer scale exhibits extreme variability, particularly in specifically designed hybrid organic/inorganic networks or composites with *in situ* formed nanoparticles. The size and distribution of these particles can be hardly probed by other experimental techniques.

X-Ray scattering is an important tool for the structural characterization of phases and the structural changes associated with phase transitions. While SAXS permits the determination of the overall structure and dimensions on a length scale ranging from a few nanometers to about 100 nm, WAXS patterns permit the investigation of molecular packing in the sub-nanometer scale.

Typically, for $2\theta > 1°$ the diffraction experiment is referred to as WAXS. For $1° > 2\theta > 0.3°$ we usually talk about medium-angle x-ray scattering (MAXS) and the region of $2\theta < 0.3°$ is typical for SAXS. As demonstrated in the recent investigation of nanocomposites based on single-wall carbon nanotube (SWCNT) and thermoplastic poly(butylene terephthalate) (PBT), the WAXS region corresponds to the diffraction of the α crystalline phase of PBT, indicating that the nanotubes do not perturb the nature of the PBT crystals. In contrast, scattering maxima in the SAXS region corresponding to the diffractions of the consecutive laminar crystals separated by amorphous layers carry information about the morphology of the prepared nanocomposites [136]. SAXS thus has been shown to be of great importance while dealing with the structural analysis of soft condensed matter because structural characterization of different building blocks of a complex material can be obtained.

Further, TEM is a very important method of "inside" characterization of complex polymer systems, including PNCs. It is very often combined with a scattering (especially WAXS) technique. However, TEM will be introduced in Section 6.3, together with other microscopy techniques.

6.3 View "on the Surface" of Polymer Nanocomposite Coatings

Surface analysis is an important factor especially for 2D systems (coatings, films, membranes, etc.). This part should briefly summarize the characterization of PNC 2D systems mainly from the surface morphology/topography aspect. Analytical methods for surface determination depend on the size of the detected formations and the magnitude of the area analyzed. The possible microscopy methods are shown in Figure 6.3.

All microscopy techniques have gone through the extensive development connected with the demands on the surface analysis of nano- and microscale

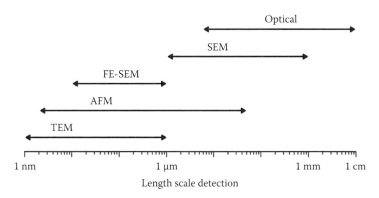

FIGURE 6.3
Instrumentation for pertinent 2D/3D surface analysis.

images of advanced multicomponent materials. Their detailed description in this chapter is inconceivable; thus, only basic techniques and their advantages and limitations for nanocomposite coating characterization will be introduced.

6.3.1 Optical Microscopy

Optical microscopy using *visible light for the sample detection* is the oldest microscopy technique of surface analysis. As the common resolution limit is ca. 250 lower than that for scanning electron microscopy (SEM) analysis, these techniques can be used for the nanocomposite coating characteristics if the large-scale images are desired, for example, the overall coating topography profile, the detections of nanofiller agglomerated into the micrometer-size formations, and so on [4,17,44,72,78,80]. Optical microscopy techniques are currently used in combination with other methods of surface material characterization, for example, with SEM, atomic force microscopy, or TEM. It can also be used for the visualization of surface damage after surface hardness tests (scratching, nanoindentation, etc.) [4,44].

6.3.2 Scanning Electron Microscopy (SEM)

SEM started to be developed approximately 75 years ago,* but SEM instrumentation began approximately 30 years later. It is an extensively used method for topography, material homogeneity, roughness, and porosity examination, and it is also common for polymer nanocomposite coating surface analysis, including film thickness determination [20,22,26,31,38,39,42,43,45,48,50, 52,59,70–72,77]. Due to the fact that magnifications from ca. 10 times to ca. 500,000 times are possible, large-scale images with a determination of nm-size

---

* Knoll [153].

formations can be very well detectable at the same time and on the same sample area. This advantage is frequently utilized in the practice. Field emission scanning electron microscopy (FESEM) offers even up to six times better resolution than conventional SEM.

The principle of SEM is *sample imaging subjected by the action of the high-energy focused beam of electrons.* Conventional SEM analyses need electrically conductive samples (in order to avoid the charge accumulations on the surface) under high vacuum. However, environmental SEM (ESEM) is the method that enables measurements demanding neither high vacuum nor sample surface conductivity. Besides biological or other sensitive coating materials, it is also used for the *in situ* nanocomposite coating surface and surface-change analysis.

6.3.3 Atomic Force Microscopy (AFM)

AFM, a member of the broad family of scanning probe microscopies, belongs to the current methods for polymer nanocomposite coating characterization [20,22,28,33,40,43,47,51,72]. The surface detection is similar to the result obtained with SEM; however, some differences (advantages or disadvantages) exist between SEM and AFM analysis: images obtained by the AFM technique can be 2D or 3D, unlike only 2D SEM projection. AFM works preferably in an air environment, unlike SEM (which mostly works under high vacuum). Samples prepared for AFM analysis do not need any further treatment and AFM analysis works for insulating materials (i.e., sample coating by conductive metal or carbon is not necessary, like in the case of SEM). The two main disadvantages of AFM compared to SEM are the limited area and height of the AFM image detection (scale x,y, are ca. 100 μm, and z in units of μm in the maximum) and the substantially lower AFM scanning speed compared to SEM. Further AFM disadvantages are thermal drift and measurement limitations for rough samples with steep walls leading to AFM artifacts, hysteresis, and nonlinearity of the piezoelectric element, for example. Current AFM instrumentation is developed minimizing the aforementioned disadvantages. The combination of AFM with other analytical techniques (e.g., optical microscopy, Raman, thermal analysis, etc.) is sometimes utilized for obtaining the most universal material information on the nanometer scale.

For the topography analysis, the tapping mode is the most used AFM image technique. The tip is not in direct contact with the sample (like in the contact mode), so that the probability of sample damage is limited. Moreover, further surface information is available (e.g., homogeneity on the nanometer scale).

6.3.4 Transmission Electron Microscopy

The principle of the TEM image is based *on the interactions of a beam of electrons passing through the ultrathin sample.* Due to the small de Broglie wavelength of electrons, the TEM analysis enables one to three orders higher resolution than SEM or light microscopes, and hence, it is one of the principal and expanded methods of nanocomposite characterization preferably on the nano- or micrometer level.

As samples for TEM have to be ultrathin, they are not often used on a real PNC coating surface; they are mainly utilized to obtain information regarding what is "inside" the material (and it should be correctly mentioned in Section 6.2). TEM analysis is almost universal for polymer nanocomposite coating characterization, but it is preferably used for detecting the quality of nanofiller dispergation in the polymer matrix. TEM is very often used for the analysis of 2D layered nanofillers (clays, MMT, graphene, etc.) [18,22,23,27,31,35–37,41,43,45–47,50,52,55,68,70,72,75,82]. TEM analysis offers *direct images* of the nanofiller organization/ordering within the polymer matrix and is a very powerful tool for a (qualitative) view of what is really inside complex PNCs. As the overall nanoscale particle dispersion is well visualized by TEM analysis, it was found to be a very efficient complementary method to x-ray analysis, which can (quantitatively) detect the degree of dispergation like the periodicity of intercalated layered nanofillers. For the correct interpretation, the combination of WAXS/TEM analyses was found to have the most suitable characteristics of polymer–nanofiller arrangement in PNCs.

Other methods commonly used for PNC coating characterization, like infrared spectroscopy [6,23,25,26,33,38–40,46,48,50,51,59,70,76], surface hardness [4,25,26,28,31,32,33,38,80], permittivity [2,23,42,74,77,82], corrosion and other resistivity determination [6,17,19,20,22,23,36,39,40,49,52,73], mechanical properties [24,26,31,32,35,37,42,44–48,50,52,77,78], and thermal stability [6,25,32,33,50,51,77] are not discussed in detail.

6.4 Practical Example: Epoxy-Based Organic–Inorganic Nanocomposite Coatings Containing 0D and 2D Inorganic Nanofillers

The variability of end-use properties of complex nanocomposite coatings will be documented in detail on epoxy-based organic–inorganic (O–I) systems containing further either colloidal silica (0D) particles or layered silicates (natural or modified MMT and bentonite), that is, 2D fillers. This chapter shows that tailor-made products could be prepared only when all starting materials are characterized, the procedure technique is controlled at all steps, intermediate products are detected in depth, and all pertinent end-use properties are completely analyzed (for further details, see Refs. [137–148]).

6.4.1 Materials and Preparation Procedure Optimization

6.4.1.1 Materials

For the O–I matrix preparation, we started with (a) two functionalized organosilicon precursors differing in the number of alkoxysilane groups, active in the sol–gel process: (3-[(glycidyloxy)propyl]trimethoxysilane (GTMS) and diethoxy[3-(glycidyloxy)propyl]-methylsilane (GMDES)); and (b) three oligo(oxypropylene) oligomers differing in chain length and structure: diamines D230 and D400 and triamine T403, all end-capped with primary amino groups [137–148]. The mixed solvent (water/alcohol, 7:3 by weight) contributed to environment-friendly reaction conditions. The nanocomposite O–I matrix preparation can be described by Equations 6.1 through 6.4:

(1) Formation of inorganic structures: sol–gel process
(a)."Acidic" step (hydrolysis), Equation 6.1

"Alkaline" step (polycondensation), Equation 6.2

(2) Formation of organic polymeric matrix: polyaddition reactions, Equations 6.3 and 6.4

For the *in situ* inorganic nanostructure formation, a sol–gel process consisting of "acidic" and "alkaline" steps (Equations 6.1 and 6.2) at room temperature was utilized first. The organic polymeric matrix formation (i.e., polyaddition reactions of epoxy and amine groups, Equations 6.3 and 6.4) at elevated temperature was the second and final step (carried out simultaneously with solvent and sol–gel low-molecular products evaporation).

Although the organosilicon precursors on the one hand and oligomeric amines on the other are relatively similar in chemical constitution (see, for example, Schemes 6.1 and 6.2), markedly different end-use properties were found. For example, the temperature of glass transition of the organic–inorganic interphase (determined from dynamic mechanical thermal analysis) varied from –16°C to +56°C [140]. This means that coatings at room temperature (23°C) in glassy, main-transition-region, or rubbery states can be prepared just by varying the material, functional group (epoxy to NH) ratio, and procedure technique [140]. In order to find the rules and principles for such markedly different end-use properties, we had to study all the steps of the preparation procedure in detail. This knowledge was necessary for controlling the complex nanocomposite preparation leading to targeted end-use properties. As the heterogeneities in the coatings were suppressed to the sub-micrometer scale, all O–I nanocomposite systems visually appear to be homogeneous, that is, they are *transparent*, colorless, or slightly yellowish.

6.4.1.2 *Preparation Procedure Optimization*

ss-NMR spectroscopy was found to be a powerful and unique method for the optimization of the preparation procedure. Recently, a fairly wide composition range of reaction mixtures involving GTMS, GMDES, D230 and T403, and colloidal silica has been studied by ^{13}C and ^{29}Si MAS NMR to follow the kinetics of the reaction process, the formation of initial building units, the determination of the rate of polycondensation, as well as to predict the final structure of the resulting materials. It was found that during the "acidic" step, hydrolysis and formation of short oligomers is predominant in systems without colloidal silica. Extensive polycondensation starts during the "alkaline" step. However, the presence of colloidal silica accelerates condensation reactions even at low pH (Figure 6.4; ^{29}Si MAS NMR spectra) as NH$_4^+$ ions stabilizing colloidal silica particles promote condensation reactions. No cleavage of oxirane rings was observed during the whole process before thermal curing, leaving these functional groups accessible for organic network buildup (Figure 6.4; ^{13}C MAS NMR spectra).

6.4.1.2.1 *Examples of the Variability of End-Use Properties Using Changes in Preparation Procedure*

This chapter shows some examples of the importance of *preparation process control*. Property variability is achieved just by changing *one* variable, keeping the others constant.

Example 1: Change of the polycondensation step duration

The duration of the polycondensation step was found to be dominant for the mechanical properties of O–I coatings [140]. The prolongation of the polycondensation step from 1 to 2.5 h in nanocomposite coatings either in a rubbery or

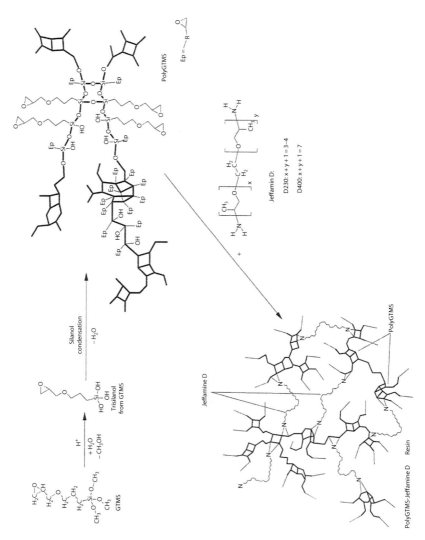

SCHEME 6.1
GTMS + Jeffamine D.

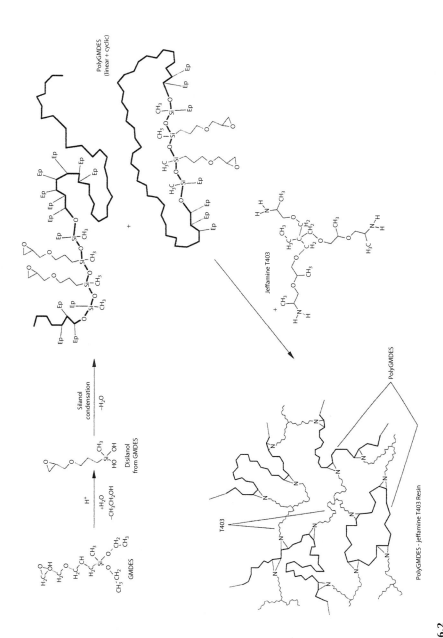

SCHEME 6.2
GMDES + Jeffamine T403.

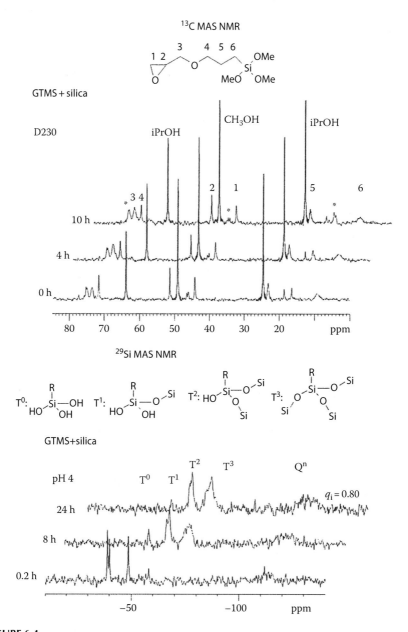

FIGURE 6.4
^{13}C and ^{29}Si MAS NMR spectra of selected reaction mixtures at various reaction times (signal assignment of basic structure units and condensation degree q_i of siloxane unit at the end of "acidic" step is presented in each spectrum; signals corresponding to oxypropylene unit [CH, CH$_2$ and CH$_3$] are marked by asterisks).

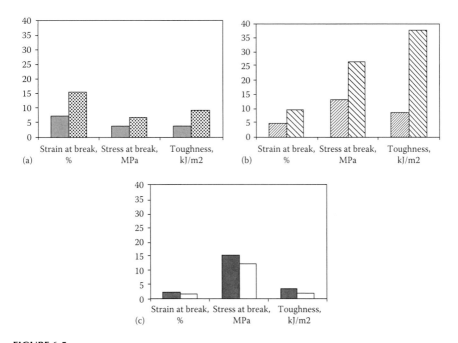

FIGURE 6.5
The influence of the duration of polycondensation step on mechanical properties of O–I nanocomposite coatings being in (a) rubbery, (b) main-transition, and (c) glassy states. Duration of polycondensation step: 1 h (left dependences) and 2.5 h (right dependences). (Data taken from Špírková, M. et al., *J. Appl. Polym. Sci.*, 92, 937, 2004): (Table III, samples 4A–6C).

main-transition state led to the improvement of tensile properties.* However, the effect for coatings in the glassy state was quite the opposite (see Figure 6.5).

Example2: Addition of nanofiller

The O–I nanocomposite coating is already made from organosilicon precursor (GTMS, GMDES) and oligomeric amine (D230, D400 or T403). However, in some cases the addition of a nanofiller is desirable for further property tuning.

ss-NMR spectroscopy revealed that if colloidal silica particles are added into the reaction mixture right at the beginning of hydrolysis, they can be covalently bonded via silanol groups with the O–I matrix. This phenomenon is important for, for example, the reinforcement of soft and elastic coatings (Ref. [140], Table V, samples 29A and 37A).

A different situation occurs if layered silicates are used as nanofillers. The order of MMT addition significantly affects not only the mechanical properties of PNCs, but also surface morphology. If layered silicates were added into the reaction mixture during (or after) hydrolysis, their 2D shape caused

---

* Toughness, i.e., the energy necessary to break the sample per volume unit was taken as the quality criterion.

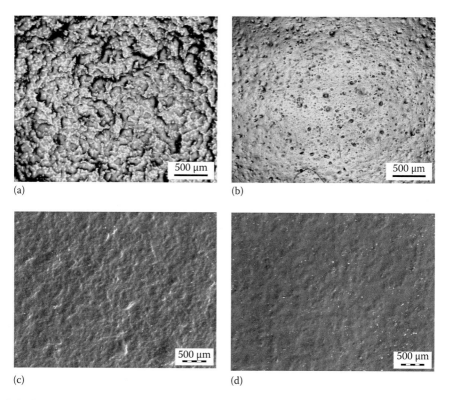

FIGURE 6.6
The influence of the order of natural montmorillonite (MNa) addition on the surface morphology detected by light microscopy (top) and SEM (bottom). (a) and (c): MNa dispersion was added before polycondensation process; and (b) and (d): MNa dispersion was added at the end of polycondensation process, just before polyaddition steps (O–I matrix is made in both cases from GTMS and D400).

a hindrance for the complete polycondensation process, which resulted in deteriorated mechanical properties (Table I in Ref. [144]) and in increased surface roughness (Figure 6.6). Hence, the addition of layered silicates is more convenient at the end of the sol–gel process and just before the beginning of the polyaddition process. The self-assembled inorganic stuctures are already built into the high degree of conversion (over 80% for GTMS systems and ca. 70% for GMDES-based analogues, as detected by ss-NMR spectroscopy).

6.4.1.2.2 Unification of Preparation Procedure of O–I Nanocomposite Coating

As the properties were found to be significantly dependent on the technique of the preparation [137,138,140–142], a *unified procedure* for O–I coating preparation was proposed and further utilized: the hydrolysis of alkoxy groups of GTMS (GMDES) proceeded at 23°C for 24 h (i.e., reaction mixture is composed from organosilicon precursor, water, 2-propanol, and diluted HCl; pH 4), followed by alkaline polycondensation at 23°C for 1 h (addition of oligomeric amine solution into the reaction mixture; pH is switched to 9).* The coating process preparation was completed by thermal curing for 2 h at 80°C and for 1 h at 105°C after casting the reaction mixture on treated polypropylene or glass sheets. If the coatings are prepared in the presence of a nanofiller, applying colloidal silica is appropriate just at the beginning of the hydrolysis during the sol–gel process but applying layered silicates (natural or modified MMT, bentonite, or Laponite) is convenient at the end of the polycondensation process, that is, just before the beginning of organic polymer network formation by poly-addition (thermal curing).

6.4.2 "Inside" Characterization of Nanocomposite Coatings

The variability of the properties of the O–I nanocomposite coatings prepared is not random; it originates from the differences in the system constitution. To identify what is "inside" the studied systems one needs to cover the whole scale characterization, that is, search the properties from the atomic up to the macroscopic levels. The techniques used and presented here are ss-NMR spectroscopy, SAXS and WAXS, and TEM (i.e., techniques analyzing the process up to micrometre size).

Similar to checking the preparation procedure, ss-NMR spectroscopy was found to be essential for the "inside" characterization, preferably on the segmental and molecular levels.

6.4.2.1 Solid-State NMR and SAXS

Trying to understand the relationships between molecular structure and material properties, in particular in characterizing both local and long-range orders in polymeric nanocomposites, a combination of ^{1}H–^{1}H spin-diffusion ss-NMR experiments with SAXS/WAXS measurements seems to be extremely promising. Because both the experimental approaches are complementary, their application can bring about a big picture into the hierarchic architecture of multicomponent solids. While various 2D homonuclear and heteronuclear solid-state correlation experiments employing ^{1}H–^{1}H spin diffusion confirm spectral assignment, detect through-space

---

* The *optimum* polycondensation degree of GTMS units before thermal curing is about $q_i = 0.80$, indicating a high portion of cyclization and formation of polycyclic clusters while their *aggregation is still prevented*. This indicates that isolated star-like epoxy-functionalized building blocks are formed before thermal curing. The optimum condensation rate at the end of the polycondensation step of GMDES units is about $q_i = 0.7$.

interatomic contacts (interactions), and thus elucidate global architecture and the extent of self-organization at nanometer scale, the established diffraction techniques provide information about the order of detecting polymer materials, for instance, about long-range periods (diffraction correlation length) between the *in situ* formed polysiloxane clusters in organic/inorganic hybrids, spatial separation of sheets in nanocomposites containing layered silicates, and so on.

Since 1985, when the first application of the 2D exchange experiment exploiting ^{1}H spin diffusion combined with ^{1}H detection using the CRAMPS technique in both time-evolution domains was used to probe the miscibility of a polymer blend polystyrene–polyvinylmethylether (PS–PVME) [149], numerous systems have been successfully described by this experimental approach. Even complicated morphology and intermolecular interactions were precisely determined for polymeric electrolytes based on complexes of poly(2-ethyl-2-oxazoline) with $AgCF_3SO_3$ [150], mixtures of semicrystalline polymers such as polethylenoxide and polycarbonates of bis-phenol A [151], and a range of pharmaceutical systems. Later, the experimental technique was applied to probe conformation and local arrangement of crystalline organic solids. Recently, a new approach providing complete reconstruction of a long-range periodic crystal structure has been developed. This procedure, currently known as NMR crystallography [152], thus opened a new way to characterize powdered organic solids. Because inorganic solids contain surprisingly high amounts of water, the ^{1}H–^{1}H correlation experiments have also been applied to study the chemical nature and distribution of surface hydroxyls in various silicate minerals and siloxane networks. Therefore, it is not surprising that hybrid organic/inorganic composites were extensively studied by ^{1}H–^{1}H spin-diffusion experiments. The success of this approach, however, strongly depends on the resolution of off-diagonal ^{1}H NMR signals in 2D ^{1}H–^{1}H correlation spectra, because the right dependencies of the correlation signal intensities on spin-diffusion mixing time reflect ^{1}H–^{1}H interatomic distances and/or the size of domains in heterogeneous systems.

Sufficient spectral resolution in ^{1}H dimensions of heterogeneous polymers, however, has been achieved very rarely. Rather the achieved spectral resolution allows for the separation of only a few signals reflecting the main function groups. This problem was recently observed in the case of epoxy networks reinforced by *in situ* formed siloxane clusters when we were able to resolve only the off-diagonal signals at 3.5 and 0.8 ppm, reflecting mainly the intramonomer magnetization exchange between CH_3 and CH–O (CH_2–O) protons in propyleneoxide units. Nevertheless, analysis of the obtained spin-diffusion built-up curves revealed a two-step behavior, the initial part of which was attributed to the fast intramonomer magnetization exchange while the slower step observed in the later phases of spin diffusion was attributed to long-range polarization transfer reflecting the segregation and ordering of oxypropylene chains and siloxane tails. To analyze the observed process in detail, the technique that nicely combines the advantage of ^{1}H nuclei (100% isotopic abundance) with a high resolution of ^{13}C NMR spectra was utilized. This 2D ^{1}H–^{13}C HETCOR experiment, in which ^{1}H–^{1}H spin exchange occurs during the mixing delay inserted after the first ^{1}H detection period, provides a significantly better resolution afforded by ^{1}H–^{13}C chemical shift separation. In the spectra, we can see two types of cross-peaks: first, signals reflecting the shortest (usually one-bond) ^{1}H–^{13}C distance, and second, cross-peaks reflecting long-range through-space polarization transfer, that is, long-range interatomic contacts. The latter signals can be used to estimate the geometry of the studied system. For instance, in a GTMS–silica system for a short mixing time of 250 μs, cross-peaks for only directly bonded ^{1}H–^{13}C pairs and the shortest intraresidual contacts evolve. For longer mixing times, cross-peaks related to interresidual transfer between oxypropylene chains and siloxane tails can be detected, although they are very weak and the contacts within the same types of monomer units still dominate. As complete equilibration of magnetization is reached after 2000 μs, the systems must be arranged into the partly separated domains whose size and extent of separation can be extracted from the analysis of cross-peaks buildups using the second Fick's law (Figure 6.7a). In case of epoxy–siloxane networks, the determined value of the spin-diffusion coefficient $D = 0.22$–0.24 nm^2 ms^{-1} was intermediate between the values estimated for highly mobile polymers (0.05–0.15 nm^2 ms^{-1}) and rigid hard segments (0.6–0.8 nm^2 ms^{-1}). Consequently, the largest polarization transfer pathway corresponded to ca. 0.8 nm, which roughly reflects the largest distance between CH$_2$–O (CH–O) in oxypropylene units and CH$_2$–Si protons on the surface of siloxane clusters (in the network, the largest distance corresponds to the middle of the CH$_2$O-chain). This finding indicated significant spatial separation of GTMS residues and oxypropylene chains, which are concentrated between siloxane cage-like clusters (Figure 6.7b).

The alternative approach represented by SAXS can not only confirm such assumptions but also provide precise structural information about the components that cannot be directly detected by NMR spectroscopy. This is particularly true for multicomponent hybrid organic–inorganic solids, the diffraction experiments of which significantly enhance precision of structure elucidations. For instance, the high scattering intensity at low scattering vector $q \to 0$ as demonstrated in Figure 6.7c reflects random distribution of silica particles in a polymer matrix. A small hump at $q = 0.4$ nm^{-1} reflects the size of silica particles which is ca. 29 nm. The long-range ordered "two-phase" structure is clearly detected through the interference maximum at the scattering vector $q = 3.5$ nm^{-1}, reflecting correlation distances of $\xi = 1.79$ nm, which can be interpreted as the average distance between siloxane cage-like clusters. The oxypropylene chains form a well-organized phase separating the organic tails of siloxane clusters, thus providing links between epoxy-functionalized cage-like units. It can be concluded that a combination

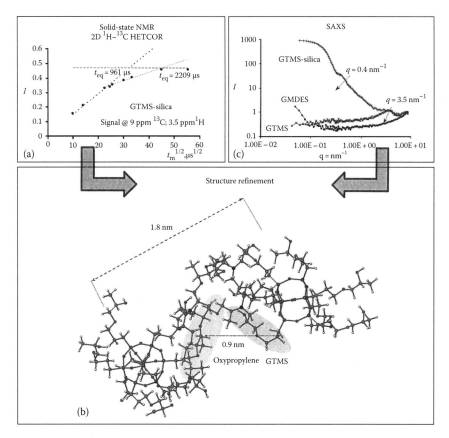

FIGURE 6.7
(a) Spin-diffusion built-up curves of the cross-peak in ^{1}H–^{13}C HETCOR spectra correlating at 9 ppm (^{13}C) and 3.5 ppm (^{1}H) obtained for GTMS–silica; (b) MM+ optimized idealized model of GTMS-based network; and (c) SAXS scattering curves obtained for the system containing *in situ* prepared siloxane cage-like clusters (GTMS, GTMS–silica, and GMDES).

of 2D ^{1}H–^{1}H CRAMPS, ^{1}H–^{13}C HETCOR NMR, and SAXS experiments further definitely confirm that GMDES-based systems are homogeneous without any regular long-range arrangement, unlike GTMS-based systems. The results obtained by ss-NMR spin-diffusion experiments are in clear agreement with x-ray scattering data.

Differences in the structural arrangement shown in Figure 6.7 for GTMS- and GMDES-based systems were used for the choice of preparation of the model O–I matrix. The preparation process is shown in Schemes 6.1 and 6.2. Further intensive detailed studies of O–I nanocomposite coatings including nanofillers were realized on two matrices, one made from GTMS and D400 (matrix "*D*") and the second from GMDES and T403 (matrix "*T*"), both at glycidyl/NH ratio equal to 1.0 [144–148].

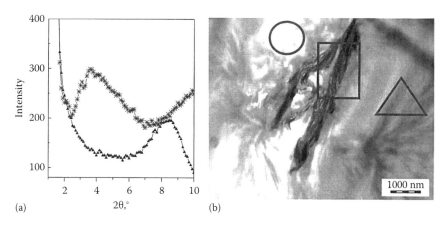

FIGURE 6.8
(a) WAXS patterns of original natural montmorillonite (MMT Na) powder (triangles; bottom) and O–I nanocomposite made from GTMS and D400 containing 1 wt.% of MMT Na (stars; top). (b) TEM image of identical O–I nanocomposite. Circle: O–I matrix, triangle: exfoliated MMT Na, and rectangle: intercalated MMT Na.

6.4.2.2 Transmission Electron Microscopy and Wide-Angle X-Ray Scattering

TEM is a powerful method for the ***direct*** visualization and determination of the internal arrangement and ordering of layered silicates in a nanocomposite matrix. Similar indirect information can be obtained by WAXS. Hence, TEM analysis was used for this purpose in combination with WAXS experiments. Figure 6.8a shows the diffraction patterns of an O–I nanocomposite containing MMT Na in comparison with the pattern of an MMT Na nanofiller powder. The lamellar periodicity of pure MMT Na is 1.21 nm and the periodicity of its mixture with a polymer increased to 2.39 nm. This change is caused by the intercalation of the O–I nanocomposite matrix between the layers of MMT. The broad diffraction maximum on the WAXS pattern of the mixture and the absence of higher orders of reflection corresponds to the not very well-ordered lamellar structure. The nanocomposite does not contain any original untreated MMT. Figure 6.8b shows MMT Na distributed in the nanocomposite. Three distinct regions are clearly visible: pure O–I matrix (circle), exfoliated region (triangle), and MMT intercalated by the O–I matrix (rectangular).

6.4.3 View "on the Surface" of Nanocomposite Coatings

As already mentioned, surface characteristics are crucial for 2D systems (films, coatings, membranes, etc.).

The presentation of the surface topography of O–I nanocomposite coatings prepared will be documented from the lowest to the highest magnification.

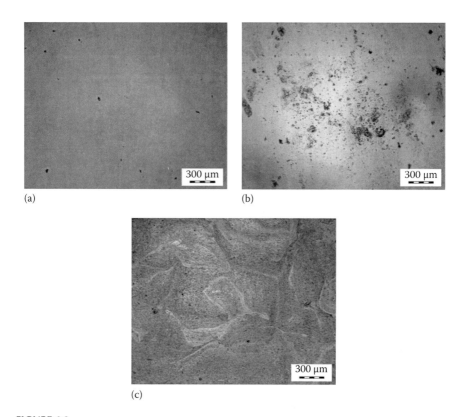

(a)   (b)

(c)

FIGURE 6.9
Surface images detected by light microscopy: O–I nanocomposite made from GTMS and D400: (a) matrix "D", (b) GTMS-D400 product containing 0.5 wt.% of bentonite for water systems, and (c) GTMS-D400 based nanocomposite containing 0.5 wt.% of bentonite modified by tris(hydroxymethyl)aminomethane (TRIS).

The information about the global relief, the arrangement, and the roughness of the coatings on the micrometer and millimeter scale obtained by light and SEM microscopies were used (Figure 6.6). As the light microscopy in the reflected light was found to be more effective than the SEM analysis tool, it was further used. Figure 6.9 shows examples of surface morphology changes due to different self-assemblies of modified nanofillers in the nanocomposite coatings.

In the condition where the surface relief on the nanometer scale is being tested (eventually units of micrometer in the maximum), atomic force microscopy was successfully used. Figure 6.10 shows differences in surface morphologies of nanocomposite coatings on the micrometer scale: 3D relief of nanocomposite coatings without any nanofiller, 2D scan of coatings containing colloidal silica particles, and 3D scans of coatings containing layered silicates: bentonite, unmodified MMT MNa, and organically modified MMT

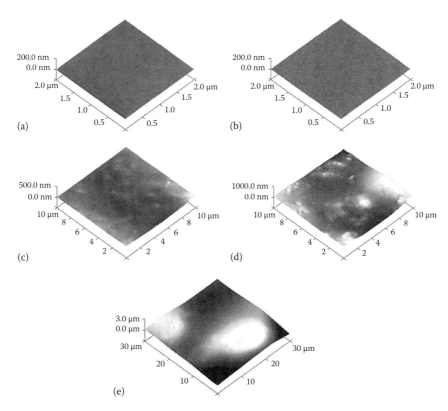

FIGURE 6.10
AFM topography images of coating surfaces: (a) O–I matrix without any additive "D"; (b) "D" with colloidal silica particles; (c) "D" with chemically modified montmorillonite; (d) "D" with bentonite for water systems; and (e) "D" with natural montmorillonite.

K30. Though the surface reliefs are different, all nanofillers are well built into the O–I nanocomposite matrix. For further details, see Refs. [137–148].

6.5 Conclusions

This brief view from inside to the surface of polymer systems shows the necessity of multidisciplinary characterization for the description of complex organic–inorganic nanocomposite coatings. The combination of "classical" and "up-to date" analytical techniques spanning from the segmental up to the macroscopic level was found to be the best method. The proper combination of analytical techniques, for example, ss-NMR spectroscopy with SAXS/WAXS analysis or WAXS/TEM, provides a deeper view "inside." Light (SEM) microscopy with atomic force microscopy offers a more precise view "on the surface" of PNCs.

Acknowledgment

The authors wish to thank institutional support RVO:61389013 and the support of the Grant Agency of the Academy of Sciences of the Czech Republic (project No. IAAX082409001). The colleagues from IMC—A. Strachota, R. Poręba, J. Hromádková, H. Vlková, and J. Baldrian—are acknowledged for their help in the graphical documentation.

References

1. Ajayan, P.M., Schadler, L.S., and Braun, P.V. 2003. *Nanocomposite Science and Technology.* Weinheim, Germany: Wiley. ISBN 3-527-30359-6, p. 2.
2. Messersmith, P.B. and Giannelis, E.P. 1995. Synthesis and barrier properties of poly(epsilon-caprolactone)-layered silicate nanocomposites. *J. Polym. Sci., Part A: Polym. Chem.,* 33:1047–1057.
3. Toselli, M., Marini, M., Fabbri, P., Messori, M., and Pilati, F. 2007. Sol-gel derived hybrid coatings for the improvement of scratch resistance of polyethylene. *J. Sol-Gel Sci. Technol.,* 43:73–83.
4. Amerio, E., Fabbri, P., Malucelli, G., Messori, M., Sangermano, M., and Taurino, R. 2008. Scratch resistance of nano-silica reinforced acrylic coatings. *Prog. Org. Coat.,* 62:129–133.
5. Duran, A., Castro, Y., Aparicio, M., Conde, A., and de Damborenea, J.J. 2007. Protection and surface modification of metals with sol-gel coatings. *Int. Mater. Rev.,* 52:175–192.
6. Rosero-Navarro, N.C., Pellice, S.A., Castro, Y., Aparicio, M., and Duran, A. 2009. Improved corrosion resistance of AA2024 alloys through hybrid organic–inorganic sol–gel coatings produced from sols with controlled polymerisation. *Surf. Coat. Tech.,* 203:1897–1903.
7. Pathak, S.S., Khanna, A.S., and Sinha, T.J.M. 2006. Sol-gel derived organic-inorganic hybrid coating: A new era in corrosion protection of material. *Corros. Rev.,* 24:281–306.
8. Wu, K.H., Chao, C.M., Yeh, T.F., and Chang, T.C. 2006. Thermal stability and corrosion resistance of polysiloxane coatings on 2024-T3 and 6061-T6 aluminum alloy. *Surf. Coat. Technol.,* 201:5782–5788.
9. He, J.Y., Zhou, L., Soucek, M.D., Wollyung, K.M., and Wesdekmiotis, C. 2007. UV-curable hybrid coatings based on vinylfunctionalized siloxane oligomer and acrylated polyester. *J. Appl. Polym. Sci.,* 105:2376–2386.
10. Grundwurmer, M., Nuyken, O., Meyer, M., Wehr, J., and Shupl, N. 2007. Sol-gel derived erosion protection coatings against damage caused by liquid impact. *Wear,* 263:318–329.
11. Sangermano, M., Gasparia, E., Vescovo, L., and Messori, M. 2011. Enhancement of scratch-resistance properties of methacrylated UV-cured coatings. *Prog. Org. Coat.,* 72:287–291.

12. Shanaghi, A., Rouhaghdam, A.S., Shahrabi, T., and Aliofkhazraei, M. 2008. Study of TiO$_2$ nanoparticle coatings by the sol-gel method for corrosion protection. *Mater. Sci.*, 44:233–247.

13. Apohan, N.K., Karatas, S., Bilen, B., and Guengoer, A., 2008. In situ formed silica nanofiber reinforced UV-curable phenylphosphine oxide containing coatings. *J. Sol-Gel Sci. Technol.*, 46:87–97.

14. Chang, K.C., Lin, H.F., Lin, C.Y., Kuo, T.H., Huang, H.H., Hsu, S.C., Yeh, J.M., Yang, J.C., and Yu, Y.H. 2008. Effect of amino-modified silica nanoparticles on the corrosion protection properties of epoxy resin-silica hybrid materials. *J. Nanosci. Nanotechnol.*, 8:3040–3049.

15. Chau, J.L.H., Tung, C.T., Lin, Y.M., and Li, A.K., 2008. Preparation and optical properties of titania/epoxy nanocomposite coatings. *Mater. Lett.*, 62:3416–3418.

16. Chiang, T.H., Liu, S.L., Lee, S.Y., and Hsieh, T.E. 2008. Preparation, microstructure, and property characterizations of fluorinated polyimide-organosilicate hybrids. *Eur. Polym. J.*, 44:3482–3492.

17. Nematollahi, M., Heidarian, M., Peikari, M., Kassiriha, S.M., Arianpouya, N., and Esmaeilpour, M. 2010. Comparison between the effect of nanoglass flake and montmorillonite organoclay on corrosion performance of epoxy coating. *Corros. Sci.*, 52:1809–1817.

18. Chen, C.G., Khobaib, M., and Curliss, D. 2003. Epoxy layered-silicate nanocomposites. *Prog. Org. Coat.*, 47:376–383.

19. Kuo, T.H., Weng, C.J., Chen, C.L., Chen, Y.L., Chang C.H., and Yeh, J.M. 2012. Electrochemical investigations on the corrosion protection effect of poly(vinyl carbazole)-silica hybrid sol–gel materials. *Polym. Composite*, 33:275–281.

20. Selvakumar, N., Jeyasubramanian, K., and Sharmila, R. 2012. Smart coating for corrosion protection by adopting nano particles. *Prog. Org. Coat.*, 74: 461–469.

21. Leroux, F., Stimpfling, T., and Hintze-Bruening, H. 2012. Relevance and performance of LDH platelets in coatings. *Rec. Pat. Nanotech.*, 6:238–248.

22. Gonzalez, E., Pavez, J., Azocar, I., Zagal, J.H., Zhou, X., Melo, F., Thompson, G.E., and Páez, M.A. 2012. A silanol-based nanocomposite coating for protection of AA-2024 aluminium alloy. *Electrochim. Acta*, 56:7586–7595.

23. Huang, T.C., Hsieh, C.F., Yeh, T.C., Lai, C.L., Tsai, M.H., and Yeh, J.M. 2011. Comparative studies on corrosion protection properties of polyimide-silica and polyimide-clay composite materials. *J. Appl. Polym. Sci.*, 119:548–557.

24. Sowntharya, L., Lavanya, S., Ravi Chandra, G., Hebalkar, N.Y., and Subasri, R. 2012. Investigations on the mechanical properties of hybrid nanocomposite hard coatings on polycarbonate. *Ceram. Int.*, 38:4221–4228.

25. Chattopadhyay, D.K., Zakula, A.D., and Webster, D.C. 2009. Organic–inorganic hybrid coatings prepared from glycidyl carbamate resin, 3-aminopropyl trimethoxy silane and tetraethoxyorthosilicate. *Prog. Org. Coat.*, 64:128–137.

26. Belon, C., Chemtob, A., Croutxe-Berghorn, C., Rigolet, S., Schmitt, M., Bistac, S., Le Houerou, V., and Gauthier, C. 2010. Nanocomposite coatings via simultaneous organic–inorganic photo-induced polymerization: Synthesis, structural investigation and mechanical characterization. *Polym. Int.*, 59:1175–1186.

27. Medda, S.M. and De, G. 2009. Inorganic-organic nanocomposite based hard coatings on plastics using in situ generated nano-SiO$_2$ bonded with ≡Si–O–Si-PEO hybrid network. *Ind. Eng. Chem. Res.*, 48:4326–4333.

28. Hwang, J.H., Lee, B.I., Klep, V., and Luzinov, I. 2008. Transparent hydrophobic organic–inorganic nanocomposite films. *Mater. Res. Bull.*, 43:2652–2657.
29. He, J., Chisholm, B.J., Mayo, B.A., Bao, H.Z., Risan, J., Christianson, D.A., and Rafferty, C.L. 2012. Hybrid organic/inorganic coatings produced using a dual-cure mechanism. *J. Coat. Technol. Res.*, 9:423–431.
30. Poovarodom, S., Hosseinpour, D., and Berg, J.C. 2008. Effect of particle aggregation on the mechanical properties of a reinforced organic-inorganic hybrid sol-gel composite. *Ind. Eng. Chem. Res.*, 47:2623–2629.
31. Hu, L. and Shi, W. 2011. UV-cured organic–inorganic hybrid nanocomposite initiated by trimethoxysilane-modified fragmental photoinitiator. *Compos. Part A-Appl. S.*, 42:631–638.
32. Amerio, E., Sangermano, M., Colucci, G., Malucelli, G., Messori, M., Taurino, R., and Fabbri, P. 2008. UV curing of organic-inorganic hybrid coatings containing polyhedral oligomeric silsesquioxane blocks. *Macromol. Mater. Eng.*, 293:700–707.
33. Chattopadhyay, D.K. and Webster, D.C. 2009. Hybrid coatings from novel silane-modified glycidyl carbamate resins and amine crosslinkers. *Prog. Org. Coat.*, 66:73–85.
34. Messersmith, P.B. and Giannelis, E.P. 1994. Synthesis and characterization of layered silicate-epoxy nanocomposites. *Chem. Mater.*, 6:1918–1925.
35. Ceccia, S., Turcato, E.A., Maffettone, P.L., and Bongiovanni, R. 2008. Nanocomposite UV-cured coatings: Organoclay intercalation by an epoxy resin. *Prog. Org. Coat.*, 63:110–115.
36. Huang, K.Y., Weng, C.J., Lin, S.Y., Yu, Y.H., and Yeh, J.M. 2009. Preparation and anticorrosive properties of hybrid coatings based on epoxy-silica hybrid materials. *J. Appl. Polym. Sci.*, 112:1933–1942.
37. Chen, C.G. and Morgan, A.B. 2009. Mild processing and characterization of silica epoxy hybrid nanocomposite. *Polymer*, 50:6265–6273.
38. Isin, D., Kayaman-Apohan, N., and Gungor, A. 2009, Preparation and characterization of UV-curable epoxy/silica nanocomposite coatings. *Prog. Org. Coat.*, 65:477–483.
39. Chen, L., Zhou, S.X., Song, S.S., Zhang, B., and Gu, G.X. 2011. Preparation and anticorrosive performances of polysiloxane-modified epoxy coatings based on polyaminopropylmethylsiloxane-containing amine curing agent. *J. Coat. Technol. Res.*, 8:481–487.
40. Singh-Beemat, J. and Iroh, J.O. 2012. Characterization of corrosion resistant clay/epoxy ester composite coatings and thin films. *Prog. Org. Coat.*, 74:173–180.
41. Tao, P., Viswanath, A., Schadler, L.S., Benicewicz, B.C., and Siegel, R.W. 2011. Preparation and optical properties of indium tin oxide/epoxy nanocomposites with polyglycidyl methacrylate grafted nanoparticles. *Appl. Mater. Interfaces*, 3:3638–3645.
42. Gopakumar, T.G., Patel, N.S., and Xanthos, M. 2006. Effect of nanofillers on the properties of flexible protective polymer coatings. *Polym. Composite*, 27:368–380.
43. Yu, Y.Y., Chien, W.C., and Chen, S.Y. 2010. Preparation and optical properties of organic/inorganic nanocomposite materials by UV curing proces. *Mater. Des.*, 31:2061–2070.
44. Mammeri, F., Teyssandier, J., Connan, C., Le Bourhis, E., and Chehimi, M.M. 2012. Mechanical properties of carbon nanotube–PMMA based hybrid coatings: The importance of surface chemistry. *RSC Adv.*, 2:2462–2468.

45. Zhang, S.W., Yu, A.X., Liu, S.L., Zhao, J., Jiang, J.Q., and Liu, X.Y. 2012. Effect of silica nanoparticles on structure and properties of waterborne UV-curable polyurethane nanocomposites. *Polym. Bull.*, 68:1469–1482.
46. Yuan, Y. and Shi, W.F. 2010. Preparation and properties of exfoliated nanocomposites through intercalated a photoinitiator into LDH interlayer used for UV curing coatings. *Prog. Org. Coat.*, 69:92–99.
47. Madbouly, S.A. and Otaigbe, J.U. 2009. Recent advances in synthesis, characterization and rheological properties of polyurethanes and POSS/polyurethane nanocomposites dispersions and films. *Prog. Polym. Sci.*, 34:1283–1332.
48. Kim, D., Jeon, K., Lee, Y., Seo, J., Seo, K., Han, H., and Khan, S. 2012. Preparation and characterization of UV-cured polyurethane acrylate/ZnO nanocomposite films based on surface modified ZnO. *Prog. Org. Coat.*, 74:435–442.
49. Heidarian, M., Shishesaz, M.R., Kassiriha, S.M., and Nematollahi, M. 2010. Characterization of structure and corrosion resistivity of polyurethane/organoclay nanocomposite coatings prepared through an ultrasonication assisted process. *Prog. Org. Coat.*, 68:180–188.
50. Jin, H., Wie, J.J., and Kim, S.C. 2010. Effect of organoclays on the properties of polyurethane/clay nanocomposite coatings. *J. Appl. Polym. Sci.*, 117:2090–2100.
51. Mishra, A.K., Narayan, R., Aminabhavi, T.M., Pradhan, S.K., and Raju, K.V.S.N. 2012. Hyperbranched polyurethane-urea-imide/o-clay-silica hybrids: Synthesis and characterization. *J. Appl. Polym. Sci.*, 125:E67–E75.
52. Ashhari, S., Sarabi, A.A., Kasiriha, S.A., and Zaarei, D. 2011. Aliphatic polyurethane–montmorillonite nanocomposite coatings: Preparation, characterization, and anticorrosive properties. *J. Appl. Polym. Sci.*, 119:523–529.
53. Iroh, J.O. and Longun, J. 2012. Viscoelastic properties of montmorillonite clay/polyimide composite membranes and thin films. *J. Inorg. Organomet. Polym.*, 22:653–661.
54. Kizilkaya, C., Dumludag, F., Karatas, S., Apohan, N.K., Altindal, A., and Gungor, A. 2012. The effect of titania content on the physical properties of polyimide/titania nanohybrid films. *J. Appl. Polym. Sci.*, 125:3802–3810.
55. Boumbimba, R.M., Bouquey, M., Muller, R., Jourdainne, L., Triki, B., Hébraud, P., and Pfeiffer, P. 2012. Dispersion and morphology of polypropylene nanocomposites: Characterization based on a compact and flexible optical sensor. *Polym. Test.*, 31:800–809.
56. Liao, S.H., Weng, C.C., Yen, C.Y., Hsiao, M.C., Ma, C.C.M., Tsai, M.C., Su, A., Yen, M.Y., Lin, Y.F., and Liu, P.L. 2010. Preparation and properties of functionalized multiwalled carbon nanotubes/polypropylene nanocomposite bipolar plates for polymer electrolyte membrane fuel cells. *J. Power Sources*, 195:263–270.
57. Lin, Y., Chen, H.B., Chan, C.M., and Wu, J.S. 2011. Effects of coating amount and particle concentration on the impact toughness of polypropylene/CaCO3 nanocomposites. *Eur. Polym. J.*, 47:294–304.
58. Lepoittevin, B., Pantoustier, N., Devalckenaere, M., Alexandre, M., Kubies, D., Calberg, C., Jerome, R., and Dubois, P. 2002. Poly(epsilon-caprolactone)/clay nanocomposites by in-situ intercalative polymerization catalyzed by dibutyltin dimethoxide *Macromolecules*, 35:8385–8390.
59. Patole, A.S., Patole, S.P., Jung, S.Y., Yoo, J.B., An, J.H., and Kim, T.H. 2012. Self assembled graphene/carbon nanotube/polystyrene hybrid nanocomposite by in situ microemulsion polymerization. *Eur. Polym. J.*, 45:252–259.

60. Siengchin, S. and Karger-Kocsis, J. 2012. Polystyrene nanocomposites produced by melt-compounding with polymer-coated magnesium carbonate nanoparticles. *J. Reinf. Plast. Comp.*, 31:145–152.
61. Jeeju, P.P. and Jayalekshmi, S. 2011. On the interesting optical properties of highly transparent, thermally stable, spin-coated polystyrene/zinc oxide nanocomposite films. *J. Appl. Polym. Sci.*, 120:1361–1366.
62. Iroh, J.O. and Rajamani, D. 2012. Synthesis and structure of environmentally friendly hybrid clay/organosilane nanocomposite coatings. *J. Inorg. Organomet. Polym.* 22:595–603.
63. Ding, X.F., Zhou, S.X., Gu, G.X., and Wu, L.M. 2011. Facile fabrication of superhydrophobic polysiloxane/magnetite nanocomposite coatings with electromagnetic shielding property. *J. Coat. Technol. Res.* 8:757–764.
64. Therias, S., Larche, J.F., Bussiere, P.O., Gardette, J.L., Murariu, M., and Dubois, P. 2012. Photochemical behavior of polylactide/ZnO nanocomposite films. *Biomacromolecules*, 13:2383–2391.
65. Silvino, A.C., de Souza, K.S., Dahmouche, K., and Dias, M.L. 2012. Polylactide/ clay nanocomposites: A fresh look into the in situ polymerization proces. *J. Appl. Polym. Sci.*, 124:1217–1224.
66. Zeng, F.L., Liu, Y.Z., Sun, Y., Hu, E.L., and Zhou, Y. 2012. Nanoindentation, nanoscratch, and nanotensile testing of poly(vinylidene fluoride)-polyhedral oligomeric silsesquioxane nanocomposites. *J. Polym. Sci. Pol. Phys.*, 50: 1597–1611.
67. Han, P., Fan, J.B., Zhu, L., Min, C.Y., Shen, X.Q., and Pan, T.Z. 2012. Structure, thermal stability and electrical properties of reduced graphene/poly(vinylidene fluoride) nanocomposite films. *J. Nanocsci. Nanotechnol.*, 12:7290–7295.
68. Kirubaharan, A.M.K., Selvaraj, M., Maruthan, K., and Jeyakumar, D. 2012. Synthesis and characterization of nanosized titanium dioxide and silicon dioxide for corrosion resistance applications. *J. Coat. Technol. Res.*, 9:163–170.
69. Stefanescu, E.A., Daranga, C., and Stefanescu, C. 2009. Insight into the broad field of polymer nanocomposites: From carbon nanotubes to clay nanoplatelets, via metal nanoparticles. *Materials*, 2(4):2095–2153.
70. Zhao, X., Sui, K.Y., Wu, W.W., Liang, H.C., Li, Y.J., Wu, Z.M., and Xia, Y.Z. 2012. Synthesis and properties of amphiphilic block polymer functionalized multi-walled carbon nanotubes and nanocomposites. *Composites*, A43:758–764.
71. Breuer, O. and Sundararaj, U. 2004. Big returns from small fibers: A review of polymer/carbon nanotube composites. *Polym. Composite*, 25:630–645.
72. Podsiadlo, P., Shim, B.S., and Kotov, N.A. 2009. Polymer/clay and polymer/carbon nanotube hybrid organic–inorganic multilayered composites made by sequential layering of nanometer scale films. *Coordin. Chem. Rev.*, 253:2835–2851.
73. Zaarei, D., Sarabi, A.A., Sharif, F., and Kassiriha, S.M. 2008. Structure, properties and corrosion resistivity of polymeric nanocomposite coatings based on layered silicates. *J. Coat. Technol. Res.*, 5:241–249.
74. Choudalakis, G. and Gotsis, A.D. 2009. Permeability of polymer/clay nanocomposites: A review. *Europ. Polym. J.*, 45:967–984.
75. Morgan, A.B. and Gilman, J.W. 2003. Characterization of polymer-layered silicate (clay) nanocomposites by transmission electron microscopy and X-ray diffraction: A comparative study. *J. Appl. Polym. Sci.*, 87:1329–1338.

76. Gianneelis, E.P. 1996. Polymer layered silicate nanocomposites. *Adv. Mater.* 8:29–35.

77. Ramanathan, T., Abdala, A.A., Stankovich, S., Dikin, D.A., Herrera-Alonso, M., Piner, R.D., Adamson, D.H. et al. 2008. Functionalized graphene sheets for polymer nanocomposites. *Nat. Nanotechnol.*, 3:327–331.

78. Kim, H.W., Abdala, A.A., and Macosko, C.W. 2010. Graphene/polymer nanocomposites. *Macromolecules*, 43:6515–6530.

79. Gong, L., Young, R.J., Kinloch, I.A., Riaz, I., Jalil, R., and Novoselov, K.S. 2012. Optimizing the reinforcement of polymer-based nanocomposites by graphene. *ACS NANO*, 6:2086–2095.

80. Sangermano, M. and Messori, M. 2010. Scratch resistance enhancement of polymer coatings. *Macromol. Mater. Eng.*, 295:603–612.

81. Emel'yanenko, A.M. and Boinovich, L.B. 2011. Analysis of wetting as an efficient method for studying the characteristics of coatings and surfaces and the processes that occur on them: A Review. *Inorg. Mat.*, 47:1667–1675.

82. Paul, D.R. and Robeson, L.M. 2008. Polymer nanotechnology: Nanocomposites. *Polymer*, 49:3187–3204.

83. Öchsner, A., Ahmed, W., and Ali, N. 2009. *Nanocomposite Coatings and Nanocomposite Materials*. Published Materials Science Foundations, Vols. 54–55, Trans. Tech. Publications, Inc. ISBN: 978-0-87849-346-3, 402p.

84. Ray, S.S., Bousmina M. 2006. *Polymer Nanocomposites and Their Applications*. ISBN: 158883-099-3

85. Friedrich, K. and Schlarb, A.K. 2008. *Tribology of Polymeric Nanocomposites. Friction and Wear of Bulk Materials and Coatings*. Oxford, U.K., Elsevier, ISBN 9780444531551. 568p.

86. Fernando, R.H. 2009. *Nanotechnology Applications in Coatings*. ACS Symposium Series 1008, Cary, NC: Oxford University Press, ISBN: 9780841274488.

87. Utracki, L.A. 2004. *Clay-Containing Polymeric Nanocomposites*. Shawbury, U.K.: Rapra Technology, ISBN1-85957-437-8 and 1-85957-482-3.

88. Rurack, K. and Martinez-Manez, R. 2010. *The Supramolecular Chemistry of Organic–Inorganic Hybrid Materials*, Hoboken, NJ: Wiley, 2010, ISBN 978-0-470-37621-8, 766p.

89. Nalwa, H.S. 2003. *Handbook of Organi–Inorganic Hybrid Materials and Nanocomposites*. 1. Hybrid materials; 381p, 2. Nanocomposites; 386p. Stevenson Ranch, CA: American Scientific Publishers. ISBN 1-58883-028-4 (Vol. 1) and 1-58883-029-2 (Vol. 2).

90. Sanchez, C., Laine, R.M., Yang, S., and Brinker, C.J. 2002. *Organic–Inorganic Hybrid Materials*. Warrendale, PA: Materials Research Society, Vol. 726, ISBN: 1-55899-662-1, 420p.

91. Chauhan, B.P.S. 2011. *Hybrid Nanomaterials. Synthesis, Characterization and Applications*. Hoboken, NJ: Wiley, ISBN 978-1-118-00349-7, 334p.

92. Merhari, L. 2009. *Hybrid Nanocomposties for Nanotechnology. Electronic, Optical, Magnetic and Biomedical Applications*. New York: Springer. ISBN 978-0-387-72398-3. 846p.

93. Mittal, V. 2012. *In-Situ Synthesis of Polymer Nanocomposites*. Weinheim, Germany: Wiley-VCH, ISBN 978-3-527-32879-6. 400p.

94. Kumar, Ch. 2011. *Polymeric Nanomaterials*. Weinheim, Germany: Wiley-VCH. ISBN: 978-3-527-32170-4. 520p.

95. Gutierrez-Wing, C., Rodriguez-Lopez, J.L., Graeve, O.A., Boeckl, J.J., and Soukiassian, P. 2011. *Nanostructured Materials and Nanotechnology*. Warrendale, PA: Materials Research Society, and Cambridge University. ISBN: 978-1-605811-348-7. 153p.

96. Mehring, M. 1983. Principles of high resolution NMR in solids, *Am. Chem. Soc.* 185:138-PHYS.

97. Schmidt-Rohr, K. and Spiess, H.W. 1994. *Multidimensional Solid-State NMR and Polymers*. London, U.K.: Academic Press, pp. 385–403.

98. Lowe, I. J. 1959. Free induction decays of rotating solids. *Phys. Rev. Lett.* 2:285–287.

99. Andrew, E.R., Bradbury, A., and Eadges, R.G. 1958. Nuclear magnetic resonance spectra from a crystal rotated at high speed. *Nature* 182:1659–1659.

100. Bennett, A.E., Rienstra, C.M., Auger, M., Lakshmi, K.V., and Griffin, R.G. 1995. Heteronuclear decoupling in rotating solids. *J. Chem. Phys.* 103:6951–6958.

101. Scholz, I., Hodgkinson, P., Meier, B.H., and Ernst, M.J. 2009. Understanding two-pulse phase-modulated decoupling in solid-state NMR. *J. Chem. Phys.* 130:114510(1)–114510(17).

102. Maricq, M.M. and Waugh, J.S. 1979. NMR in rotating solids, *J. Chem. Phys.*, 70:3300–3316.

103. Burum, D.P. and Rhim, W.K. 1979. Analysis of multiple pulse NMR in solids 3. *J. Chem. Phys.*, 71:944–956.

104. Lee, M. and Goldburg, W.I., 1965. Nuclear-magnetic-resonance line narrowing by a rotating RF field. *Phys. Rev. A*, 140:1261–1271.

105. Bielecki, A., Kolbert, A.C., de Groot, H.J.M., Griffin, R.G., and Levitt, M.H. 1990. Frequency-switched Lee-Goldburg sequences in solids, *Adv. Magn. Reson.*, 14:111.

106. Vinaogradov, E., Madhu, P. K., and Vega, S. 1999. High-resolution proton solid-state NMR spectroscopy by phase-modulated Lee-Goldburg experiment. *Chem. Phys. Lett.*, 314:443–450.

107. Sakellariou, D., Lesage, A., and Emsley, L. 2001. Proton-proton constrains in powdered solids from H-1-H-1 and H-1-H-1-C-13 three-dimensional NMR chemical soft correlation spectroscopy. *J. Am. Chem. Soc.*, 123:5604–5605.

108. Brus, J., Petříčková, H., and Dybal, J. 2002. Potential and limitations of 2D H-1-H-1 spin-exchange CRAMPS experiments to characterize structures of organic solids. *Monatshefte Fur Chemie*, 133:1587–1612.

109. Pines, A., Gibby, M.G., and Waugh, J.S. 1973. Proton-enhanced NMR of dilute spins in solids. *J. Chem. Phys.*, 59:569–590.

110. Schaefer, J. and Stejskal, E.O.J. 1976. C-13 nuclear magnetic-resonance of polymers spinning at magic angle. *J. Am. Chem. Soc.*, 98:1031–1032.

111. Kobera, L., Slavik, R., Koloušek, D., Čubová Urbanová, M., Kotek, J., and Brus, J. 2011. Structural stability of aluminosilicate inorganic polymers: Influence of the preparation procedure. *Ceramics-Silikáty*, 55:343–354.

112. Goldman, M. and Shen, L. 1966. Spin-spin relaxation in LAF3. *Phys. Rev.*, 144:321–331.

113. Lesage, A., Steuernagel, S., and Emsley, L. 1998. Carbon-13 spectral editing in solid-state NMR using heteronuclear scalar couplings. *J. Am. Chem. Soc.*, 120:7095–7100.

114. Ramamoorthy, A., Gierasch, L.M., and Opella, S.J. 1996. Three-dimensional solid-state NMR correlation experiment with H-1 homonuclear spin exchange. *J. Mag. Reson. Ser. B*, 111:81–84.

115. van Rossum, B.J., de Groot, C.P., Ladizhansky, V., Vega, S., and de Groot, H.J.M. 2000. A method for measuring heteronuclear (H-1-C-13) distances in high speed MAS NMR. *J. Am. Chem. Soc.*, 122:3465–3472.
116. Ladizhansky, V. and Vega, S. 2000. Polarization transfer dynamics in Lee-Goldburg cross polarization nuclear magnetic resonance experiments on rotating solids. *J. Chem. Phys.*, 112:7158–7168.
117. Schmidt-Rohr, K., Clauss, J., and Spiess, H.W. 1992. Correlation of structure, mobility, and morphological information in heterogeneous polymer materials by 2-dimensional wideline-separation NMR-spectroscopy. *Macromolecules*, 25:3273–3277.
118. Clauss, J., Schmidt-Rohr, K., and Spiess, H.W. 1993. Determination of domain sizes in heterogeneous polymers by solid-state NMR. *Acta Polym.*, 44:1–17.
119. Munowitz, M., Aue, W.P., and Griffin, R.G. 1982. Two-dimensional separation of dipolar and scaled isotropic chemical-shift interactions in magic angle NMR-spectra. *J. Chem. Phys.*, 77:1686–1689.
120. Brus, J., Urbanová, M., and Strachota, A. 2008. Epoxy networks reinforced with polyhedral oligomeric silsesquioxanes: Structure and segmental dynamics as studied by solid-state NMR. *Macromolecules*, 2:372–386.
121. Brus, J. and Urbanová, M. 2005 Selective measurement of heteronuclear H-1-C-13 dipolar couplings in motionally heterogeneous semicrystalline polymer systems. *J. Phys. Chem. A*, 23:5050–5054.
122. Brus, J., Urbanová, M., Kelnar, I., and Kotek, J. 2006. A solid-state NMR study of structure and segmental dynamics of semicrystalline elastomer-toughened nanocomposites. *Macromolecules*, 16:5400–5409.
123. Brus, J., Dybal, J., Sysel, P., and Hobzová, R. 2002. Mobility, structure, and domain size in polyimide-poly(dimethylsiloxane) nettworks studied by solid-state NMR spectroscopy. *Macromolecules*, 35:1253–1261.
124. Matějíček, P., Brus, J., Zhigunov, A., Pleštil, J., Uchman, M., Procházka, K., and Gradzielski, M. 2011. On the structure of polymeric composite of metallcarbo-rane with poly(ethylene oxide). *Macromolecules*, 44:3847–3858.
125. Brus, J. 2002. Solid-state NMR study of phase separation and order of water molecules and silanol groups in polysiloxane network. *J. Sol-Gel Sci.Technol.*, 25:17–28.
126. Hong, M., Yao, X., Jakes, K., and Huster, D. 2002, Investigation of molecular motions by Lee-Goldburg cross-polarization NMR Spectroscopy. *J. Phys. Chem. B*, 106:7355–7364.
127. Dvinskikh, S.V., Zimmermann, H., Maliniak, A., and Sandstrom, D. 2003. Heteronuclear dipolar recoupling in liquid crystals and solids by PISEMA-type pulse sequences. *J. Magn. Reson.*, 164:165–170.
128. Huster, D., Xiao, L.S., and Hong, M. 2001. Solid-state NMR investigation of the dynamics of the soluble and membrane-bound colicin la channel-forming domain. *Biochemistry*, 40:7662–7674.
129. Palmer, A.G., Williams, J., and McDermott, A. 1996. Nuclear magnetic resonance studies of biopolymer dynamics. *J. Phys. Chem.*, 100:13293–13310.
130. Zhang, R., Chen, Y., Chen, T., Sun, P., Li, B., and Ding, D. 2012. Accessing structure and dynamics of mobile phase in organic solids by real-time T-1C Filter PISEMA NMR spectroscopy, *J. Phys. Chem. A.*, 116:979–984.
131. Ohgo, K., Niemczura, W.P., Muroi, T., Onizuka, A.K., and Kumashiro, K.K. 2009. Wideline Separation (WISE) NMR of Native Elastin, *Macromolecules*, 42:8899–8906.

132. Yan, B. and Stark, R.E., 1998. A WISE NMR approach to heterogeneous biopolymer mixtures: Dynamics and domains in wounded potato tissues, *Macromolecules*, 31:2600–2665.

133. Demco, D.E., Johanson, A., and Tegenfeldt, J. 1995. Proton spin-diffusion for spatial heterogeneity and morphology investigations of polymers, *Solid State Nucl. Magn. Reson.*, 4:13–38.

134. García-Gutiérrez, M.C., Nogales, A., Hernádez, J.J., Rueda, D.R. and Ezquerra, T.A. 2007. X-ray scattering applied to the analysis of carbon nanotubes, polymers and nanocomposites, *Opt. Pura Apl.*, 40:195–205.

135. Bell, J.L., Sarin, P., Provis, J.L., Haggerty, R.P., Driemeyer, P.E., Chupas, P.J., van Deventer, J.S.J., and Kriven, W.M. 2008. Atomic structure of a cesium aluminosilicate geopolymer: A pair distribution function study, *Chem. Mater.*, 20:4768–4776.

136. Nogales, A., Broza, G., Roslaniec, Z., Schulte, K., Sÿics, I., Hsiao, B.S., Sanz, A. et al. 2004. Low percolation threshold in nanocomposities based on oxidized single wall carbon nanotubes and poly (butylene terephthalate, *Macromolecules*, 37:7669–7672.

137. Špírková, M., Brus, J., Hlavatá, D., Kamišová, H., Matějka, L., and Strachota, A. 2003. Preparation and characterization of hybrid organic–inorganic coatings and films. *Surf. Coat. Int. B: Coat. Trans.*, 86:187–193.

138. Brus, J., Špírková, M., Hlavatá, D., and Strachota, A. 2004. Self-organization, structure, dynamic properties and surface morphology of silica/epoxy films as seen by solid-state NMR. *Macromolecules*, 37:1346–1357.

139. Špírková, M., Stejskal, J., and Prokeš, J. 2004. Hybrid organic–inorganic coatings and films containing conducting polyaniline nanoparticles. *Macromol. Symp.*, 212:343–348.

140. Špírková, M., Brus, J., Hlavatá, D., Kamišová, H., Matějka, L., and Strachota, A. 2004. Preparation and characterization of hybrid organic–inorganic epoxide-based films and coatings prepared by sol-gel process. *J. Appl. Polym. Sci.*, 92:937–950.

141. Brus, J. and Špírková, M. 2005. NMR spectroscopy and atomic force microscopy characterization of hybrid organic–inorganic coatings. *Macromol. Symp.*, 220:155–164.

142. Špírková, M., Brus, J., Baldrian, J., Šlouf, M., Kotek, J., and Bláhová, O. 2005. Preparation and surface characterization of novel epoxy-based organic–inorganic nanocomposite coatings. *Surf. Coat. Int., Part B: Coat. Trans.*, 88:237–242.

143. Špírková, M., Šlouf, M., Bláhová, O., Farkačová, T., and Benešová, J. 2006. Submicrometer characterization of surfaces of epoxy-based organic-inorganic nanocomposite coatings. A comparison of AFM study with currently used testing techniques. *J. Appl. Polym. Sci.*, 102:5763–5774.

144. Špírková, M., Brus, J., Brožová, L., Strachota, A., Baldrian, J., Urbanová, M., Kotek, J., Strachotová, B., and Šlouf, M. 2008. A view from inside onto the surface of self-assembled nanocomposite coatings. *Prog. Org. Coat.*, 61:145–155.

145. Špírková, M., Strachota, A., Strachotová, B., and Urbanová, M. 2008. Effect of montmorillonite on properties of nanocomposite coatings. *Surf. Eng.*, 24:268–271.

146. Špírková, M., Strachota, A., Brožová, L., Brus, J., Urbanová, M., Baldrian, J., Šlouf, M., Bláhová, O., and Duchek, P. 2010. The influence of nanoadditives on surface, permeability and mechanical properties of self-organized organicinorganic nanocomposite coatings. *J. Coat. Technol. Res.*, 7:219–228.

147. Špírková, M., Duchek, P., Strachota, A., Poręba, R., Kotek, J., Baldrian, J., and Šlouf, M. 2011. The role of organic modification of layered nanosilicates on mechanical and surface properties of organic-inorganic coatings. *J. Coat. Technol. Res.*, 8:311–328.

148. Špírková, M., Duchek, P., Strachota, A., Brus, J., Poręba, R., Baldrian, J., and Šlouf, M. 2011. Organic-inorganic coatings. Preparation and characterization. pp. 313–323. In *Advances in Applied Surface Engineering*, published by Research Publishing, Singapore. ISBN 978-981-087922-8.

149. Caravatti, P., Neuenschwander, P., and Ernst, R.R. 1985. Characterization of heterogeneous polymer blends by 2-dimensional proton spin diffusion spectroscopy. *Macromolecules*, 18:119–122.

150. Spěváček, J., Brus, J., Dybal, J., and Kang, J. 2005. Solid-state C-13 NMR and DFT quantum-chemical study of polymer electrolyte poly(2-ethyl-2-oxazoline)/AgCF3SO3. *Macromolecules*, 38:5083–5087.

151. Brus, J., Dybal, J., Schmidt, P., Kratochvíl, P., and Baldrian, J. 2000. Order and mobility in polycarbonate-poly(ethylene oxide) blends studied by solid-state NMR and other techniques. *Macromolecules*, 33:6448–6459.

152. Elena, B. and Emsley, L. 2005. Powder crystallography by proton solid-state NMR spectroscopy. *J. Am. Chem. Soc.*, 127:9140–9146.

153. Knoll, M. 1935. Aufladepotentiel und Sekundäremission elektronenbestrahlter Körper. *Zeitschrift für technische Physik* 16:467–75.

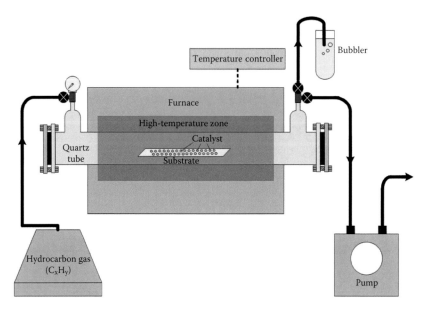

FIGURE 3.1
Schematic diagram of simplified CVD system for growing CNTs.

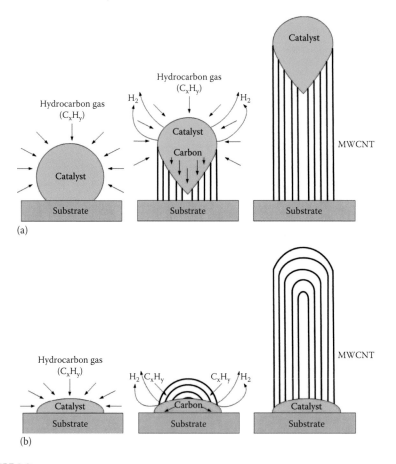

FIGURE 3.2
Growth mechanisms of CNTs in CVD system: (a) tip-growth model and (b) base-growth model.

FIGURE 5.11
Appearance of PPN-coated tablets with (a) CS film and CS-MAS films in the ratios of (b) 1:0.2, (c) 1:0.6, and (d) 1:1 at 4.3 mg cm^{-2} coating level.

FIGURE 7.4
Photograph depicting the plasma treatment process using atmospheric air.

FIGURE 9.11
Laspra treated with (a) Tegosivin HL 100 and (b) TMSPMA. Repedea treated with (c) Tegosivin HL 100 and (d) TMSPMA.

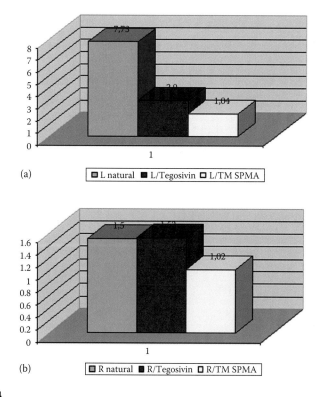

FIGURE 9.14
(a) Laspra—weight loss after 30 salt mist aging cycles and (b) Repedea—weight loss after 60 salt mist aging cycles.

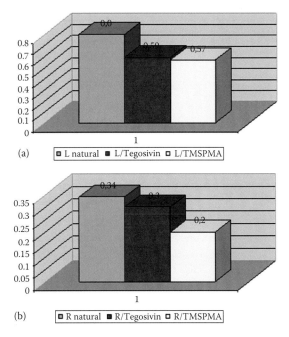

(a)
☐ L natural ■ L/Tegosivin ☐ L/TMSPMA

(b)
☐ R natural ■ R/Tegosivin ☐ R/TMSPMA

FIGURE 9.19
Weight loss of (a) Laspra after exposure to SO_2 saturated atmosphere and (b) Repedea after exposure to SO_2 saturated atmosphere.

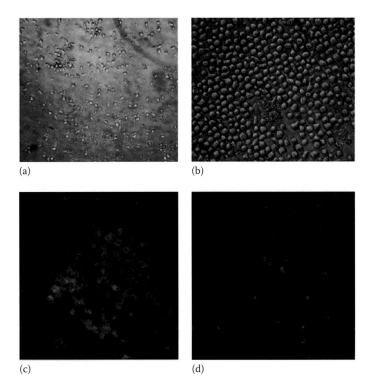

(a)

(b)

(c)

(d)

FIGURE 10.3
Proliferation of endothelial progenitor cells on POSS–PCU: (a) spindle-shaped at day 7, (b) cobblestone-shaped at day 21, (c) cells stained with von Willebrand factor, and (d) cells stained with vascular endothelial growth factor receptor-2 (VEGFR2). (Reproduced from Ghanbari, H., de Mel, A., and Seifalian, A.M., *Int. J. Nanomed.*, 6, 775, 2011. With permission from Dove Press. Copyright 2011.)

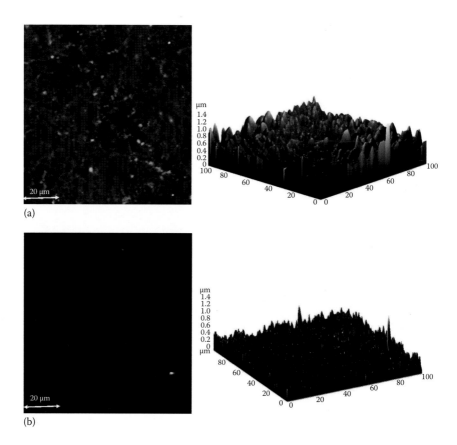

(a)

(b)

FIGURE 10.4
Atomic force microscopy of PCU and POSS–PCU. Addition of POSS into PCU (making POSS–PCU) results in nanotopography patterns on the surface of the nanocomposite. (Reproduced with permission from Springer+Business Media: *Applications of Polyhedral Oligomeric Silsesquioxanes*, Biomedical application of polyhedral oligomeric silsesquioxane nanoparticles, 2011, 363, Ghanbari, H., Marashi, S.M., Rafiei, Y., Chaloupka, K., and Seifalian, A.M., Copyright 2011.)

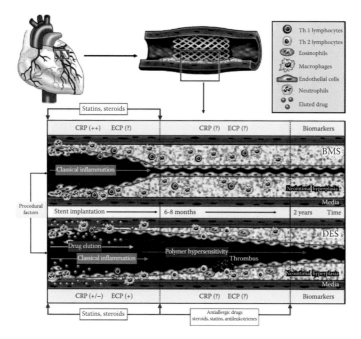

FIGURE 10.6
Problems associated with bare-metal stents (BMSs) and drug-eluting stents (DESs). In-stent restenosis (ISR) is observed with BMS, necessitating repair procedures. Although ISR was largely circumvented in DES, late stent thrombosis (ST) was seen in DES, which is a potentially fatal complication.(Reproduced from Niccoli, G., Montone, R.A., Ferrante, G., and Crea, F., The evolving role of inflammatory biomarkers in risk assessment after stent implantation, *J. Am. Coll. Cardiol.*, 56, 1783–1793, 2010, Copyright 2010. With permission from American College of Cardiology Foundation.)

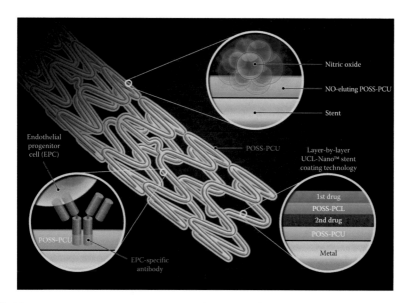

FIGURE 10.7
Convergence of biotechnology and nanotechnology. We envisage that next-generation coronary stent coatings would incorporate nanotechnology, like endothelial progenitor cell capture, biofunctionalization via nitric oxide (NO), and layer-by-layer self-assembly for sustained and controlled drug release. (Reproduced from *Trends Biotechnol.*, 30, Tan, A., Alavijeh, M.S., Seifalian, A.M., Next generation stent coatings: Convergence of biotechnology and nanotechnology, 406, Copyright 2012, with permission from Elsevier.)

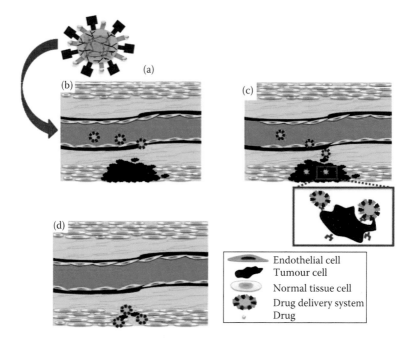

FIGURE 10.10
Therapeutic drug delivery system using POSS: (a) POSS-based nanocarrier, (b) penetration of nanocarriers through capillary walls toward cancer cells, (c) targeted drug delivery using specific receptors, and (d) destruction of cancer cells via targeted POSS-based nano-drug delivery.

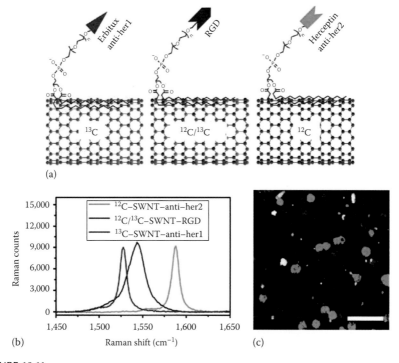

FIGURE 10.11
Carbon nanotubes (CNTs) for theranostic applications: (a) CNTs can be conjugated with antibodies and peptide sequences, (b) Raman spectroscopy of different antibody-conjugated CNTs, and (c) deconvoluted Raman image of cancer cells, which internalized different antibody-conjugated CNTs. (Copyright © 2009 Springer.)

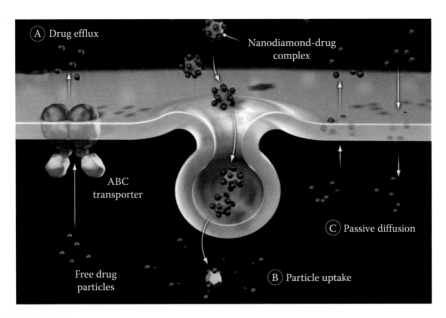

FIGURE 10.14
Nanodiamond drug delivery. Enhancement of drug delivery via nanodiamonds can be achieved with coating and functionalization of POSS–PCU. (From Merkel, T.J. and DeSimone, J.M., Dodging drug-resistant cancer with diamonds, *Sci. Trans. Med.*, 3, 73–78, 2011, Copyright 2011. Reproduced with permissions of AAAS.)

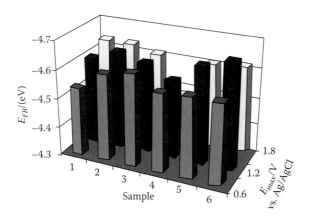

FIGURE 11.7
E_{FB} of different samples as a function of potential at which oxide was obtained. 1-CAT, 2-RES, 3-HQ, 4-HQS, 5-BDS, and 6-DHBDS.

7

Properties of Nanocomposite Hard Coatings on Polycarbonate

Raghavan Subasri

CONTENTS

7.1 Introduction

The demand for transparent plastics has been on the increase in many industrial sectors. Due to its lightweight, lower density, and excellent optical clarity than glasses, many of the transparent polymeric materials such as polycarbonate (PC), polymethylmethacrylate (PMMA), and diallyl diglycol carbonate (CR-39) are being utilized to make windows, lenses, or other optical devices [1]. PC, which belongs to the class of engineering thermoplastics, possesses superior toughness and finds application in many fields. The high impact strength and transparency of PC renders it an ideal material for applications in optical and automobile industries, replacing glass. Unfortunately, the poor abrasion and scratch resistance and low solvent resistance cause a faster deterioration of the optical quality of the uncoated plastic [2]. Improvement of such properties is a major requirement in industries and it is a scientific challenge [3]. These properties can be improved by loading suitable additives during polymer processing itself, but this method has its own problems like leaching of the additives, change in mechanical properties, and it is also not economical. An alternative solution for this problem is to give a functional protective coating above the substrates [2]. Although coating deposition on polymeric substrates is a big challenge because of their thermo-sensitive nature, and poor wettability/adhesion due to low surface free energy, in order to overcome the wear in plastics and to capitalize on their bulk properties, abrasion-resistant hard coatings have been developed over the past few years using different techniques like gas-phase, vacuum deposition (physical vapor deposition [4] and chemical vapor deposition [5,6]), and sol–gel method [7–10]. Purely inorganic coating materials were used in the earlier process, which had a problem of high-cost and poor adhesion due to the difference in thermal expansion coefficient between the substrate and the coating material. Preparation of organic–inorganic hybrid coatings by sol–gel method is an alternative to gas-phase or vacuum deposition techniques [7–12]. These hybrid coating materials are developed to achieve specific properties that organic or inorganic materials cannot provide independently [13,14]. The low-cost and low-temperature sol–gel process is the most interesting and viable alternative when compared to gas-phase and vacuum-based techniques for coating deposition on plastics.

7.2 Hybrid Sol–Gel Nanocomposite Hard Coatings

Hybrid coatings can be prepared over a continuous compositional range from almost organic to almost inorganic. The properties of these coatings can be changed continuously to form an optimum coating. The inorganic components contribute to an increase in scratch resistance, durability, and adhesion. The organic component increases thickness, density, and flexibility and makes the coatings amenable to be cured using alternate low-temperature curing methods like ultraviolet (UV) curing, etc. Organically modified sol–gel coatings can be classified into two groups based on the type of the organic part in the organoalkoxysilanes. They are nonfunctional organoalkoxysilanes and organofunctional alkoxysilanes. Alkyl or aryl groups (like methyl or phenyl) are commonly used as organic parts in the preparation of hybrids with nonfunctional component. Organofunctional precursors are epoxy, methacrylic, acrylic, allyl, alkyd, and pyridine, amino, or vinyl-functional organosilanes that can get polymerized using UV/IR radiation.

Some of the examples of commonly used organically modified silanes are listed as follows:

1. Methacryloxy propyl trimethoxysilane (MPTMS)
2. Glycidoxypropyl trimethoxy silane (GPTMS)

3. Vinyltriethoxysilane (VTES)

4. Allyltriethoxysilane (ATES)

5. Trimethoxysilylpropylmethacrylate

6. Bis-methacryloyloxypropyl aminopropyltriethoxysilane

7. γ-Aminopropyltriethoxysilane

Incorporation of nanostructured inorganic particles, like silica, boehmite, zirconia, alumina, titania, or silver, into a hybrid sol–gel matrix to improve the modulus, scratch, abrasion resistance, or antibacterial property of the coating material has been reported [15–23]. The nanoparticles can be incorporated using two methods, that is, either *in situ* building up within the hybrid matrix by an addition of fully hydrolysable silane precursors like tetraethoxysilane (TEOS) [24,25]/tetramethoxysilane or by external addition and homogeneous dispersion of nanoparticles into the hybrid matrix [26]. The properties of the resulting nanocomposites can be tailored by controlling the degree of interaction between the hybrid matrix phase and the nanoparticles. It is expected that when strong interfacial adhesion between the two phases is formed, the hybrid materials exhibit a stable homogeneous microstructure without macro-phase separation that in turn helps in achieving improved material properties. The successful application of nanoparticles depends upon both the synthesis methodology and the nature of surface of these nanoparticles. Jeon et al. reported on the dispersion of boehmite nanoparticles in an organosilane matrix (MPTMS; phenyl trimethoxy silane; vinyl trimethoxy silane; GPTMS), coatings of which when deposited on PMMA substrates yielded a pencil scratch hardness of 8H [27]. Kasemann et al. also investigated the effect of homogeneous dispersion of boehmite nanoparticles (<50 nm) in a sol-based on TEOS and GPTMS [28]. A systematic study of the structure of the nanocomposite of boehmite and hybrid sol–gel matrix was carried out using ^{27}Al, ^{13}C, and ^{29}Si NMR spectroscopy. They interestingly reported that boehmite nanoparticles act as catalysts for epoxy polymerization in the GPTMS, and a special structure is formed at the particle–matrix interface, which yields unusually high mechanical properties. GPTMS is known to undergo polymerization to polyethylene oxide in the presence of catalysts like titanium or zirconium alkoxides [29]. A schematic of different reactions that GPTMS can undergo is shown in Figure 7.1 [29]. The effect of the morphology of boehmite nanoparticles in GPTMS-based hybrid coatings was also investigated [30]. Chen et al. [30] dispersed boehmite nanoparticles and nanorods in GPTMS matrix and found that nanorods greatly improved the crack toughness of the coatings, though the modulus and hardness of the nanorod filled coatings were slightly lower than coatings of the same composition filled with boehmite nanoparticles.

In some of the previous investigations, the effect of the addition of different concentrations of colloidal silica nanoparticles in the hybrid matrix on the mechanical properties of coatings have been reported [7,8,15,16,26]. The

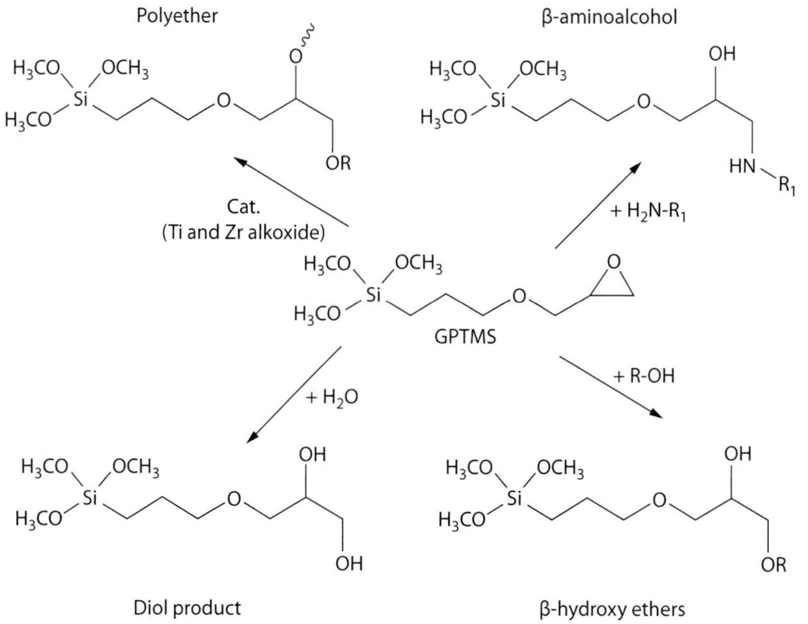

FIGURE 7.1
Schematic showing of how GPTMS can undergo different reactions to form the corresponding diol or β-hydroxy ethers or polyether linkages respectively. (From Schottner, G., *Chem. Mater.*, 13, 3422, 2001.)

major problem in dispersing the nanoparticles into an organic–inorganic hybrid matrix is the aggregation, which results in poor mechanical properties of the coatings. For example, when silica nanoparticles with no surface modification are directly loaded into hybrid matrix comprising MPTMS, it has been reported that there will be phase separation and agglomeration of the particles [31]. This is due to the incompatibility of the silica and acrylic groups from MPTMS.

Moreover, an increase in the concentration of the nanoparticle rapidly increases the sol viscosity, which affects the critical thickness of the coating [32]. In order to overcome these problems, the surface of the nanoparticles has to be modified so as to improve their interaction with the matrix. Sometimes, even though the nanoparticles used are surface modified, as shown in Figure 7.2, if the nanoparticles are not homogeneously distributed and dispersed in the hybrid matrix, this may lead to phase separation resulting in poor scratch resistance [26]. Recent investigations by Sowntharya et al. have shown that better mechanical properties can be achieved when nanoparticles are generated *in situ*, when compared to the direct addition of nanoparticles in the hybrid sol–gel matrix [33].

The processing steps for generating a hybrid sol–gel coating on PC substrate is shown in Figure 7.3.

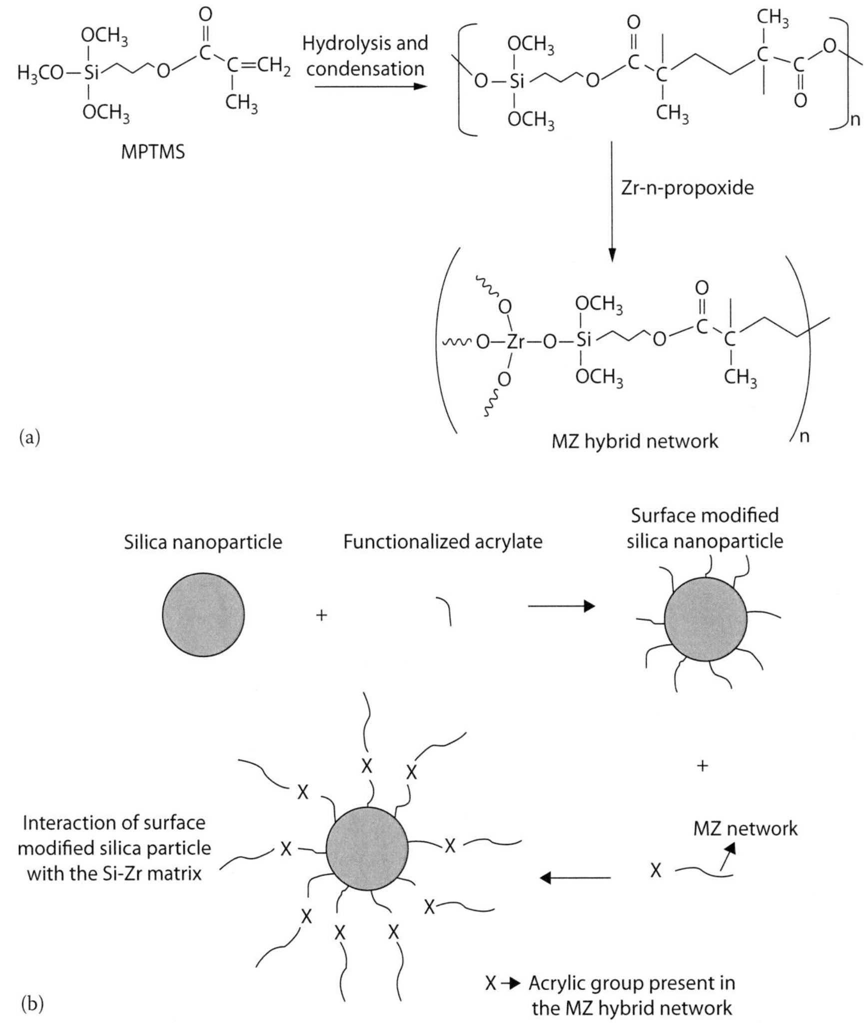

(a)

(b)

FIGURE 7.2
Schematic representation of (a) a hybrid network formed between methacryloxypropyltri-methoxy silane (M) and zirconium-n-propoxide (Z), and (b) hybrid MZ matrix with surface-modified silica nanoparticles dispersed. (Sowntharya, L. et al., *Ceram. Int.*, 39, 4245, 2013.)

7.3 Surface Cleaning/Activation

Since sol–gel coatings are in general very thin in thicknesses ranging from submicron to a few micrometers only and are necessarily transparent on transparent plastic substrates like PC, even a single dust particle or a small

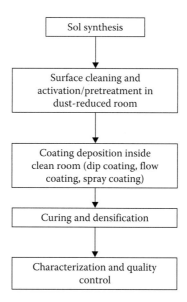

FIGURE 7.3
Schematic showing the processing steps undertaken in the sol–gel coating on polycarbonate substrates.

amount of greasy organic matter on the substrate surface can leave an uncoated area or coating with poor adhesion upon curing. Hence, cleaning of the substrate surface or activation becomes very important in ensuring good-quality coatings not only for coating continuity, but also in increasing the adhesion to the substrate [34] since the adhesion between the sol–gel coating and substrate plays a major role in improving the properties of the coating. Surface activation methods such as polymer base interlayer [34,35] have been reported to improve the adhesion of coating and the substrate. Low-pressure soft plasma treatment has been studied to modify the surface properties of polymer substrates so as to improve the adhesion between the substrate and the coating [36]. Here, the surface modification takes place due to the energetic photons generated from the plasma, which perform multiple activities like cleaning, surface activation, and surface chemical functionalization, due to which improved bonding between coating and substrate could be achieved. The effect of plasma treatment depends on the kind of plasma (DC, radio frequency [RF], or microwave [MW]); the discharge power density; the pressure; flow rate of the gas/gas mixture; and the treatment time. Accordingly, plasma activation in vacuum chamber, inert gas, and oxygen plasma are also used to activate the surface [8,21,37–39] in order to improve the wettability and bonding between the coating and substrate. Kitova et al. [38] studied the effect of Ar, Ar/H$_2$O, and Ar/C$_2$H$_5$OH RF plasma treatments on PC and PMMA substrates for treatment times varying from 1 to 30 min. They found that short, that is, 1–5 min of, Ar/C$_2$H$_5$OH and Ar/H$_2$O plasma treatments are more effective for increasing the polar part of surface free energy, and hence for improving the wettability, than the pure Ar treatment. In addition, a short Ar/C_2H_5OH treatment has the same effectiveness as Ar/H_2O treatment, but it results in more uniform polymer surfaces, without defects or damage, when compared to Ar- and Ar/H_2O-modified surfaces. In case of oxygen plasma etching [39], it was postulated that various oxygen-containing species were introduced by oxygen plasma etching into the plastic surface, including carboxylates and hydroxyl groups possibly leading to the formation of Si–O–C bonds at the interface between the coating and the plastic substrate. As a result, very good film adhesion was achieved. Recently, Gururaj et al. [40] and Soma Raju et al. [41] have reported on the use of atmospheric air plasma pretreatment of PC prior to sol–gel coating deposition of a hybrid silica–zirconia sol, as shown in Figure 7.4. The contact angles after atmospheric air plasma treatment were measured to be $43° \pm 1°$ when compared to $80° \pm 2°$ for the as-cleaned substrates, which showed improved wettability of the surface after the plasma pretreatment. However, they reported that the effect of plasma pretreatment on the PC substrates was retained for only up to 1 h after the treatment and, thereafter, slowly decreased with time and, hence, concluded that the optimum time lag between plasma treatment and coating deposition should be less than or equal to 10 min. The coating adhesion was ranked as 4B after plasma pretreatment when compared to 2B prior to treatment, which clearly indicated the improved adhesion after plasma pretreatment. Plasma treatment removes organic contaminants on the surface of the work piece by the bombardment of reactive species of plasma on the surface and thereby causes simultaneous surface oxidation [42]. Subsequently, these free radicals couple with active species from the plasma environment to form polar groups such as –(C–O)–, –(C=O)–, and –(C=O)–O– on the surface

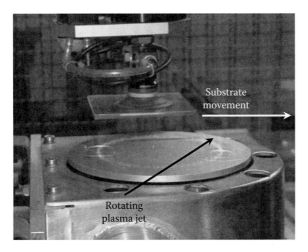

FIGURE 7.4
(See color insert.) Photograph depicting the plasma treatment process using atmospheric air.

of work piece [43]. Water molecule has a property to form hydrogen bond with polar functional groups created on the polymer surface leading to the decrease in contact angle measurements, which is a reflection of increase in surface free energy. Higher the surface free energy, lower the water contact angle, and hence the adhesive strength between polymer and coating is higher [44,45].

7.4 Alternate Curing Techniques

Plastic substrates are thermo-sensitive, and hence, the curing of coatings has to be done at temperature lower than their glass transition temperature. Conventionally, these coatings are cured thermally, which have longer cycle times and high-energy requirements. Thermal curing occurs by the conduction of heat, which results in thermal gradient leading to uneven curing. In fact, a fast curing step is a major requirement from a technology development point of view. In view of this, UV and MW curing, which offer several advantages over conventional thermal curing, have drawn attention. UV-initiated polymerization is versatile and very fast when compared to thermal curing [46]. MW energy can also be used for curing coatings, since MWs have been used in the processing of polymer composites [47,48] and curing of adhesives based on epoxy resins [49]. MW-assisted curing carried out on inorganic SiO_2 and TiO_2 coatings on PC substrates and the study of their mechanical properties have been reported recently by Dinelli et al. [50]. Most reports focus on MW curing of epoxy-based systems and very few reports on MW curing of acrylic-based systems [51]. From the technology point of view, it would be beneficial to study the effect of MW curing on acrylic groups based on organic–inorganic hybrid coatings on PC substrates, since acrylic-based coatings are more stable toward UV degradation. Recently, Sowntharya and Subasri studied and compared the mechanical properties of hybrid sol–gel coatings on PC substrates cured using MW and UV radiations [52]. Organic–inorganic hybrid nanocomposite coatings were prepared using an acrylic-based organically modified silane, namely, MPTMS (M) along with titanium tetraisopropoxide (TIPO) (T), abbreviated as MT, and coatings were cured using MW and UV radiations. The purpose of using TiO_2 in the multimetal oxide sol–gel network was that TiO_2 is a good absorber of MWs [53], and hence, there was no requirement to separately add an MW absorber in the coatings, to facilitate MW curing. Moreover, TiO_2 is also a good absorber of UV radiation. Hence, the hybrid nanocomposite coatings could be UV cured also without the addition of any photoinitiator. The authors compared the mechanical properties of the coatings cured conventionally and used MW and UV radiations. The MW and UV curing processes were of very short duration, that is, 2 and 4 min respectively, when compared to 4 h curing time for conventional curing. They also studied UV followed by MW curing. They reported that the coatings cured by only MW and only UV radiation were found not to have dried completely and exhibited poor scratch resistance lower than that of the substrate. MW followed by UV curing showed marked improvement in pencil scratch hardness when compared to that of only MW and only UV curing. Conventional curing, however, yielded the maximum scratch hardness, 3H, when compared with 2H obtained for MW- plus UV-cured coatings. They reported that fine tuning in the dual curing process (MW + UV) would further enhance the mechanical properties and will provide substantial time saving for the curing of the coatings. From a technology point of view, MW and UV curing show good promise to be very good alternatives to thermal curing, though it still requires several research investigations prior to commercialization.

7.5 Mechanical Property Studies on Hard Coatings on Polycarbonate

7.5.1 Scratch Hardness and Adhesion

The scratch hardness of hybrid sol–gel coatings is usually reported in terms of its pencil scratch resistance; the pencil test uses constant pressure and variable hardness of the test tool as its fundamental principal. The degree of hardness of the pencil that damages the surface is taken as a measurement for scratch hardness. According to the usual testing procedure, carried out according to ASTM D3363-05 [54], there are different pencils varying from grades 9B to 9H, and the order of hardness increases as given in Figure 7.5. The pencil scratch test is well accepted by the industry. However, the damage caused on coatings on soft plastic substrates, by the scratch generated using a pencil lead, is a complex phenomenon [8]. A precise understanding of the pencil scratch failure mode along with the factors affecting the pencil hardness is vital. Two categories of scratch behavior were put forth by Douce et al. [55], namely, (1) fine scratches due to very small particles and (2) visible and wide cracks due to bigger particles. The latter is closely related to the pencil scratch hardness, which was discussed in detail by Wu et al. [8]. The substrate and coating properties contribute to the total scratch resistance. They reported that the failure mode during pencil scratching contains more than one failure mechanism, which is not considered in the classical scratch

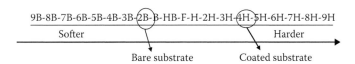

FIGURE 7.5
Schematic showing the gradient in pencil scratch hardness.

mechanism. The pencil scratch hardness of a nanocomposite coating with GPTMS/TEOS mixture along with colloidal silica nanoparticles as filler was studied. Different filler concentrations were used, and coatings of different thicknesses were generated. It was found that the pencil scratch hardness increased with increase in coating thickness. If the coating thickness were increased to such an extent so as to accommodate the plastic zone of the substrate, no scratches would occur. An effective way of increasing the scratch resistance would be to increase the coating thickness, which can be achieved by way of multiple layer coating or by increasing the withdrawal speed, if the coating deposition is done by dip coating. In case of multilayered coating, it has to be ensured that the interface between the two layers is adherent.

The scratch hardness of coatings is also measured using a scratch tester, which normally uses a diamond indenter tip. This is a simple and rapid method to characterize the coatings, but the results obtained are influenced by various factors such as coating thickness, substrate mechanical properties, interfacial bond strength, and test conditions such as scratch speed, load, and indenter tip radius. Scratch tester has a tip (normally with a radius of 200 μm), which is placed with a controlled scratch load F_z on the surface to be tested. The scratch tester tip is brought in contact with the coated surface, and the sample is moved at a constant speed while the tip normal load is progressively increased to the set maximum value. The output is measured in terms of acoustic emission, penetration depth, and tangential frictional force. When the sample is scratched, the tip is stationary and the sample moves. The resulting frictional force F_x is monitored while scratch track is generated. The onset load for coating cracking is commonly referred to as the critical load and is associated with the rise in the value of the coefficient of friction and acoustic emission.

Scratch resistance is evaluated on the basis of critical load, crack formation, and recovery [32]. The coatings with excellent scratch resistance give evidence to high critical load, low cracking tendency, and high recovery. In one of the investigations where mechanical properties of hybrid nanocomposite coatings on PC were studied [33], for one set of coatings, an *in situ* silica network was generated by making use of TEOS in the hybrid sol–gel matrix and another set was generated by adding surface-modified silica nanoparticles to the hybrid sol–gel matrix. Scratch testing revealed that the coatings with *in situ* generated silica network showed enhanced scratch resistance when compared to other coatings where nanoparticles were added externally.

Adhesion of hybrid sol–gel nanocomposite hard coatings is usually measured using cross hatch cut and a tape peel test according to ASTM D3359-02 method [56].

7.5.2 Nanoindentation Hardness

The nanoindentation hardness is an intrinsic property and gives an indication of the hardness of the coating despite the coating thickness as shown in

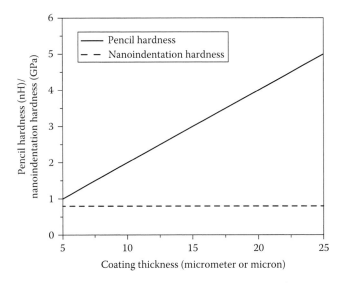

FIGURE 7.6
Schematic showing variation of pencil hardness and invariation of nanoindentation hardness with thickness. (From Wu, Y.L. et al., *Thin Solid Films*, 516, 1056, 2008.)

Figure 7.6. Usually, the hardness and elastic modulus are measured using the continuous stiffness measurement techniques in the displacement-control mode. The load at the final depth of indentation is typically held for a duration of 30 s, and the samples are indented to depths of 1/10th of the coating thickness, since it is generally accepted that when the depth of indentation is 1/10th or less than the coating thickness, the substrate effect is avoided. The abrasion resistance of the coatings depends on the hardness and brittleness index of the coating material. Brittleness index is the ratio of hardness to the Young modulus [57]. For the coating to have better abrasion resistance, the nanoindentation hardness has to increase while keeping a low value of brittleness index. The brittleness index and Vickers hardness of PC are reported as 0.06 and 0.15 GPa respectively in the literature [57]. Any coating with hardness of more than 0.15 GPa and brittleness index within the range of 0.06–0.08 will have good mechanical properties and it can be judged as useful for hard coating applications on PC. This property is further enhanced by the addition of nanoparticles like silica or zirconia.

7.5.3 Abrasion Resistance

The abrasion resistance of transparent coatings on transparent plastic substrates like PC is measured using ASTM D4060-10 [58] and ASTM D1044-08 [59]. The abrasion of coated substrates is carried out using an abrader with a fixed load and specific abrading wheels (e.g., CS17, CS10, CS10F), and for a fixed number of revolutions, as per ASTM D4060-10. The haze of the abraded specimens is measured using a hazemeter employing the method as per ASTM D1044-08. The change in haze for the uncoated and coated substrates is evaluated as the abrasion resistance for substrate and coating respectively and compared. Usually, a change in haze of <2% for coatings after abrasion test using a load of 2×250 g with CS10F wheels, for 1000 abrading cycles, is considered as useful for automotive applications. Abrasion resistance also varies with the functional group in the organically modified silanes used for the synthesis of the hybrid matrix. Sowntharya et al. [60] reported that when hybrid matrices synthesized from MPTMS or GPTMS along with TIPO deposited on PC substrates are investigated for their mechanical properties like nanoindentation hardness, abrasion resistance, scratch resistance, etc., coatings derived from MPTMS exhibit higher intrinsic nanoindentation hardness than those derived from GPTMS, whereas coatings derived from GPTMS show higher abrasion and pencil scratch hardness than those coatings derived from MPTMS due to higher coating thickness. TIPO catalyzes the epoxy polymerization of GPTMS [61] and rapidly increases the coating thickness, which in turn increases the scratch and abrasion resistance [8].

7.6 Possible Applications

7.6.1 Retroreflective Lenses of Pavement Markers

The retroreflective lenses used in pavement markers have a reduced lifetime on the roads due to scratches caused by dust particles or due to sharp metallic objects lying on the roads that get attached to the wheels, which subsequently scratch the lenses during unexpected lane crossing. Such lenses require a protective coating, in order to increase their life when installed on roads. Multilayered hybrid sol–gel coating stacks with a bottom impact-resistant layer and a top scratch-resistant coating can provide very good abrasion resistance and increase the life of lenses [62]. Together, such a system is expected to provide sufficient scratch and abrasion resistance. The first layer (A) is obtained by acid-catalyzed hydrolysis and polycondensation of two UV-curable silanes (MPTMS and VTES) along with TEOS. This layer is superimposed by (B), product of acid-catalyzed hydrolysis and condensation of MPTMS and Zr-n-propoxide. Figure 7.7 shows a schematic of the functionally gradient bilayered coating and the comparison of the abrasion resistance of a bilayered coating, where the inner layer adjacent to the substrate is mostly organic and, hence, impact-resistant, and the outer layer is mostly inorganic (zirconia dispersed in a hybrid sol–gel matrix), which provides the necessary scratch resistance.

Other important applications of hard coatings on PC include (a) windshields of automotives and (b) biaspheric lenses used in indirect ophthalmoscopy.

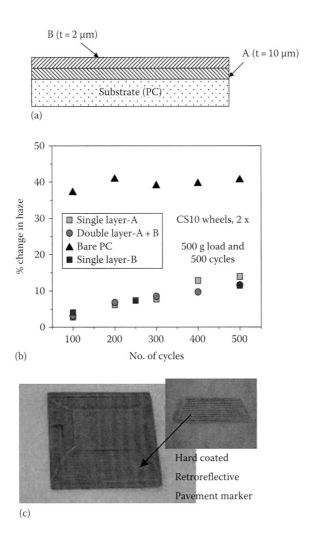

FIGURE 7.7
(a) Schematic of a bilayered functionally graded coating; layer A—impact-resistant coating and layer B—scratch-resistant coating, (b) abrasion resistance as a function of change in haze for single and bilayered coatings on PC substrate; (c) photograph of pavement marker with a protective coated retroreflective lens.

7.7 Conclusions

Hybrid sol–gel coatings provide a viable method of improving the mechanical properties of transparent plastics like PC. Nanocomposite hybrid sol–gel coatings, where the nanoparticles are generated *in situ*, improve the mechanical properties better than the dispersion of nanoparticles in the hybrid sol–gel matrix. A plasma surface pretreatment of the PC substrates prior to coatings greatly improves the adhesion between the substrate and coating. Alternate curing using MW or UV radiation substantially reduces the time taken for curing when compared to conventional thermal curing employed for coating densification. Depending on the end use of the PC components, different configurations of the hybrid nanocomposite coatings are designed, which can be single layered or multilayered. Sol–gel coating technology is a cost-effective and easy method of improving the mechanical properties of transparent plastics like PC.

Acknowledgments

The author gratefully acknowledges the support and constant encouragement provided by Dr G. Sundararajan, Director, and Dr G. Padmanabham, Associate Director, ARCI, Hyderabad, India. The author also sincerely acknowledges all present and past team members of Centre for Sol–Gel Coatings, ARCI, Hyderabad, India, who have contributed substantially to the research activities on the topic of protective hybrid sol–gel coatings on polycarbonate, some of which have been cited in this chapter.

References

1. C. Li, K. Jordens, and G.L. Wilkes, Abrasion-resistant coatings for plastic and soft metallic substrates by sol-gel reactions of a triethoxysilylated diethylenetriamine and tetramethoxysilane, *Wear* 242 (2000) 152–159.
2. N. Nakayama and T. Hayashi, Synthesis of novel UV-curable difunctional thiourethane methacrylate and studies on organic–inorganic nanocomposite hard coatings for high refractive index plastic lenses, *Prog. Org. Coat.* 62 (2008) 274–284.
3. P. Fabbri, C. Leonelli, M. Messori, F. Pilati, M. Toselli, P. Veronesi, S.M. Thérias, A. Rivaton, and J. L. Gardette, Improvement of surface properties of polycarbonate by organic–Inorganic hybrid coating, *J. Appl. Polym. Sci.* 108 (2008) 1426–1436.
4. S.V. Fortuna, Y.P. Sharkeev, A.J. Perry, J.N. Matossian, and I.A. Shulepov, Microstructural features of wear-resistant titanium nitride coatings deposited by different methods, *Thin Solid Films* 377 (2000) 512–517.
5. R.J. Martin-Palma, R. Gago, V. Torres-Costa, P. Fernandez-Hidalgo, U. Kreissig, and J. M. Martinez Duart, Optical and compositional analysis of functional SiO_xC_y: H coatings on polymers, *Thin Solid Films* 515 (2006) 2493–2496.
6. T. Schmauder, K.-D. Nauenberg, K. Kruse, and G. Ickes, Hard coatings by plasma CVD on polycarbonate for automotive and optical applications, *Thin Solid Films* 502 (2006) 270–274.

7. L.Y.L.Wu, G.H. Tan, X.T. Zeng, T.H. Li, and Z. Chen, Synthesis and characterization of transparent hydrophobic sol-gel hard coatings, *J. Sol–Gel Sci. Technol.* 38 (2006) 85–89.

8. L.Y.L. Wu, E. Chwa, Z. Chen, and X.T. Zeng, A study towards improving mechanical properties of sol-gel coatings for polycarbonate, *Thin Solid Films* 516 (2008) 1056–1062.

9. H. Schmidt, Transparent inorganic/organic copolymer by sol- gel process, *J. Sol-Gel Sci. Technol.* 1 (1994) 217–231.

10. S. Sepeur, N. Kunze, B. Werner, and H. Schmidt, UV curable hard coatings on plastics, *Thin Solid Films* 351 (1999) 216–219.

11. P.G. Romero and C. Sanchez, (eds.) *Functional Hybrid Materials*, Weinheim, Germany: Wiley VCH Verlag GmbH & Co., 2004

12. G. Kickelbick, (ed.) *Hybrid Materials: Synthesis, Characterization and Applications*, Weinheim, Germany: Wiley VCH Verlag GmbH & Co., 2007.

13. Y.-H. Han, A. Taylor, M.D. Mantle, and K.M. Knowles, UV curing of organic–inorganic hybrid coating materials, *J. Sol-Gel Sci. Technol.* 43 (2007) 111–123.

14. R. Kasemann and H. Schmidt, Coatings for mechanical and chemical protection based on organic-inorganic sol-gel nanocomposite, *New J. Chem.* 18 (1994) 1117–1123.

15. M. Spirkova, A. Strachota, L. Brozova, J. Bruz, M. Urbanova, J. Baldrian, M. Slouf, O. Blahova, and P. Duchek, The influence of nanoadditives on surface, permeability and mechanical properties of self-organized organic–inorganic nanocomposite coatings, *J. Coat. Technol. Res.* 7 (2010) 219–228.

16. H. Schmidt, Nanoparticles by chemical synthesis, processing to materials and innovative applications, *Appl. Organometal. Chem.* 15 (2001) 331–343.

17. S.K. Medda and G. De, Inorganic-organic nanocomposite-based hard coatings on plastics using in-situ generated nano-SiO_2 bonded with $\equiv$Si–O–Si-PEO hybrid network, *Ind. Eng. Chem. Res.* 48 (2009) 4326–4333.

18. G. De and D. Kundu, Silver-nanocluster-doped inorganic–organic hybrid coatings on polycarbonate substrates, *J. Non-Cryst. Solids* 288 (2001) 221–225.

19. S.K. Medda, D. Kundu, and G. De, Inorganic–organic hybrid coatings on polycarbonate, spectroscopic studies on the simultaneous polymerizations of methacrylate and silica networks, *J. Non-Cryst. Solid* 318 (2003) 149–156.

20. L.Y.L. Wu, A.M. Soutar, and X.T. Zeng, Increasing hydrophobicity of sol-gel hard coatings by chemical and morphological modifications, *Surf. Coat. Technol.* 198 (2005) 420–424.

21. D.K. Hwang, J.H. Moon, Y.G. Shul, K.T. Jung, D.H. Kim, and D.W. Lee, Scratch resistant and transparent UV- protective coating on polycarbonate, *J. Sol-Gel Sci. Technol.* 26 (2003) 783–787.

22. M.S. Lee and N.J. Jo, Coating of Methyltriethoxysilane-modified colloidal silica on polymer substrates for abrasion resistance, *J. Sol-Gel Sci. Technol.* 24 (2002) 175–180.

23. M.E.L. Woulters, D.P. Wolfs, M.C. van der Linde, J.H.P. Hovens, and A.H.A. Tinnemans, Transparent UV curable antistatic hybrid coatings on polycarbonate prepared by the sol-gel method, *Prog. Org. Coat.* 51 (2004) 312–320.

24. B. Ramezanzadeh, M. Mohseni, and A. Karbasi, Preparation of sol–gel-based nanostructured hybrid coatings; part 1: Morphological and mechanical studies, *J. Mater. Sci.* 47 (2012) 440–454.

25. Y. Ma, M. Kanezashi, and T. Tsuru, Synthesis and characterization of hydrophobic silica using methyltriethoxysilane and tetraethorthosilicate as a co-precursor, *J. Sol-Gel Sci. Technol.* 53 (2010) 93–99.

26. M. Sangermano and M. Messori, Scratch resistance enhancement of polymer coatings, *Macromol. Mater. Eng.* 295 (2010) 603–612.

27. S.J. Jeon, J.J. Lee, W. Kim, T.S. Chang, and S.M. Koo, Hard coating films based on organosilane-modified boehmite nanoparticles under UV/thermal dual curing, *Thin Solid Films* 516 (2008) 3904–3909.

28. R. Kasemann, H.K. Schmidt, and E. Wintrich, A new type of a sol-gel-derived inorganic–organic nanocomposite, *Proc. Mater. Res. Soc.*, 346 (1994) 915–921.

29. G. Schottner, Hybrid sol-gel derived polymers: Applications of multifunctional materials, *Chem. Mater.* 13 (2001) 3422–3435.

30. Q. Chen, J.G.H. Tan, S.C. Shen, Y.C. Liu, W.K. Ng, and X.T. Zeng, Effect of boehmite nanorods on the properties of glycidoxypropyl-trimethoxysilane (GPTS) hybrid coatings, *J. Sol-Gel Sci. Technol.* 44 (2007) 125–131.

31. D.L. Chandler, A new approach to scratch resistance, MIT News office, Cambridge, MA, (August 17, 2011). http://web.mit.edu/newsoffice/2011/scratch-resistance-0817.html.

32. E. Amerio, P. Fabbri, G. Malucelli, M. Messori, M. Sangermano, and R. Taurino, Scratch resistance of nano-silica reinforced acrylic coatings, *Prog. Org. Coat.* 62 (2008) 129–133.

33. L. Sowntharya, R.C. Gundakaram, K.R.C. Soma Raju, and R. Subasri, Effect of addition of surface modified nanosilica into silica-zirconia hybrid sol-gel matrix, *Ceram. Int.* 39 (2013) 4245–4252.

34. A. Hozumi, Y. Kato, and O. Takai, Two-layer hard coatings on transparent resin substrates for improvement of abrasion resistance, *Surf. Coat. Tech.* 82 (1996) 16–22.

35. Y.J. Shin, M.H. Oh, Y.S. Yoon, and J.S. Shin, Hard coatings on polycarbonate plate by sol-gel reactions of melamine derivative, PHEMA and silicates, *Polym. Eng. Sci.* 48 (2008) 1289–1295.

36. E.M. Liston, L. Martinu, and M.R. Wertheimer, Plasma surface modification of polymers for improved adhesion: A critical review, *J. Adhesion Sci. Technol.* 7 (1993) 1091–1127.

37. Z. Chen, L.Y.L. Wu, E. Chwa, and O. Tham, Scratch resistance of brittle thin films on complaint substrates, *Mater. Sci. Eng. A* 493 (2008) 292–298.

38. S. Kitova, M. Minchev, and G. Danev, Soft plasma treatment of polymer surfaces, *J. Optoelectron. Adv. Mater.* 7 (2005) 249–252.

39. C.M. Chan, G. Z. Cao, H. Fong, and M. Sarikaya, Nanoindentation and adhesion of sol-gel- derived hard coatings on polyester, *J. Mater. Res.* 15 (2000) 148–154.

40. T. Gururaj, R. Subasri, K.R.C. Soma Raju, and G. Padmanabham, Effect of plasma pre-treatment on adhesion and mechanical properties of UV-curable coatings on plastics, *Appl. Surf. Sci.* 257 (2011) 4360–4364.

41. K.R.C. Soma Raju, L. Sowntharya, S. Lavanya, and R. Subasri, Effect of plasma pre-treatment on adhesion and mechanical properties of sol-gel nanocomposite coatings on polycarbonate, *Compos. Interfaces.* 19 (2012) 259–270.

42. C.H. Yi, Y.H. Lee, and G.Y. Yeom, The study of atmospheric pressure plasma for surface cleaning, *Surf. Coat. Technol.* 171 (2002) 237–240.

43. S.K. Koh, J.S. Cho, K.H. Kim, S. Han, Y.W. Beag, Altering a polymer surface chemical structure by an ion-assisted reaction, *J. Adhes. Sci. Technol.* 16 (2002) 129–142.

44. C.-S. Ren, K. Wang, Q.-Y. Nie, D.-Z. Wang, and S.-H. Guo, Surface modification of PE film by DBD plasma in air, *Appl. Surf. Sci.* 255 (2008) 3421–3425.

45. G. Borcia, C.A. Anderson, and N.M.D. Brown, The surface oxidation of selected polymers using an atmospheric pressure air dielectric barrier discharge-part I, *Appl. Surf. Sci.* 221 (2004) 203–214.

46. C.F. Tsang, Investigation into the curing and thermal behaviour of an epoxy-based UV curable coating in microelectronics assembly, *Microelectron. Int.* 7 (2000) 27–41.

47. R. Yusoff, M.K. Aroua, A. Nesbitt, and R.J. Day, Curing of polymeric composites using microwave resin transfer moulding (RTM), *J. Eng. Sci. Technol.* 2 (2007) 151–163.

48. N. Uyanik, M. Karadayi, and C. Erbil, Studies on thermal and morphological characteristics of siloxane containing epoxy resins cured by microwave radiation, *Surf. Coat. Int. Part. B-Coat. Trans.* 87 (2004) 71–148.

49. B. Soesatyo, A. Blicblau, and E. Siores, Effect of rapid curing doped epoxy adhesive between two polycarbonate substrates on the bond tensile strength, *J. Mater. Process. Technol.* 89–90 (1999) 451–456.

50. M. Dinelli, E. Fabbri, and F. Bondioli, TiO$_2$–SiO$_2$ hard coating on polycarbonate substrate by microwave assisted sol–gel technique, *J. Sol-Gel Sci. Technol.* 58 (2011) 463–469.

51. C.P. Lai, M.-H. Tsai, M. Chen, H.-S. Chang, and H.-H. Tay, Morphology and properties of denture acrylic resins cured by microwave energy and conventional water bath, *Dent. Mater.* 20 (2004) 133–141.

52. L. Sowntharya and R. Subasri, A comparative study of different curing techniques for SiO$_2$–TiO$_2$ hybrid coatings on polycarbonate, *Ceram Int.* 39 (2013) 4689–4693.

53. D. Di Claudio, A.R. Phani, and S. Santucci, Enhanced optical properties of sol–gel derived TiO$_2$ films using microwave irradiation, *Opt. Mater.* 30 (2007) 279–284.

54. ASTM D3363-05, *Standard Test Method for Film Hardness by Pencil Test*, West Conshohocken, PA: ASTM International 2000.

55. J. Douce, J.-P. Boilot, J. Biteau, L. Scodellaro, and A. Jimenez, Effect of filler size and surface condition of nano-sized silica particles in polysiloxane coatings, *Thin Solid Films* 466 (2004) 114–122.

56. ASTM D3359-02, *Standard Test Methods for Measuring Adhesion by Tape Test*, West Conshohocken, PA: ASTM International 2002.

57. L. Hu, X. Zhang, Y. Sun, and R.J.J. Williams, Hardness and elastic modulus profiles of hybrid coatings, *J. Sol–Gel Sci. Technol.* 34 (2005) 41–46.

58. ASTM D4060-10, *Standard Test Method for Abrasion Resistance of Organic Coatings by the Taber Abraser*, ASTM International 2010.

59. ASTM D1044-08, *Standard Test Method for Resistance of Transparent Plastics to Surface Abrasion*, ASTM International 2008.

60. L. Sowntharya, S. Lavanya, G. Ravi Chandra, N.Y. Hebalkar, and R. Subasri, Investigations on the mechanical properties of hybrid nanocomposite hard coatings on polycarbonate, *Ceram. Int.* 38 (2012) 4221–4228.

61. G. Philipp and H. Schmidt, The reactivity of TiO$_2$ and ZrO$_2$ in organically modified silicates, *J. Non-Cryst. Solids* 82 (1986) 31–36.

62. T. Gururaj, K.R.C. Soma Raju, R. Subasri, and G. Padmanabham, Indian patent application entitled. *Improved Scratch and Abrasion Resistant Compositions for Coating Plastic Surfaces, A Process for Their Preparation and a Process for Coating Using the Compositions* filed as no. 2427/DEL/2010 on 12/10/2010.

8

UV-Cured Polymer: Boehmite Nanocomposite Coatings

Carola Esposito Corcione and Mariaenrica Frigione

CONTENTS

8.1 Introduction

Polymer composites reinforced with inorganic fillers of dimensions in the nanometer range, known as nanocomposites, have recently attracted a great interest due to unexpected synergistic properties derived from two components. The most studied polymer nanocomposites are composed of thermoplastic matrix and organically modified montmorillonite [1–5].

The studies focused on thermosetting nanocomposites are fewer than those carried out on thermoplastic matrices. Even if the thermosets cover a wide range of industrial applications, in fact, their use as nanocomposites is less common than thermoplastics. Adhesives and coatings, especially based on epoxies and acrylates, are two widespread families of products that are under continuous development. Due to the intrinsic fragility of thermosetting resins, the use of reinforcing agents for such applications is often mandatory: this represents the main reason for the recent popularity of studies on nanocomposites.

The formation of an epoxy nanocomposite through montmorillonite exfoliation is one of the most common routes to obtain better performance [6,7], for instance, increased Young's modulus [8] and fracture toughness [9]. Nanocomposites prepared with nano-carbon fibers generally display higher temperature capability and mechanical performance, extremely high environmental corrosion resistance, and good dimensional control compared to neat epoxy [10]. Only a few examples related to the incorporation of boehmite into polymers to prepare nanocomposite materials have been reported in literature [11–15].

The boehmite particles are colloidal plate-like crystals with a high anisotropy. They consist of double layers of oxygen octahedra partially filled with Al cations [16]. Their aqueous dispersions exhibit flow birefringence and thixotropy [17]. A valuable advantage of boehmite nanoparticles is their availability on a large industrial scale coupled with a tailorable interface (either hydrophobic, hydrophilic, or silane treated), which is able to promote their dispersion in a large number of resins. A proper surface modification is required, in fact, to obtain a good compatibility with different matrices still using the conventional preparation techniques for nanocomposites [18].

Nanocomposites based on acrylic resins have been studied in the last 10 years [19], especially due to the growth of UV-curable acrylic-based coatings in different industries. UV-cured coatings display important advantages for the finishing technology, in terms of reduced curing times, absence of solvent emissions, good scratch and wear resistance, and improvements in final performance [11–27].

In this study, starting from a photopolymerizable siloxane-modified acrylic formulation previous developed [15,27–29], UV-cured boehmite nanocomposites were prepared as potential protective coatings for wood elements. Photo-DSC and fourier transform infrared (FTIR) spectroscopy were used to study the effect of the inclusion of nanofiller on the kinetics of polymerization. The rheological behavior of formulations produced was studied as a function of the shear rate using a parallel plate rheometer. On the formulations coated on a glass substrate and photo-cured in air by using a medium-pressure Hg UV-lamp, the following properties were measured: glass transition temperature, gel content, transparency, scratch and surface hardness, contact angle, and accelerated weathering resistance. The water absorption behavior of walnut wood elements coated with nanocomposite was finally analyzed.

8.2 Experimental

8.2.1 Materials

Trimethylolpropane trimethacrylate (TMPTMA) was chosen as the main polymeric component of the coating, due to its high reactivity. The product, whose formula is schematized in Scheme 8.1, was supplied by Cray Valley.

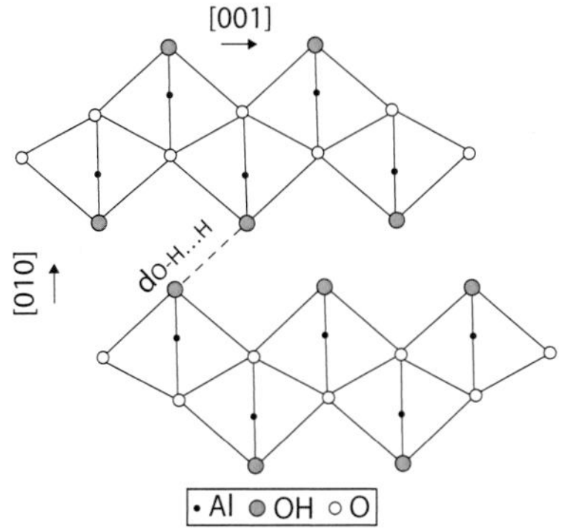

UB

$[H_2C{=}C(CH_3)CO_2CH_2]_3CC_2H_5$

TMPTMA

$CH_2{=}C(CH_3){-}COO{-}(CH_2)_3{-}Si(OCH_3)_3$

MEMO

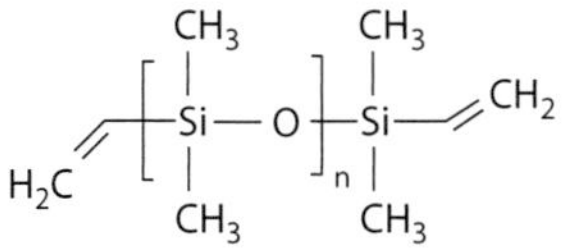

VT PDMS

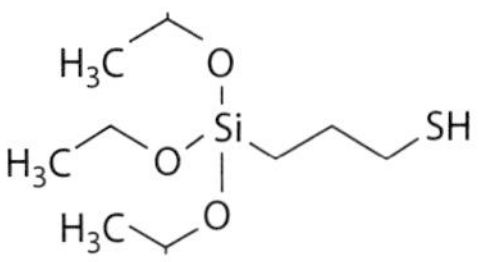

(MPTS)

IRGACURE 819 and IRGACURE 819 DW

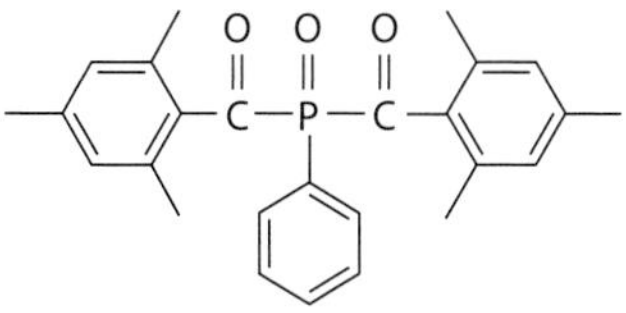

SCHEME 8.1
Chemical formula of all the materials employed.

Organically modified boehmite (OMB), supplied by Sasol, commercially available as Disperal-MEMO was selected as a nanofiller for the methacrylic resin. The organic stabilizer for OMB nanoparticles is the trimethoxypropyl silane metacrylate (MEMO). MEMO was also used as a coupling agent to enhance the dispersion of the nanofiller into organic coatings. It was supplied by Dow

Corning as Z6030. The projection of the crystalline structure of unmodified boehmite and the chemical formula of MEMO is reported in Scheme 8.1.

Vinyl-terminated polydimethylsiloxane (VT PDMS), supplied by Aldrich, was added to the methacrylic mixture to enhance the water repellence of the coating. The VT PDMS used has a number-average molecular weight in the region of 25,000. Its chemical formula is again reported in Scheme 8.1.

3-Mercaptopropyltriethoxysilane (MPTS), supplied by Aldrich, was added to the siloxane-modified methacrylic resin in order to reduce the effect of the inhibition of oxygen towards radical photopolymerization, as previously demonstrated [29]. MPTS has a number-average molecular weight of about 238.42 g/mol. The functionalization of VT PDMS with MPTS was performed by mixing the two components at 100°C in 1:1 molar ratio in the presence of 1% wt. of Diethylamine (DTA, supplied by Aldrich).

The photoinitiator, Irgacure 819, was purchased from Ciba. It was chosen for its broad absorption characteristics and its capability to catalyze the reaction under the UV-radiation. Both unfilled and filled mixtures contained one part per hundred (pph) of the photoinitiator. The chemical formula of Irgacure 819 is reported in Scheme 8.1.

8.2.2 Preparation of Nanocomposite Mixture

In order to improve the dispersion of OMB in the methacrylic-based mixture, and in turn to obtain a true nanocomposite, two different modifications of the conventional "solvent dispersion method" were proposed in a recent paper [15]. Following these procedures, OMB was initially dispersed in MEMO silane, due to its high compatibility with modified boehmite, and subsequently added to TMPTMA resin and VT PDMS, the latter functionalized with MPTS (in the following text named as PDMS$_m$). It was found that the method employing a magnetic stirring for long times was more efficient to disperse the boehmite than the simple sonication procedure. With the first procedure, in fact, a more homogeneous and more stable dispersed nanofilled system was achieved (average dimension of the particles: about 80 nm), both in liquid and cured states [15]. Starting from these findings, the optimized procedure employing a magnetic stirring was chosen to prepare the nanocomposite.

Details of the composition of organic control and the nanofilled suspensions produced are reported in Table 8.1.

8.2.3 Experimental Techniques

The UV-curing process was monitored for the filled and unfilled mixtures by a differential scanning calorimeter (Mettler Toledo DSC1 StareSystem), which is able to allow the irradiation of the sample by means of a UV/visible-lamp. This technique is known as photo-DSC (p-DSC). The light source, produced by a 300 W Xenon lamp Hamamatsu LC8, was limited to a wavelength of 370 nm, corresponding to UV-radiation. Small-size samples (0.9–1.1 mg) were

TABLE 8.1

Details of Composition of the Systems Produced

Sample	Weight Composition
$100T_{819}$	100% TMPTMA + 1% Irgacure 819
$97T$-$3PDMS_{819}$	97% TMPTMA + 3% VT PDMS + 1% Irgacure 819
$97T$-$3PDMS_{m819}$	97% TMPTMA + 3% VT PDMS/MPTS + 1% Irgacure 819
$90T$-$10PDMS_{m819}$	90% TMPTMA + 10% VT PDMS/MPTS + 1% Irgacure 819
$80T$-$20PDMS_{m819}$	80%TMPTMA + 20% VT PDMS/MPTS + 1% Irgacure 819
$50T$-$50PDMS_{m819}$	50% TMPTMA + 50% VT PDMS/MPTS + 1% Irgacure 819
$85T$-$10M$-$5PDMS_{m819}$	85% TMPTMA, 10% MEMO, + 5% VT PDMS/MPTS + 1% Irgacure 819
$85T$-$10M$-$5PDMS_m$-$3OMB_{819}$	85% TMPTMA, 10% MEMO, 5% VT PDMS/MPTS + 1% Irgacure 819 + 3pph OMB

used in order to achieve isothermal conditions and a uniform degree of cure through the sample thickness. Isothermal scans were run at 25°C in nitrogen atmosphere and with a radiation intensity of 9.6 µW/mm². The experiment was continued until no residual exothermal signal could be detected. In all the DSC experiments, each sample was irradiated after 30 s from the beginning of the test. During the first 30 s, in fact, the sample was maintained under dark conditions. The tangent to the heat flow curve during the test under dark conditions was used as a baseline for peak integration. This procedure reduced the possibilities that the slow recovering of the baseline used for kinetic analysis is inside the experimental errors. The photo-calorimetric experiments were repeated at least three times to check the accuracy of results.

FTIR spectrometer (FT/IR-6300 Jasco) was used to monitor the consumption of the constituents in liquid mixtures, and in turn to analyze the crosslinking reaction mechanism. The tests were carried out in air, in order to simulate the true curing conditions. Preliminary FTIR analyses were performed on a thin layer of uncured mixtures, using a NaCl support. The same sample was then irradiated using the same xenon lamp used in the p-DSC experiments (with an irradiation intensity of 9.6 µW/mm²) every 10 s, up to 12 h. After each cross-linking reaction step, the sample on the support was placed in the FTIR instrument, and its characteristic peaks and their intensities were determined.

The rheological characterization of all the formulation realized was carried out in a strain-controlled rheometer (Ares Rheometric Scientific). All the formulations were tested after their preparation, avoiding any aging process. The tests were performed with a plate and plate flow geometry (radius of the plate 12.5 mm) in steady-state mode. The experimental parameters were chosen to simulate the possible operative conditions. To this end, the tests were conducted at room temperature (25°C) using a shear rate range of 0.05 to 200 s⁻¹. A first sweep experiment was always followed by a second experiment performed on the same sample, and using the same conditions, in order to avoid any aggregation processes. The rheological experiments were repeated at least three times to check the repeatability of results. No settling phenomena occurred in any formulation produced.

Both filled and unfilled liquid mixtures were applied on a glass substrate and UV-cured for 12 h using a UV-lamp with an irradiation intensity of 9.6 $\mu W/mm^2$.

The morphology and size of the boehmite particles, dispersed in UV-cured films, were analyzed by scanning electron microscopy (SEM) using a Jeol JSM-6550F.

The transmittance spectra of the cured coatings (of 100 μm thickness) were recorded at normal light incidence in the spectral range of wavelength 200–1500 nm with a Varian Cary 500 UV–Vis–NIR double-beam spectrophotometer.

The glass transition temperature and residual heat of cross-linking reactions, when present, of all UV-cured systems, removed from the glass support, were measured by using differential scanning calorimetry (Mettler Toledo DSC1 StareSystem), heating 10–20 mg of each sample from 20°C to 250°C, at a heating rate of 10°C/min in nitrogen atmosphere. Three specimens for each formulation were analyzed in DSC, and the calorimetric results were averaged.

Dynamic contact angle measurements of the UV-cured films coated on glass substrate were performed with a COSTECH instrument, equipped with a video camera and an image analyzer. The analyses were performed at room temperature by means of the sessile drop technique. The measuring liquid was distilled water. Twenty measurements were performed on each sample, and the values were averaged.

Pencil hardness measurements were used to determine the hardness of 100 μm thickness coating films, according to the standard test method for film hardness by pencil test, that is, ASTM D 3633 [30].

The Hardness Shore D was measured on all the coatings using a digital Durometer Sh D, Gibitre Instruments, according to ASTM D2240 [31]. The thickness of each tested sample was 1 mm. In this case, liquid mixtures were poured in polycarbonate molds and subsequently UV-cured, using the same conditions previously described (lamp intensity, temperature, and time). Five tests were performed for each system.

The photo degradation of the unfilled and filled films was analyzed by QUV accelerated weathering tester, which was able to reproduce the damage caused by sunlight, rain, and dew. To this aim, the UV-cured films coated on aluminum foils were subjected to alternating cycles (100 h each) consisting of exposure to UV-light and moisture (60%), at a controlled temperature (25°C). The test parameters were settled according to EN ISO 11507 [32]. Two subsequent cycles were performed on each specimen. After each cycle, the Tg, the contact angle, the pencil and Shore D hardness, and the color of the films were evaluated, in order to estimate the degradation of each system upon exposure to accelerated weathering.

The evaluation of the color change of films, coated on glass substrates before and after QUV exposure, was carried out by means of a Konica Minolta CR-410 in total reflectance and double channel mode, using a xenon lamp light source. The determinations were carried out according to the corresponding normal protocol 43/93 [33]. The color changes were evaluated by the L*a*b* system (ASTM D-1925, CIE 1976) and expressed as ΔE. Twenty color determinations were carried out on different spots of each glass-coated specimen before and after each cycle was performed. The instrument was calibrated on a standard white plate reference before each measurement. On the same aged films, the transmittance analysis was also performed, as previously described.

Chemical modifications brought about by the QUV cycles were also quantitatively measured by attenuated total reflection infrared spectroscopy (ATR) analysis performed on the UV-cured films before and after the accelerated aging. A PerkinElmer Spectrum One instrument was employed to this aim, for performing the scan from 4000 to 500 cm^{-1}.

Gel content was determined on the UV-cured films by measuring the weight loss after a 48 h extraction at 80°C, according to the standard test method, ASTM D2665-84 [34].

Finally, the efficacy as water repellence coating of the acrylic-based nanocomposite was proved by applying the new product on walnut wood specimens. The procedure used for the UV-curing of coatings after their application was 12 h using a UV lamp with an irradiation intensity of 9.6 μWatt/mm^2. Sets of 10 coated and uncoated specimens ($20 \times 20 \times 20$ mm^3) of wood were oven dried at 105°C ± 2°C until a constant weight was reached and then placed in a tank, filled with deionized water for 21 days. Fresh water was replaced daily during the immersion period. The samples were periodically removed from the tank, shacked to remove residual water, weighed, and their dimensions measured. Then, water absorption values (WA$_t$) were calculated according to the following equation:

$$WA_t = \frac{W_t - W_0}{W_0} * 100 \tag{8.1}$$

where
   W$_t$ was the weight of the specimen soaked in water for time t(g)
   W$_0$ was the initial dry weight of the specimen (g)

8.3 Results

8.3.1 Kinetic Characterization

The values of the heat of reaction (H$_{max}$) and the time to reach peak (t$_{peak}$) measured in p-DSC for each formulation are reported in Table 8.2. The addition of either a silane coupling agent or a high-molecular-weight polysiloxane monomer to the methacrylic resin is found to reduce the heat of reaction, both

TABLE 8.2

Calorimetric Results from p-DSC (the Values of Heat of Reaction, H_{max}, Have Been Normalized to the Effective Resins' Mixture Content)

Sample	H_{max} (J/g)	t_{peak} (s)	H_{max} (J/g)	t_{peak} (s)
	Nitrogen		Air	
$100T_{819}$	768.5 ± 43.1	14.0 ± 1.4	321.8 ± 32.1	16.0 ± 1.6
97 T-3 PDMS$_{819}$	263.2 ± 2.6	12.0 ± 0.9	212.9 ± 1.2	12.0 ± 0.9
97 T-3 PDMS$_{m819}$	313.7 ± 3.0	12.0 ± 0.8	298.8 ± 1.1	12.1 ± 0.9
90 T-10 PDMS$_{m819}$	289.9 ± 4.4	13.1 ± 2.1	274.2 ± 7.0	14.9 ± 1.2
80 T-20 PDMS$_{m819}$	244.2 ± 6.5	17.4 ± 3.7	223.4 ± 6.6	29.4 ± 5.1
50 T-50 PDMS$_{m819}$	232.7 ± 5.8	31.1 ± 3.7	211.0 ± 2.8	43.1 ± 3.5
85T-10M-5 PDMS$_{m819}$	272.1 ± 4.5	12.0 ± 1.1	228.9 ± 1.6	12.9 ± 0.2
85T-10M-5PDMS$_m$-3OMB$_{819}$	275.0 ± 5.6	12.1 ± 0.5	232.4 ± 0.9	13.1 ± 1.7

in an inert and in air atmosphere. This effect could be attributed to the low reactivity of the silane MEMO and VT PDMS, which was confirmed by the observation that the calculated heat of reaction of MEMO is much lower than that of the acrylic resin TMPTMA, employing the same experimental conditions, as previously reported [24]. Similar results were found by other authors for an acrylate polydimethylsiloxane (AF-PDMS) added as a reactive additive in UV-curable coating formulations based on mixtures of trimethylol propane triacrylate and polyester acrylate, although in that case the photopolymerization was carried out only in air [35]. Moreover, the possibility of a small effect arising from the poor solubility of the UV-photoinitiator in the liquid mixture with VT PDMS cannot be excluded, as previously observed [26].

Referring to the value of the time to reach the peak, a t_{peak} similar to the value calculated for the TMPTMA resin is obtained when VT PDMS is also present in the mixture. The reactivity, expressed in terms of both heat developed and rate of reaction, is generally found to decrease when the UV-photopolymerization is carried out in air, always at room temperature and with a radiation intensity of 9.6 μW/mm², due to the inhibiting action of the oxygen towards free radical polymerization [26]. It is well known that oxygen inhibition of free radical photopolymerization causes numerous deleterious effects on free-radically cured products, including slow polymerization rates, long induction periods, low conversion, short polymeric chain length, and tacky surface properties [36–38]. The reaction mechanism of free radical photopolymerization can be represented with three subsequent steps: initiation, propagation, and termination. In the absence of oxygen, the initiation reactions can be schematically described by the following:

$$I + h\nu \xrightarrow{k_d} R^* \tag{8.2}$$

$$R^* + M \xrightarrow{k_i} M_1^* \tag{8.3}$$

In the first step, Equation 8.2 represents the photolysis of initiator, I, to give two primary radicals, R*, while Equation 8.3 is the chain initiation process. In the latter reaction, a primary radical reacts with monomer, M, to form the first repeat unit of the growing polymer chain, M_1^*. The rate of both reactions is controlled by the kinetic constants for the photolysis of initiator, k_d, and the chain initiation, k_i, respectively.

When the curing reaction takes place in the presence of oxygen, in addition to reactions schematized with Equations 8.1 and 8.2, the chain initiation process comprises an additional reaction [36–38]:

$$R^* + O_2 \xrightarrow{k_0} RO_2^*$$ (8.4)

where k_0 is the kinetic constant of the radical scavenging by O_2 molecules. When the reaction is conducted in air, then a part of radicals, R*, are subtracted by reacting with the monomer since they can also react with oxygen molecules present in air. As a consequence, the reaction slows down and proceeds with a lower conversion.

The data reported in Table 8.2 confirms that the heat of reaction of 97T-3PDMS$_{819}$ mixture is decreased by about 19% when the reaction is carried out in air instead of nitrogen. Under air atmosphere, the values of the heat of reaction calculated for all mixtures can be primarily attributed to the reaction of the acrylic resin. Indeed, when the reaction is performed in air, it is not possible to register any heat for the reaction of MEMO or VT PDMS in the presence of the UV-photoinitiator, when using a low-power UV-lamp, such as that operating in p-DSC. The presence of MPTS in 97T-3PDMS$_{819}$ mixture allows to limit the decreasing of the heat of reaction, due to the oxygen effect, to only 5%. Moreover, the presence of MPTS is able to increase the conversion of the reactive species even in the presence of nitrogen, as can be inferred by the increase in the heat of reaction. This result is explained with an increased reactivity of the VT PDMS functionalized with MPTS. In fact, while the heat of reaction of VT PDMS in isolation is about 329 J/g [24], it increases to about 504 J/g when VT PDMS is functionalized with MPTS. The trend observed for the time to reach the exothermic peak of all the mixtures is similar to that observed when the reaction is carried out in an inert atmosphere.

The effect of the presence of the OMB nanoparticles on the kinetic behavior of the siloxane-modified methacrylate resin mixture 80T-10M-5PDMS$_{m819}$ is analyzed by p-DSC. The maximum heat of reaction, H_{max}, normalized to the weight of the organic compound (TMPTMA/MEMO/mPDMS), and the time to reach the exothermic peak are again reported in Table 8.2. The presence of boehmite nanoparticles in the siloxane-modified acrylic resin does not appreciably affect the maximum heat of reaction, nor the time to reach the exothermic peak. The latter result confirms that, even in the presence of small amount of nanoparticles, the polymerization induced in nitrogen or air atmosphere by UV-radiation is still very fast and compatible with the potential use of the experimented products as protective coatings.

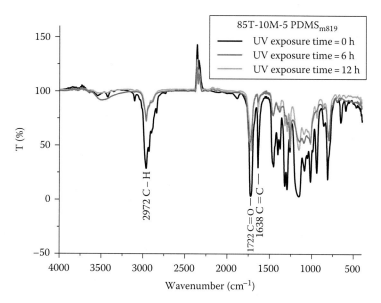

FIGURE 8.1
FTIR curves relative to 85T-10M-5 PDMS$_{m819}$ system.

FTIR analysis was performed on two systems, 85T-10M-5 PDMS$_{m819}$ and 85T-10M-5PDMS$_m$-3OMB$_{819}$, before and after the photopolymerization process at different times, up to 12 h, in air atmosphere, in order to deeply analyze the effect of the presence of OMB nanoparticles on the kinetic photopolymerization mechanism of the methacrylic-based mixture. The FTIR spectra for the organic and nanofilled systems before and during (after 6 and 12 h) the photopolymerization reactions are reported in Figures 8.1 and 8.2, respectively.

The presence of the nanofiller in the 85T-10M-5PDMS$_m$-3OMB$_{819}$ mixture is confirmed by the peak at 500 cm^{-1}, present only in the spectra reported in Figure 8.2, and that corresponding to AlOOH boehmite particles, as reported in literature [39].

The presence of the peak at 1638 cm^{-1}, related to the double C=C bond of the organic monomer, can be still noticed in all the spectra reported in Figures 8.1 and 8.2 (better evidenced in the frame shown in Figure 8.2b). This result suggests that the UV-curing of the organic unfilled and nano-filled mixtures was not complete in air, even after 12 h of exposure to the UV-radiation. This means that the inhibition effect of oxygen towards free radical photopolymerization, even if reduced by the presence of MPTS [30], is not completely eliminated. The identification of the main peaks with the associated chemical bonds and wave numbers is reported in Table 8.3.

The maximum extent of reaction (α) for both systems is, then, determined as follows [40]:

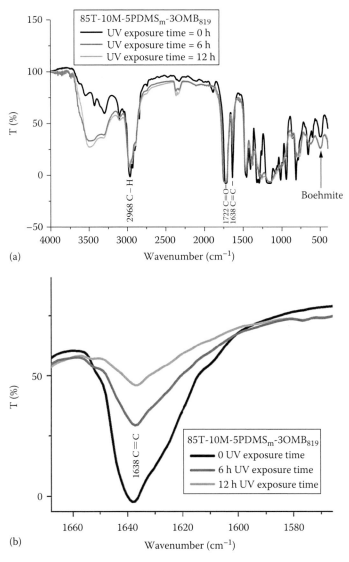

FIGURE 8.2
(a) FTIR curves related to 85T-10M-5PDMS$_m$-3OMB$_{819}$ system; and (b) frame of the peak at 1638 cm^{-1}, related to the double C=C bond of the organic monomer.

$$\alpha = \frac{(A_{C=C})_{ti} - (A_{C=C})_{t0}}{(A_{C=C})_{t0}} \tag{8.5}$$

where

$(A_{C=C})_{ti}$ is the absorbance related to C=C bond at time ti

$(A_{C=C})_{t0}$ is the absorbance of the invariant related to C=C bond of the initial time t_0

TABLE 8.3

Identification of the Main FTIR Peaks with the Associated
Chemical Bonds and Wave Numbers

Chemical Bond	Wave Numbers (cm^{-1})	
	85T-10M-5 PDMS$_{m819}$	85T-10M-5PDMS$_m$-3OMB$_{819}$
CH $_{stretching}$	2972	2966
CH $_{bending}$	1470	1470
C=O $_{stretching}$	1722	1722
C=C $_{stretching}$	1638	1638
C–O $_{stretching}$	1160	1160
P=O $_{stretching}$	1297	1297
Si–O–C $_{stretching}$	1017	1017
Si–H$_3$ $_{bending\ asimm.}$	935	935
AlOOH	— —	500

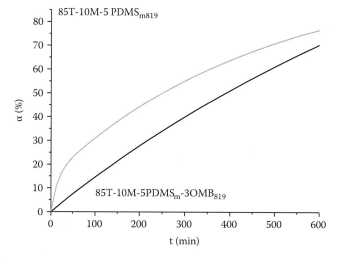

FIGURE 8.3
Maximum extent of reaction (α) for 85T-10M-5PDMS$_m$-3OMB$_{819}$ and 85T-10M-5PDMS$_m$-3OMB$_{819}$
systems, obtained from FTIR results.

The results obtained are reported in Figure 8.3.

It can be noticed that, after a 600 min exposure to UV-radiation in air
atmosphere, both the systems achieve a final α value not higher than 70%.
Moreover, the nanofilled system shows a slightly lower extent of reaction,
even though the difference in reactivity is reduced for longer exposure times.

As a final consideration, the UV-curing times required to the new systems
under consideration, in the order of 10 h, are still much lower than the typical
curing times required to the systems thermally cured at room temperature,
which are in the order of weeks [41].

8.3.2 Rheological Characterization

In Figure 8.4, the viscosities of neat methacrylic resin photo-initiated with Irgacure 819 (100 T_{819}) and of the photo-initiated resin in the presence of 3% wt. of VT PDMS (97T-3PDMS$_{819}$) are reported as a function of the shear rate. Both mixtures display a fairly Newtonian behavior, and no differences in the rheological curve were observed from the two consecutive steps performed on the same specimen. The presence of 3% wt. of VT PDMS in the photo-initiated TMPTMA resin allows to slightly increase the viscosity of the mixture, that is, from 38 to 46 mPa*s.

In order to further increase the viscosity of the protective mixtures to a more appropriate value, as required for the specific application selected, the VT PDMS content in the mixtures with TMPTMA was increased from 3% to 50% wt. Moreover, VT PDMS was functionalized with MPTS monomer (as previously described), in order to reduce the negative effect of the presence of oxygen. The rheological curves of all the mixtures containing VT PDMS/MPTS are reported in Figure 8.5.

All the mixtures reported in Figure 8.5 exhibit a pseudo-plastic rheological behavior, that is, the viscosity decreases with shear rate. The presence of MPTS determines, therefore, a change in the rheological behavior from Newtonian (Figure 8.4) to pseudo-plastic (Figure 8.5) in the mixture composed of 97% TMPTMA + 3% VT PDMS + 1.5% Irgacure 819. This feature could be attributed to the reaction of functionalization between VT PDMS and MPTS monomers. It is also evident that the viscosity of the liquid mixtures increases by increasing the content of VT PDMS functionalized with MPTS. The viscosity achieves the adequate value for the selected application, [42], that is, protective coatings for porous stones elements, in the

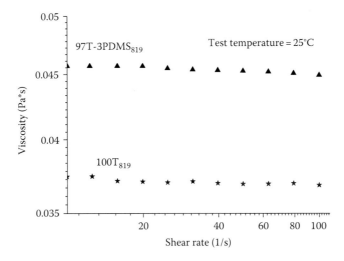

FIGURE 8.4
Experimental viscosity curves for some of the mixtures realized as a function of shear rate.

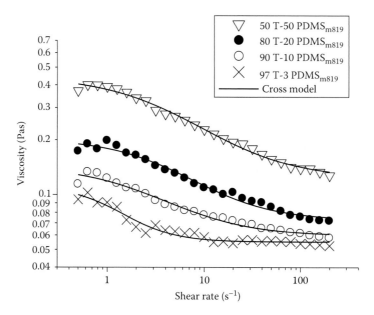

FIGURE 8.5
Experimental viscosity curves for some of the mixtures realized as a function of shear rate with the predictions of cross model (full lines).

presence of 50% wt. of VT PDMS/MPTS (i.e., about 0.3–0.5 Pa*s). Moreover, the occurrence of a pseudo-plastic behavior at the lowest shear rate values would exclude any particle settling during the preparation stage; a lower value of viscosity at medium shear rate values would be an advantage during the application stage, in particular by spreading.

The viscosity curves obtained are fitted according to the cross model [26,43]

$$\eta = \eta_\infty + \frac{\eta_0 - \eta_\infty}{1 + \left(\lambda\dot{\gamma}\right)^m} \tag{8.6}$$

where
η_0 and η_∞ are the lower and upper Newtonian limit viscosities
λ is the reciprocal of the shear rate at which the calculated value of η
    equals η_0
the parameter m is related to the power law index, n, by the expression:
$m = 1 - n$

The fitting curves are reported in Figure 8.5, as full lines, and the relative parameters are listed in Table 8.4. A good agreement can be observed between the experimental results and model predictions.

TABLE 8.4

Model Parameters of the Cross Model Fitting Experimental
Viscosity Data

Mixture	η_0 (Pa*s)	η_∞ (Pa*s)	λ (s)	m
97 T-3 PDMS$_{m819}$	0.44 ± 0.01	0.12 ± 0.01	0.22 ± 0.01	0.88 ± 0.12
90 T-10 PDMS$_{m819}$	0.20 ± 0.01	0.07 ± 0.01	0.22 ± 0.04	0.96 ± 0.13
80 T-20 PDMS$_{m819}$	0.14 ± 0.00	0.06 ± 0.00	0.39 ± 0.13	1.01 ± 0.18
50 T-50 PDMS$_{m819}$	0.11 ± 0.01	0.06 ± 0.00	0.85 ± 0.26	1.60 ± 0.38

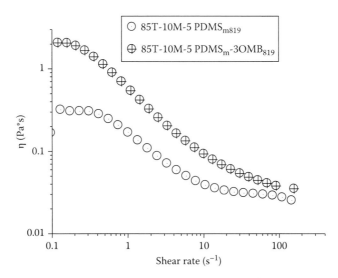

FIGURE 8.6
Rheological curves of 85T-10M-5PDMS$_{m819}$ and 85T-10M-5PDMS$_m$-3OMB$_{819}$ mixtures.

In order to quantify the effect of the presence of OMB nanofiller in 85T-10M-5PDMS$_{m819}$ formulation, and in turn to have an indirect measure of the dimensions of the nanoparticles dispersed in the same suspension, steady rheological measurements were also performed on 85T-10M-5PDMS$_{m819}$ and 85T-10M-5PDMS$_m$-3OMB$_{819}$ systems. The results, reported in Figure 8.6, show that both mixtures display a pseudo-plastic behavior and that the viscosity of the unfilled system increases in the presence of 3% wt. of OMB, in particular at low shear rates.

The rheology of multiphase systems, and more specifically of solid–liquid suspensions, was the object of numerous investigations, both theoretical and experimental, starting from the work of Einstein [44,45]. It was found, in particular, that the viscosity of a suspension depends on the dimensions of the solid particles: by increasing the dimension of the particles, an increase in the viscosity of the suspension is obtained. The Einstein equation can be applied to very dilute suspensions (solid volume fraction $\Phi < 0.02$) of rigid spheres in a Newtonian field, that is,

$$\eta_r = 1 + k_1{}^* \Phi \tag{8.7}$$

where η_r is the relative viscosity of the suspension, calculated as the ratio between the viscosity of the filled suspensions, η, and the viscosity of the suspending medium, η_s. The parameter k_1 takes into account the shape of the particles in suspension and equals 2.5 for spherical particles, as in Einstein equation. Guth found that, in the case of nonspherical particles, k_1 depends on the aspect ratio, p, that is, the ratio between the highest and lowest diameters of the ellipsoid particles, according to equation [45]

$$k_1 = \frac{p}{2\ln(2p) - 3} + 2 \tag{8.8}$$

By substituting the Equation 8.7 in Equation 8.8 and by calculating the relative viscosity of the two systems analyzed, starting from the experimental data of Figure 8.6, a p value of 2.3 was calculated for the nanofilled suspension, irrespective of the shear rate investigated. This value was in accordance with that obtained from different techniques (light transmittance, dynamic light scattering analysis, SEM) and reported in previous work [15].

8.3.3 Surface Characterization of UV-Cured Boehmite Nanocomposites

In Figure 8.7, the SEM image of the nanofilled system, 85T-10M-5PDMS$_m$-3OMB$_{819}$, applied and photo-cured on a glass substrate, is reported. It reveals the presence of boehmite particles randomly dispersed in the polymeric matrix. A statistical analysis was performed on 20 SEM images, acquired in different parts of the samples with a scan size of 10 μm, employing a dedicated program that allows the estimation of the average dimension of the particles formed during the UV-curing process. It was found that the smallest particles have the major axis ranging between 20 and 70 nm.

An elementary analysis (EDX) of the chemical elements present in the film analyzed by SEM was, then, performed, and the results are reported in Figure 8.8. The EDX analysis confirms the presence of boehmite nanoparticles in the filled siloxane-modified methacrylate system.

In order to verify if the presence of boehmite nanoparticles affects the transparency of the polymeric matrix, light transmittance measurements were performed on the photo-cured films coated on a glass support. In Figure 8.9, the comparison between the light transmittance curve for the system filled with 3% wt. of OMB nanoparticles (85T-10M-5PDMS$_m$-3OMB$_{819}$) and that for the unfilled UV-cured siloxane-modified methacrylate system (85T-10M-5 PDMS$_{m819}$) is shown. An appreciable increase in the optical transparency of the potential coating is obtained when OMB nanoparticles were

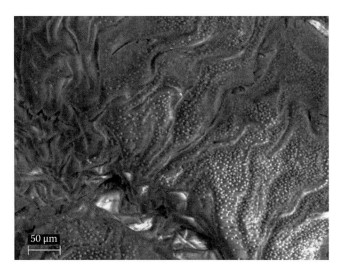

FIGURE 8.7
SEM image of the nanofilled OMB/siloxane-modified methacrylate coating (85T-10M -5PDMS$_m$-3OMB$_{819}$).

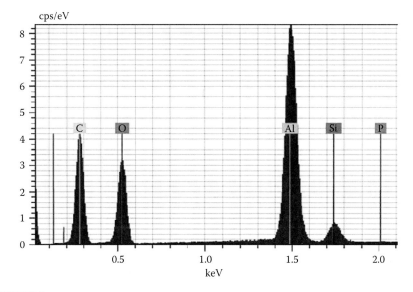

FIGURE 8.8
Elementary analysis performed by EDX on the nanofilled OMB/siloxane-modified methacrylate system (85T-10M-5PDMS$_m$-3OMB$_{819}$).

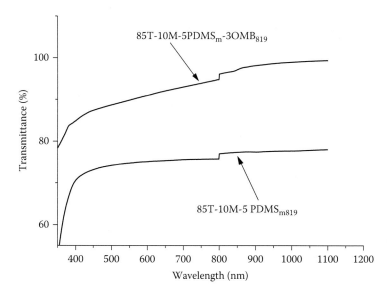

FIGURE 8.9
Light transmittance of the UV-cured films realized with the mixtures 85T-10M-5 PDMS$_{m819}$ and 85T-10M-5PDMS$_m$-3OMB$_{819}$.

added to the modified methacrylic-based mixture. The unfilled coating, that is, 85T-10M-5PDMS$_{m819}$, achieves, in fact, a light transmittance value up to 78%, while for the parent nanocomposite (i.e., 85T-10M-5PDMS$_m$-3OMB$_{819}$), a greater value was found, up to 99%.

The glass transition temperature and the residual heat of reaction of the samples photo-cured for 12 h were calculated using DSC dynamic tests. The values of Tg, summarized in Table 8.5, show that the nanofilled system possesses a glass transition temperature only slightly higher (about 3°C) than that of the organic parent system. Moreover, the significant values found for $\Delta H_{residual}$ would suggest that a 12 h exposure to UV-lamp in air atmosphere was not sufficient to complete the radical photopolymerization reactions in both unfilled and filled systems. This aspect, however, will be discussed in detail later in the text.

The photo-cured films containing boehmite nanoparticles exhibit excellent final properties, as reported in Table 8.5. The results of the contact angle

TABLE 8.5

Surface Properties of Photo-Cured Films

Sample	Tg (°C)	$\Delta H_{residual}$ (J/g)	Contact Angle with Water (°)	Pencil Hardness	Shore D Hardness (ShD)
85T-10M-5PDMS$_{m819}$	56.9 ± 2.6	85.2 ± 4.8	92 ± 1	7H	13.6 ± 0.2
85T-10M-5PDMS$_m$-3OMB$_{819}$	59.8 ± 1.4	97.5 ± 3.6	120 ± 1	>9H	22.7 ± 0.3

measurements with water suggest, in fact, that the presence of the nanofiller is able to greatly raise the hydrophobic properties of the organic film, since the contact angle of the unfilled film increases to 28° in the presence of the nanofiller.

The results of the scratch hardness tests and Shore D hardness measurements performed on UV-cured films applied on a glass substrate, reported in Table 8.2, further confirm the effectiveness of boehmite nanoparticles to improve the surface properties of the siloxane-modified methacrylate system.

The gel content of all the photo-cured systems, measured according to the procedure previously explained, was about 94%, demonstrating the general formation of a highly cross-linked network, irrespective of the presence of OMB nanoparticles.

The UV-cured films coated on aluminum foils were, finally, exposed to alternating cycles consisting of UV light radiation and moisture. After each cycle, the Tg, the contact angle, the pencil and Shore D hardness, and the color change (ΔE) of the films were evaluated. The results obtained are summarized in Table 8.6.

By comparing the results obtained from the experimental measurements performed on the photo-cured films before (Table 8.5) and after (Table 8.6) the QUV accelerated weathering procedures, it can be noticed that the first UV-exposition cycle caused a certain increase in all the properties measured, which remained almost unchanged after the second UV-cycle. In particular, the contact angle of both unfilled and filled systems increased by about 10 degrees after the first cycle, improving further the hydrophobic properties of films. This result was attributed to the completion of the free radical photopolymerization reactions, due to the additional 100 h UV exposure during the first QUV accelerated cycle. This hypothesis would be confirmed by the DSC analysis performed on the films after 1 UV-cycle: no residual reactivity was observed, since a null $\Delta H_{residual}$ was measured for both systems. As a further confirmation, the contact angle measurements, performed on

TABLE 8.6

Surface Properties of the Films after QUV Treatments

Sample	Tg (°C)		Contact Angle (°)		Pencil Hardness 1 and 2 UV Cycle	Shore D Hardness (ShD)		ΔE	
	1 UV Cycle	2 UV Cycles	1 UV Cycle	2 UV Cycles		1 UV Cycle	2 UV Cycles	1 UV Cycle	2 UV Cycles
85T-10M-5PDMS$_{m819}$	58.6±2.2	59.0±1.8	102±5	102±6	6H	44.9±1.3	44.8±1.1	16.44	26.21
85T-10M-5PDMS$_{m}$-3OMB$_{819}$	61.4±3.0	61.5±1.2	130±7	130±9	>9H	89.4±3.5	89.6±2.1	27.90	32.78

the photo-cured films after the second UV-cycle, evidenced the same value of this property, suggesting that the first UV-cycle was enough to achieve the best final properties in the coatings. The Shore D hardness seems to follow the same trend of the other properties measured, since its values highly increased after the first UV-cycle and did not further change after the second one. Almost unchanged values were found for pencil hardness after both QUV cycles.

However, the Tg of both coatings increased only by about 2°C after the first UV-exposure and remained unchanged after the second UV-cycle. Since the latter result would not justify the measurement of a null $\Delta H_{residual}$ after QUV cycles, further spectroscopic analyses are performed on some of the aged films. In Figure 8.10a, the ATR spectra measured on photopolymerizable siloxane-modified methacrylic-based unfilled film, before and after the aging procedure, are compared. In Figure 8.10b, an enlargement (limited to 1350–1200 cm^{-1}) of both spectra is shown. The main peaks with the associated chemical bonds and wave numbers are summarized in Table 8.7. As can be observed by the spectra of Figure 8.10, in both unaged and aged films, the C=C peak (at about 1640 cm^{-1}) does not appear, suggesting that the conversion of methacrylic monomers was roughly completed after a 12 h exposure to UV-lamp. On the other hand, in both spectra, the peak of Si–O–C$_{stretching}$ is visible (enlarged in Figure 8.10b), which is much less evident after the second QUV cycle treatment. This latter observation suggests that the silane components of the mixture (VT PDMS and MEMO) achieved an almost complete homopolymerization reaction only after the QUV exposure, decreasing the height of the characteristic peak to about 90%. As a conclusion, the $\Delta H_{residual}$ reported in Table 8.5 can be attributed nearly completely to the unreacted silane monomers. As a further confirmation, in the DSC thermogram obtained for the siloxane-modified methacrylic-based film after two QUV cycles, a Tg related to the silane components was measured, at about 22°C. The same film analyzed before the QUV treatment did not show any Tg at similar temperature, since the typical Tg of VT PDMS is about –127°C [46], well below the range of temperatures analyzed in this study.

Passing to analyze the color measurement results, also reported in Table 8.6, the analyses performed after the QUV exposure show that this procedure causes a severe yellowing of both unfilled and boehmite-filled films. Transmittance measurements, reported in Figure 8.11a and b for unfilled and boehmite-filled siloxane-modified methacrylic-based films, respectively, before and after the QUV cycles, confirm a severe reduction in the transparency of films, even worse in the presence of nanofiller. However, it is well recognized that accelerated weathering generally provides an overestimation of the degradation effects occurring in real outdoor applications. Accelerated tests, in fact, are performed by exposing the materials to environments more aggressive than those generally encountered in practice, that is, very high temperatures, prolonged immersion periods, and higher loads. In standardized procedures, proposed to achieve accelerated weathering, one or more

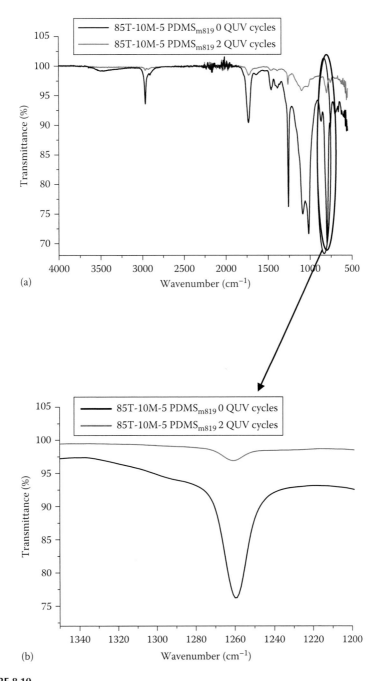

FIGURE 8.10

(a) ATR spectra measured on photopolymerizable siloxane-modified methacrylic-based films, before and after the aging procedure; and (b) spectra limited to 1350–1200 cm⁻¹.

TABLE 8.7

Identification of the Main ATR Peaks with the
Associated Chemical Bonds and Wave Numbers

Chemical Bond	85T-10M-5PDMS$_{m819}$ (0 QUV Cycles)	85T-10M-5PDMS$_{m819}$ (2 QUV Cycles)
CH$_{stretching}$	2967	2967
CH$_{bending}$	1473	1473
C=O$_{stretching}$	1725	1725
C–O$_{stretching}$	1100	1100
P=O$_{stretching}$	1248	1248
Si–O–C$_{stretching}$	1250	1250
Si–H$_{3\ bending\ asimm.}$	805	805

weather-like conditions are intensified to levels greater than those occurring naturally. As a consequence, the results of the accelerated weathering, even if carried out in accordance with the current standards (ISO, ASTM), are mainly regarded as qualitative indicators and used as a reference limit value [47].

The mixtures 85T-10M-5PDMS$_{m819}$ and 85T-10M-5PDMS$_m$-3OMB$_{819}$ were, finally, applied on walnut wood specimens ($20 \times 20 \times 20$ mm³) and photocured in air by using a medium-pressure Hg UV lamp.

The water absorption behavior of both coated and uncoated specimens of walnut wood, determined by using the procedure described in experimental section, is shown in Figures 8.12. Each point represents the average weight measurements of each set of 10 specimens. As expected, the water absorption capacity of the coated samples was reduced with respect to uncoated specimens. In particular, the samples coated with nanocomposite coating absorb much less water not only with respect to the uncoated wood but also with respect to wood coated with unfilled coating, especially in the first immersion period. This different behavior was quantified by determining the final water absorption values (WA$_t$), calculated according to Equation 8.1 for treated and control wood specimens and reported in Table 8.8.

8.4 Conclusions

Several novel photopolymerizable siloxane-modified acrylic formulations, both in the presence and without OMB nanoparticles, were prepared and characterized, in order to select the most suitable composition to be proposed as protective coating for different substrates. To this aim, a deep analysis of the time, the rate of photopolymerization reaction, and the viscosity of the experimental UV formulations was performed.

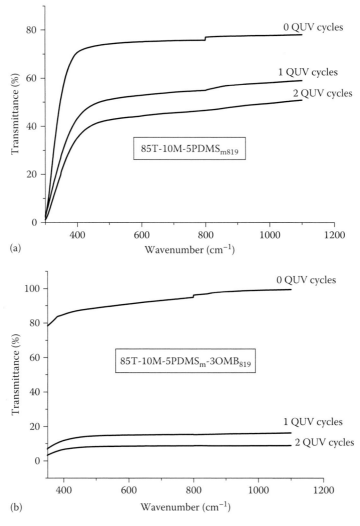

FIGURE 8.11

Transmittance measurements for (a) unfilled and (b) boehmite-filled siloxane-modified methacrylic-based films, before and after the QUV cycles.

The kinetic results obtained from p-DSC measurements, evidenced that, in nitrogen atmosphere, the photopolymerization of all the formulations produced was very fast, even in the presence of small amount of nanoparticles. On the other hand, in air atmosphere, the effect of oxygen inhibition towards free radical photopolymerization reaction was reduced by introducing a proper thiol (MPTS). However, FTIR analysis showed that the systems achieved a final α value below 70%. Moreover, the nanofilled system displayed a slightly lower extent of reaction, even though the difference in

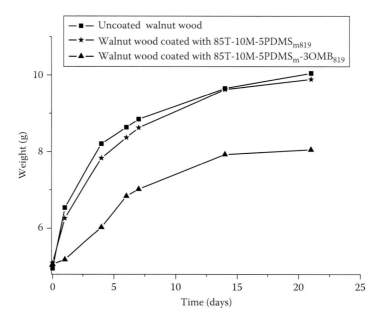

FIGURE 8.12

Water absorption behavior of both coated and uncoated specimens of walnut wood as a function of time.

TABLE 8.8

Results of Water Absorption Measurements

Wood Sample	WA$_t$ (%)
Uncoated	108
Coated with **85T-10M-5PDMS$_{m819}$**	94
Coated with **85T-10M-5PDMS$_m$-3OMB$_{819}$**	59

reactivity was reduced for longer exposure times. It must be underlined that the UV-curing times required for the development of experimental, in the order of 10 h, were still much lower than the typical curing times required for the thermally induced curing at room temperature, in the order of weeks.

Referring to the viscosity of the photo-curable methacrylic siloxane-based formulations, a pseudo-plastic behavior was generally found, well fitted by the cross relationship. The mixtures analyzed possessed a viscosity suitable for the proposed application in all ranges of shear rate investigated, even in the presence of a small amount of boehmite nanoparticles.

The characterization of the physical and surface properties of photopolymerizable siloxane-modified methacrylic-based/boehmite nanocomposite was also performed in order to quantify the modifications in properties brought about by the presence of nanoparticles finely dispersed in the thermosetting matrix. In order to verify their possible application as a protective coating for different substrates, they were first applied and photo-cured on a glass substrate and subsequently on walnut wood samples. The transparency, glass transition temperature, scratch and Shore D hardness, contact angle, and color change were measured before and after QUV accelerated weathering cycles. The presence of boehmite nanoparticles in the photo-cured films, revealed by SEM observations and EDX analysis, allows to appreciably increase the surface properties and the hydrophobicity of the film, without affecting the transparency of the unfilled coating. Moreover, the glass transition temperature of the filled film slightly increased with respect to the unfilled one. The presence of a residual reactivity before the QUV treatment was explained by the completion of homopolymerization reactions of the unreacted silane monomers. After the first QUV accelerated weathering cycle of 100 h, the final properties of the coatings further increased, achieving a plateau value that remained unchanged after the second UV cycle. The presence of OMB nanoparticles in the siloxane-modified methacrylate film was not able to limit the yellowing effect due to long UV-exposure time. Further studies are in progress to evaluate the natural weathering effect on both systems, by the outdoor exposure of coatings. The water absorption behavior of walnut wood coated samples was finally analyzed, confirming that the wood elements coated with nanocomposites absorb much less water than those uncoated or coated with the unfilled coating.

References

1. Indennidate L, Cannoletta D, Lionetto F, Greco A, Maffezzoli A. *Polymer International* 2010, 59, 486–491.
2. Greco A, Esposito Corcione C, Strafella A, Maffezzoli A. *Journal of Applied Polymer Science* 2010, 118, 3666–3672.
3. Esposito Corcione C, Cavallo A, Pesce E, Greco A, Maffezzoli A. *Polymer Engineering and Science* 2011, 51, 1280–1285.
4. Greco A, Gennaro R, Rizzo M. *Polymer International* 2012, 61, 1326–1333.
5. Greco A, Rizzo M, Maffezzoli A. *Thermochimica Acta* 2012, 534, 226–231.
6. Chen C, Curliss D. *Nanotechnology* 2003, 14, 643–648.
7. Chin IJ, Thurn-Albrecht T, Kim H-C, Russell TP. *Polymer* 2001, 42, 5947–5952.
8. Yu ZZ, Yan C, Yang M, Mai Y-W, *Polymer International* 2004, 53, 1093–1098.
9. Liu W, Hoa SV, Pugh M. *Polymer Engineering and Science* 2004, 44, 1178–1186.
10. Nayak GC, Sahoo S, Rajasekar R, Das CK. *Composites Part A* 2012, 43, 1242–1251.
11. Esposito Corcione C, Fasiello A, Maffezzoli A. *Journal of Nanostructured Polymers and Nanocomposites* 2007, 3, 82–89.
12. Esposito Corcione C, Frigione M, Maffezzoli A, Malucelli G. *European Polymer Journal* 2008, 44, 2010–2023.
13. Esposito Corcione C, Frigione M, Acierno D. *Journal of Applied Polymer Science* 2009, 112, 1302–1310.

14. Mathieu Y, Rigolet S, Valtchev V, Lebeau B. *Journal of Physical Chemistry C* 2008, 112, 18384–18392.
15. Esposito Corcione C, Frigione M. In press on: *Journal of Applied polymer Science.* DOI: 10.1002/app.38639.
16. Kloprogge JT, Ruan HD, Frost RL. *Journal of Material Science* 2002, 37, 1121–1129.
17. Kornmann X, Lindberg H, Berglund LA. *Polymer* 2001, 42, 4493–4499.
18. Glasel H-J, Bauer F, Ernst H, Findeisen M, Hartmann E, Langguth H, Mehnert R, Schubert R. *Macromolecular Chemistry and Physics* 2000, 201, 2654–2659.
19. Van den Branden S. Paint and coatings industry, *Proceedings of Radtech Europe* 2000, pp.147–149.
20. Licciulli A, Esposito Corcione C, Greco A, Amicarelli V, Maffezzoli A. *Journal of the European Ceramic Society* 2005, 25, 1581–1589.
21. Esposito Corcione C, Greco A, Maffezzoli A. *Polymer* 2005, 46, 8018–8027.
22. Esposito Corcione C, Greco A, Maffezzoli A. *Journal of Applied Polymer Science* 2004, 92, 3484–3491.
23. Esposito Corcione C, Frigione M, Maffezzoli A, Malucelli G. *Polymer Testing* 2009, 28, 157–164.
24. Esposito Corcione C, Previderio A, Frigione M. *Thermochimica Acta* 2010, 509, 56–61.
25. Esposito Corcione C, Frigione M. *Progress in Organic Coatings* 2011, 72, 522–527.
26. Esposito Corcione C, Frigione M. *Thermochimica Acta* 2012, 534, 21–27.
27. Esposito Corcione C, Frigione M. *Progress in Organic Coatings* 2012, 74, 781–787.
28. Frigione M, Esposito Corcione C. *The Open Materials Science Journal* 2012, 6, 68–76.
29. Frigione M. Esposito Corcione C. submitted to *Polymer Composites.*
30. ASTM D 3633, *Standard Test Method for Film Hardness by Pencil Test,* West Conshohocken, PA, reapproved 2006.
31. ASTM D2240, *Standard Test Method for Rubber Property—Durometer Hardness,* West Conshohocken, PA, reapproved 2010.
32. EN ISO 11507:2007, *Paints and Varnishes—Exposure of Coatings to Artificial Weathering—Exposure to Fluorescent UV Lamps and Water.*
33. Normal Protocol 43/93 *Colour Determinations of Opaque Surfaces;* ICR-CNR: Rome, Italy, 1993.
34. ASTM D2665-84, *Standard Specification for Poly(Vinyl Chloride) (PVC) Plastic Drain, Waste, and Vent Pipe and Fittings,* reapproved 2009.
35. Kim HK, Ju HT, Hong JW. Characterization of UV-cured polyester acrylate films containing acrylate functional polydimethylsiloxane. *European Polymer Journal* 2003, 39, 2235–2241.
36. Lee TY, Guymon CA, Sonny Jonsson E, Hoyle CE. The effect of monomer structure on oxygen inhibition of (meth)acrylates photopolymerization. *Polymer* 2004, 45, 6155–6162.
37. Goodnera MD, Bowman CN. Development of a comprehensive free radical photopolymerization model incorporating heat and mass transfer effects in thick films. *Chemical Engineering Science* 2002, 57, 887–900.
38. Studer K, Decker C, Beck E, Schwalm R. Overcoming oxygen inhibition in UV-curing of acrylate coatings by carbon dioxide inerting, Part I. *Progress in Organic Coating* 2003, 48, 92–100.
39. Chen Q, Udomsangpetch C, Shen SC, Liu YC, Chen Z, Zeng XT. The effect of AlOOH boehmite nanorods on mechanical property of hybrid composite coatings. *Thin Solid Films* 2009, 517, 4871–4874.

40. Kerbouch P, Lebaudy P, Lecamp L, Bunel C. Numerical simulation to correlate photo-polymerization kinetics monitoring by RT-NIR spectroscopy and photo-calorimetry. *Thermochimica Acta* 2004, 410, 73–78.

41. Lettieri M, Lionetto F, Frigione M, Prezzi L, Mascia L. Cold-cured epoxy-silica hybrids: Effects of large variation in specimen thickness on the evolution of the Tg and related properties. *Polymer Engineering and Science* 2011, 51, 358–368.

42. Kim EK, Won J, Do J, Kim SD, Kang YS, *Journal of Cultural Heritage* 2009, 10, 214–221.

43. Esposito Corcione C, Frigione M. A novel procedure able to predict the rheological behavior of trifunctional epoxy resin/hyperbranched aliphatic polyester mixtures. *Polymer Test* 2009, 28, 830–835.

44. Dealy JM, Wissbrun KF, Reinhold V. *Melt Rheology and Its Role in Plastics Processing Theory and Applications.* Van Nostrand Reinhold: New York; 1990.

45. Utracki L, *Polymer Alloys and Blends: Thermodynamics and Rheology.* Hanser Publishers: New York; 1989.

46. Dollase T., Spiess H.W., Gottlieb M., Rozen Y. *Europhysics Letters* 2002, 60(3), 390–396.

47. Russo P., Acierno D., Marinucci L., Greco A., Frigione M. *Journal of Applied Polymer Science* 2013, 127, 2213–2219.

9

Protective Coatings Based on Silsesquioxane Nanocomposite Materials

Bogdana Simionescu, Irina-Elena Bordianu, Magdalena Aflori,
Florica Doroftei, Corneliu Cotofana, and Mihaela Olaru

CONTENTS

9.1 State of the Art

Various polymer systems (consolidants and coatings) are nowadays being used for reducing or delaying the weathering of monuments and historic buildings. These coatings include different common systems such as alkoxysilanes, acrylics, epoxies, as well as other not-so-used organic systems, such as polyurethanes, isocyanates, and so on. Recently, polyfluorinated compounds possessing water repellent abilities have been tested, mostly in Italy. It is however, not very well known and understood how these synthetic materials respond to the problems given by different environmental conditions.

The sol–gel process showed a good efficiency when used in applications in the stone conservation field. Hybrid organic–inorganic composites resulted from sol–gel processing are materials that may be designed and destined for a large application range. The structural diversity can be achieved by controlling the relative ratio of organic *vs.* inorganic content, the chemical composition of the inorganic precursor molecule, the structural complexity of the organic part, as well as its nature, and the reaction conditions used in the synthesis. Sol–gel precursors, such as alkoxysilanes, or "silanes," particularly methyltrimethoxysilane and tetraethoxysilane (TEOS), have been commonly used in stone conservation as consolidants for limestones, sandstones, and marble due to their specific properties, such as low viscosity and chemical and light stabilities [1]. Silanes are hydrolyzed by water and they form silanols, giving birth to a silicone polymer through the polycondensation reaction. The polycondensation and hydrolysis reactions take place after the chemical product has penetrated into the building stone structure, the resulting polymer network imparts the required strength to the stone. The advantage shown by these products is their decreased viscosity leading to a deeper penetration into the porous stone substrate. After polymerization upon contact with the environmental moisture, a stable gel having a silicon–oxygen backbone is formed. The bonding, which takes place between a stone substrate and the consolidant, remains a subject of interest. For example, it is widely argued that alkoxysilanes form primary chemical bonds with the Si–OH groups from sandstone surfaces, but they will not be able to form primary bonds with a limestone substrate. The lack of bonding doesn't necessarily prove a failure; the unbonded network of consolidant leads, in some cases, provide strengthening properties. In the 1990s, various authors reported the decay of certain stones following the treatment with TEOS, unless the stones were treated with a water-repellent treatment, as well [2,3]. The surface coating has to be renewed at regular intervals, but the initial consolidation should last much longer.

The epoxy resins used as consolidants are also, subject for debate. Many researchers/conservators consider them to be brittle and viscous yellowing materials acting as efficient adhesives in some circumstances, but cannot be seriously taken into account as consolidants. Other reviews summarized the usage of epoxies as consolidants, describing their advantages and weaknesses [4], presenting, as well, two different paths adopted to treat relatively small objects and large facades, respectively. The choice of the solvent, application techniques, and post-application procedures are considered to be important for a successful result. Other materials used in stone conservation science include acrylics, silicones, alkoxysilanes, polyesters, urethanes, and other organic-inorganic polymers [5,6].

Water is involved in most aspects of building stone deterioration; therefore, water-repellency is a targeted property in surface coating. Water-repellency is generally provided by silicones, alkoxysilanes, and fluoropolymers. Water repellent coatings prevent the penetration of liquid water, but they do not prevent the passing through of water vapor. Moreover, many surface coatings are also permeable to harmful gases such as sulfur dioxide [7].

In the field of science for conservation of cultural heritage, the stone conservation, nanotechnology, and nanomaterials are being nowadays used for many purposes [8–11], most of them being related to the use of organic–inorganic hybrid nanocomposites, as well as composite materials comprising different types of polymers and silica nanoparticles. The amount of publications in the field of stone conservation is very high and they still continue to grow. While most of them are focused on the same materials or topics and procedures/techniques, the synthesis, design, and development of new chemical products prove to be a scientific and technologically evolving process.

Considering the methyl methacrylate and other acrylic/methacrylic monomers, their *in situ* polymerization has some advantages, although the high glass transition temperatures and rigidity of the resulting polymers are diminishing their interest as stone consolidants. In the 1970s and 1980s, the research in stone conservation has been focused on finding a single treatment for both the consolidation and protection of the stone-based monuments. 3-(Trimethoxysilyl)propyl methacrylate is one of the most significant sol–gel precursors presenting a dual-network forming capability, used worldwide in the synthesis of organic–inorganic hybrid nanomaterials [8,9]. This compound possesses a polymerizable methacrylate group at one end and alkoxy silane groups capable to form inorganic networks *via* sol–gel route at the other end.

The silsesquioxanes are a family of compounds possessing a ratio of 1.5 between the silicon and the oxygen atoms [12], their structures being expressed in a general formula: $(R–SiO_{1.5})_n$ (n = even number) [13]. The silsesquioxane family is nowadays internationally recognized to have a great potential as building blocks for multiple advanced materials, its applications as organic–inorganic nanocomposites being found in coordination chemistry, catalysis, and material science, a.s.o. The hydrolysis and condensation reactions of substituted alkoxysilanes, $R–Si(OR')_3$, possessing a nonhydrolyzable Si–C bond, are known to give birth to a variety of silsesquioxanes having multiple substituent groups and cage structures, as well [12,14]. By self-assembling 3-(trimethoxysilyl)propyl methacrylate-based block copolymers in solution, multiple organic–inorganic hybrid morphologies, mainly spheres [15], vesicles [16], and compound vesicles [17] with silica oxide network, have been formed by a gelation process.

The discovery of the surfactant-templated synthesis in the early 1990s allowed the formation of ordered mesoporous silica gels. The surfactant is used in a concentration below the critical micellar concentration (CMC) and has the role of a structure-directing agent during polymerization. These new materials, called *molecular sieves*, have ordered mesopores with uniform size, determined by the size of the surfactant aggregate as the main advantage. This type of synthesis prevents the cracking and the possible fissures of the gel, which may appear during the drying phase for two reasons: (1) the decrease in the surface tension provided by the surfactant reduces the capillarity pressure; and (2) the surfactant leads to a coarsening of the gel network, thus reducing the capillarity pressure, as well.

The evaluation of the long-lasting results of the chemical treatments is based on a range of standardized tests [18,19], but the applied evaluation procedure should be tailored to suit a particular stone and environment [20]. While many stone conservation approaches proved to be efficient for specific problems, limestone conservation continues to represent a challenge, due to the chemical composition and pore structure of the limestone made monuments. The characteristics desired by the conservation community for a protective/conservation product are a higher chemical resistance against dissolution, hydrophobicity accompanied by permeability to water vapor, reversibility and environmental friendliness, and strengthening or consolidation of the weakened limestone. Generally, the main drawbacks of the classical water-repellents, strengtheners, and consolidants are related to the weakness of the chemical bond between the limestone and the chemical treatment, UV degradation, and superficial penetration [6].

As an alternative to the well-known chemical products used worldwide in stone conservation, a newly designed compound, that is, a hybrid nanocomposite containing silsesquioxane and methacrylate units, was synthesized and characterized using multiple methods. The challenge was to create a crack-free gel using a surfactant (in a concentration below CMC) as a template starting from 3-(trimethoxysilyl)propyl methacrylate precursor that contains three different components, namely, methacrylate, hydroxyl functionality, and silicon–alkoxy groups. The hydrolyzable silicon alkoxide at one end may be condensed in order to form an inorganic Si–O–Si network, and simultaneous polymerization of the methacrylate group at the other end can lead to a highly crosslinked and dense structure. The present study underlines the use of silsesquioxane-based compounds as water-repellents imparting a higher durability of the treated stone samples as compared to the ones treated with commercial products, being among the first ones mentioning the use of this class of compounds in stone conservation.

9.2 Hybrid Nanocomposite with Silsesquioxane Units: Synthesis and Characterization

The hybrid nanocomposite with silsesquioxane and methacrylate units (TMSPMA) was obtained by radical polymerization of 3-(trimethoxysilyl) propyl methacrylate precursor in the presence of a primary amine surfactant (dodecylamine) and 2,2'-azobis(2-methylpropionitrile) (AIBN) [10,11]. The surfactant was added in the system in a concentration below its CMC, playing a structure-directing role during polymerization. The reaction was acid catalyzed in water/hydrochloric acid solution, possessing the HCl/precursor molar ratios equal to $1.85 \cdot 10^{-2}$ and the H_2O/precursor ratio smaller than 2 (stoichiometrically deficient amount of water). AIBN (2 wt% of precursor mass) was used as a thermal initiator of the organic monomer polymerization. In the system, the mole ratios were 1 precursor: 2 H_2O: 11 ethanol: 0.004 HCl: 0.003 dodecylamine. 3-(Trimethoxysilyl)propyl methacrylate precursor, containing dodecylamine in an ethanol solution, was mixed with AIBN and mechanically stirred for 30 min, then added to a hydrochloric acid solution (HCl 37%, Aldrich) and distilled water, then stirred for another 30 min. The polymerization process was carried out for 48 h at 60°C. After the solvent evaporation in vacuum, the designed chemical product was a transparent solid at room temperature and dried in vacuum at 60°C for 3 days. The proposed structure for TMSPMA is given in Scheme 9.1.

The TMSPMA X-ray diffraction diagram is presented in Figure 9.1. The X-ray diffraction diagram shows the presence of a narrow crystalline peak at 26.640 (3.343 Å), which can be ascribed to polyhedral silsesquioxanes (normally crystallizable), as opposite to the ladder amorphous ones.

Moreover, the broadening of the diffraction peak is probably due to the crystallite size in nanometer scale. The reflection from 26.640 (3.343 Å) is related to the hexagonal crystalline structure from 101 diffraction plane, with a hexagonal unit cell of a = 4.91 and c = 5.4 Å, besides a value of 8.14 nm of the crystallite size.

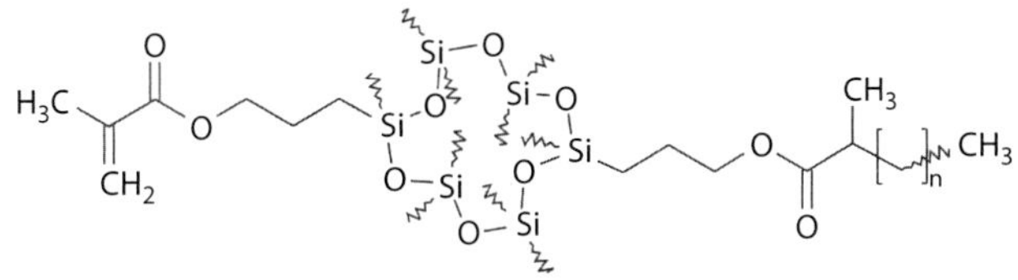

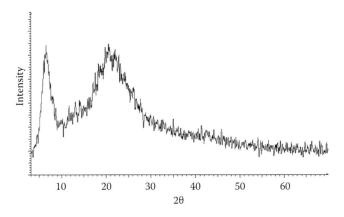

SCHEME 9.1
Proposed structure of TMSPMA hybrid nanocomposite.

FIGURE 9.1
TMSPMA X-ray diffraction diagram.

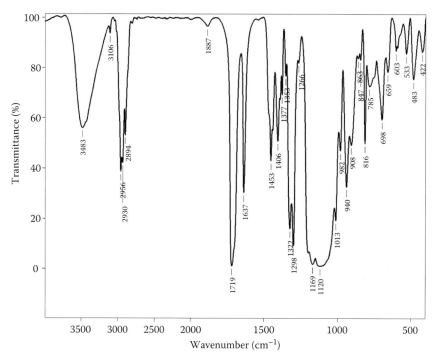

FIGURE 9.2
FT-IR spectrum of TMSPMA.

The TMSPMA FT-IR spectrum is evidenced in Figure 9.2. A characteristic band corresponding to the O–H stretching vibrations of the hydroxyl groups are evidenced around 3485 cm^{-1}. The hydrogen bonds play a significant role in such conformational arrangements, giving birth to hydrophobic associated domains. Two small peaks at 3750 and 3106 cm^{-1}, attributed to silanols and hydrogen-bonded silanol groups, respectively, may be distinguished, as well, while the vicinal ones and the physically adsorbed water are characterized by the adsorption bands in the 3000–3750 cm^{-1} range. The presence of these peaks agrees with Sommer's hypothesis stating that silane bonds are converted to silanol bonds during the process of gelation. The C–H alkyl stretching band is evidenced in the 2894–2956 cm^{-1} range. The absorption peak from 1719 cm^{-1} may be attributed to the stretching vibration of carbonyl groups, while the one around 1637 cm^{-1} may be attributed to acrylic double bonds (C=C), stating that not all of these have reacted.

The silicate vibrational spectra may be divided into two areas. The first one covers the 4000–1600 cm^{-1} frequency range, where the stretching and the bending vibrations of water molecules are found. The second region (below 1300 cm^{-1}) consists in the vibrations due to the silicate layer and charge-balancing cations. Usually, the γ_{as}(Si–O–Si) modes appear in the 1200–1000 cm^{-1}

area, whereas γ_s(Si–O–Si) appears in the 700–400 cm^{-1} region. By increasing the silica content of the hybrid composites, there is an intensity increase in some characteristic peaks of the silica phase such as the Si–O–Si stretching, the presence of such bonds confirm the existence of covalent linkages between the organic groups and the silica, this leads to a better compatibility and to a crosslinked network between the organic and inorganic components. For TMSPMA, Si–O–Si bands may be observed at 1169 and 1120 cm^{-1}. The silanol groups are capable of continuing the reactions with other molecules and producing new siloxane bridges. Additional peaks at around 698–785 cm^{-1} (for symmetric Si–O–Si stretching vibration) may be observed as well.

TMSPMA possesses self-assembling properties due to the trimethoxysilane groups. Usually, the trialkoxysilyl end groups of different alkyltrialkoxysilanes (R–Si(OR)$_3$), when hydrolyzed, enable polycondensation into crystalline or amorphous silsesquioxanes with complex (T cube-like or ladder-like) structures. Such types of silsesquioxane structures generated during the polycondensation, either polyhedra or ladders, may be inferred from the FT-IR spectroscopy observing the localization of the bands characterizing the antisymmetric Si–O–Si stretching vibrations. A single band near 1120–1130 cm^{-1} may be considered as a solid evidence for the presence of a polyhedral structure, while two bands located around 1040 and 1120–1130 cm^{-1} are characteristic to a *cis*-syndiotactic ladder configuration. The cage- or ladder-like silsesquioxane configurations are highly influenced by the class of the organic group bonded to the silicon atom, considering a steric effect promoting the intramolecular condensation. In this research, the presence of one band in both FT-IR spectra, located near 1120–1130 cm^{-1}, marks out the polyhedral structures formed by the Si–O–Si bonds.

Silica possesses a complex morphology lying down from nanometer to micron size range. The manipulation of the morphology at each structural level determines the properties of the product, the silica physical properties essentially depend on the synthesis procedure. In the silica structure, three types of particle organization may be evidenced. At the smallest size scale, quasi-spherical primary particles (ranging from 3 to 500 nm in diameter) are found, these ones control the specific surface area of the powder. Under the effect of colloidal forces, the primary particles are clustered to form disordered aggregates at 0.1 μm size scale. Typically, aggregates are moreover linked to form agglomerates that extend up to hundreds of μm in size. The dimension of the primary particles, as well as the density, aggregation, and agglomeration degrees, have an influence on the silica porosity and specific surface. SEM characterization is performed to evaluate the hybrid composite morphology (Figure 9.3).

The hybrid compound's SEM examination emphasized the presence of small silsesquioxane aggregates ranging from 100 to 800 nm. The surfactant directed the silsesquioxanes into 3-D spherical arrangements, which, at a

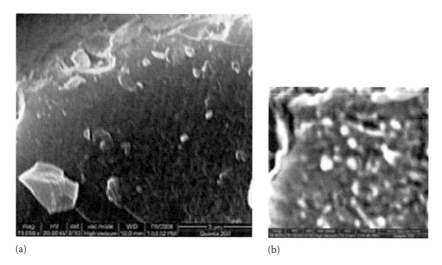

(a)   (b)

FIGURE 9.3
SEM micrographs for TMSPMA: (a) 3 µm and (b) 500 nm.

certain scale, are connected to tubular and continuous structures, while in the polymeric matrix, some ordered regions consisting in a self-assembly of semicylindrical shells may be observed.

9.3 Treated Stone Characterization

The property that has been most wanted and searched for in surface coatings is water-repellency. This approach is based on the idea that since water is involved in most forms of stone decay, a treatment preventing the ingress of water should help to reduce the decay. It has also been argued that water-repellents should help to prevent resoling, although this claim has not been adequately substantiated.

To test the newly designed nanocomposite with silsesquioxane units, two different monumental stones were selected, as well a worldwide-used water repellent, namely, Tegosivin HL 100 (DEGUSSA GOLDSCHMIDT), for comparison [21]. The first stone, Laspra, is a micritic dolomitic limestone, characteristic of the Spanish region of Asturias. This monumental stone may be widely found in Asturias, especially in Oviedo, being used, for example, as one of the three main building materials of the San Salvador Cathedral of Oviedo (1293–1587) (Figure 9.4a).

Oviedo Cathedral is one of the most significant stone built monuments of the late Gothic architectures in Asturias. Historically and artistically, the cathedral presents a special interest, reflecting the Gothic-style evolution in

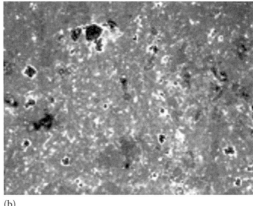

(a)   (b)

FIGURE 9.4
(a) San Salvador Cathedral of Oviedo, Spain, origin: ninth century, period: sixteenth century, Gothic style and (b) POL micrographs NCX2 of Laspra stone.

Spain. San Salvador Cathedral was the first Gothic monumental building in Asturias. Laspra was in quite common use over the centuries, and nowadays no quarry to extract this stone is available. For our study, a small block of Laspra from the Oviedo Cathedral's portico was used. Laspra presents a microcrystalline texture, with micritic and microsparitic grain sizes along the pores (Figure 9.4b). Laspra is an isotropic stone with the same values of anisotropy coefficient in all directions, obtained by the propagation velocity of ultrasonic wave test between 1.0 and 1.12. The micrographs of Laspra demonstrate the existence of small pieces of bioclasts in all mass, the observed structures attribute to not only fossils like mollusks (mainly bivalves and gastropods) and foraminifers, but also the existence of small traces of iron oxides that are easy to identify due to their reddish color. Quartz grains of spherical or ellipsoidal shapes were occasionally found, with the spherical index and roundness of 0.3–0.5 and grains sizes of 0.2–1.0 mm. Laspra can be classified as a mudstone [22] or micrite [23,24], a microporous stone presents a moldic porosity. From mercury intrusion porosimetry perspective, Laspra has an open porosity of 30%, small pore radii of 0.15 μm, and large specific surface area of 4.2 (m^2/g).

The second monumental stone, originating from Romania, is a bioclastic oolitic limestone, named Repedea, and may be found in the eastern part of Romania, along the Moldavian Platform, where an oolitic limestone slate covers a huge surface of approximately 3000 km^2. In Moldova, this limestone was and still is the main building material for the construction of basically all important churches and monasteries since antiquity until the present time. One of the most representative monasteries, from both historic and cultural points of view, situated in the northeastern part of Romania is

(a)

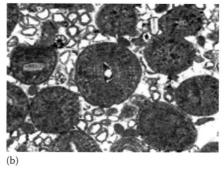
(b)

FIGURE 9.5
(a) Dobrovat Monastery, sixteenth century, Medieval Moldavian style and (b) POL micrographs NCX2 of Repedea stone.

Dobrovat monastery (Figure 9.5a), the whole monastery assembly being built (1503–1504) from Repedea limestone.

Dobrovat monastery from Iasi, Romania, is one of the most important buildings entirely built (1503–1504) using Repedea limestone. Repedea's texture (Figure 9.5b) can be described as an oolitic grainstone. From the anisotropy's point of view, Repedea exhibits large differences in the anisotropy coefficient values in two directions, that is, 4.52–6.08 for x and 0.99–1.03 for y and z because of the difference in the oolite grain sizes and because of their distribution within the stone's structure. The quartz grains are surrounded by the cement of micritic and microsparitic grain size; in some areas, pieces of fossils like mollusks, brachiopods, and gastropods are present, while in other regions, intraclasts can be identified. Repedea can be classified as a grainstone [22] and as an oomicrite–oosparite [23,24]. The mercury intrusion porosimetry test revealed that Repedea has an open porosity of 10%, large pore radii of 3.3 μm, and small specific surface area of 0.6 (m^2/g).

As for the commercial products, according to the producer, Tegosivin HL 100 is a low-molecular weight-modified polysiloxane resin (methyl-ethoxy-polysiloxane); its application requiring the dissolution in white spirit. The NMR and FT-IR spectra of Tegosivin HL 100 revealed the following chemical structure: [–Si(CH$_3$)$_2$–O–]–[– Si(CH$_3$)(OCH$_2$CH$_3$)–O–], solventless.

The different behaviors of the widely used and newly synthesized nanocomposite material when applied onto the stone surfaces are shown in Figure 9.6. These differences are more evident in Laspra's case, where one may observe the formation of a more or less compact and continuous polymeric film on the stone sample surfaces treated with Tegosivin HL 100, a film that shows no visible cracks at ×1000 order of magnification (even for higher orders of magnification, no cracks or fissures being detected). For the stones treated with TMSPMA, the samples don't show the same surface morphology. More visible in Laspra's case, the stone surface is not covered with a compact film, TMSPMA polymerizing through a sol–gel process. Due to

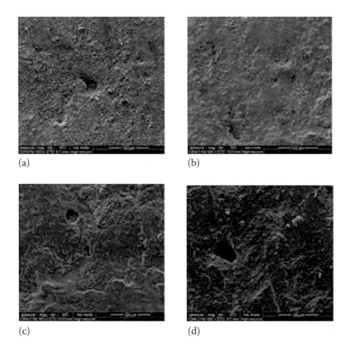

FIGURE 9.6
(a) Laspra treated with TMSPMA. (b) Tegosivin HL 100 (×1000). (c) Repedea treated with TMSPMA and (d) Tegosivin HL 100 (×1000).

its nanosized particles, the nanocomposite material is able to penetrate the limestone substrates deeper than the commercial one, not forming a compact polymeric film on top of the stone surfaces.

The ESEM micrographs of the monumental stone samples brushed with the two water-repellents allowed the determination of parameters in relation to the type of coating, film-forming capacity, adherence to material, spalling or cracking, etc.

To investigate the hydrophobicity given by the two products, the treated stones were submitted to contact angle measurements using a Kruss Easydrop Standard Goniometer. The stones were treated with Tegosivin HL 100 and TMSPMA, and submitted to contact angle measurements after solvent evaporation and complete drying (Figure 9.7) [25].

Regarding the contact angle measurements, TMSPMA behaves differently; for Laspra it shows values between 120° and 104°, and a slight decrease in the contact angle is noticed during the 30 s after water drop deposition (Table 9.1). This product can be considered less adequate from the contact angle point of view. In the case of Repedea treated with TMSPMA, a constant 105° contact angle value was registered.

One of the most important attributes for a water-repellent coating consists in its ability to allow the treated stone to "breathe," making the monumental

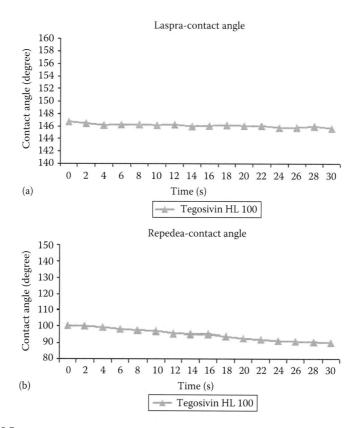

FIGURE 9.7
Contact angle values for (a) Laspra and (b) Repedea.

TABLE 9.1

Laspra and Repedea—Contact Angle
Values (Degrees)

	Tegosivin HL 100	TMSPMA
Laspra	147	120 ↓
Repedea	100 ↓	105

stone surface water-repellent, not water proof. Figure 9.8 evidences the water vapor permeability graphs for both stone samples treated with TMSPMA and Tegosivin HL 100. Following the application of the siloxane-based water-repellents, the values of water vapor permeability coefficient show a decrease. The lower the decrease in the values of water vapor permeability coefficient after the stones treatment, the better the result. Over the curing period, TMSPMA nanoparticles self-assemble and bond with the surface; this nanocoating not only forming as a continuous film on the surface, since

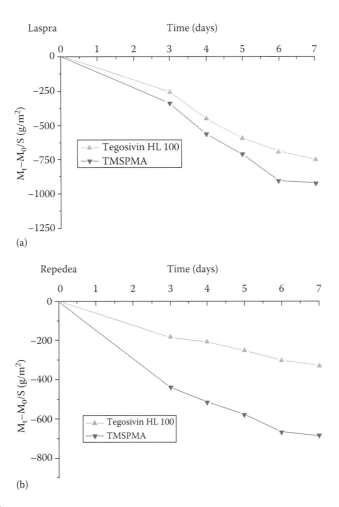

FIGURE 9.8
Water vapor permeability graphs of (a) Laspra and (b) Repedea.

nanoparticles are able to penetrate deeply inside porous stones without blocking the small pores, but also allows the stone substrate to "breathe" and leads to a better water vapor permeability.

Aesthetics is one of the most significant aspects in cultural heritage conservation. The changes in the monumental stone color that might intervene after their treatments have to be reduced to a maximum extend in order to change as little as possible as compared to the untreated stone sample. The color measurements were performed on samples of Laspra (Table 9.2) and Repedea (Table 9.3), before and after the treatment with two products, and were evaluated using CIE L* a* b* and CIE L* C* h systems, where L* is the variable lightness, which can vary from 0 (black) to 100 (white), a* and b* are the chromatic coordinates, that is, +a is red, −a is green, +b is yellow,

TABLE 9.2

Laspra—Total Color Change after
Treatment Application

Laspra	Tegosivin HL 100	TMSPMA
ΔE^*	0.75	2.87

TABLE 9.3

Repedea—Total Color Change after
Treatment Application

Repedea	Tegosivin HL 100	TMSPMA
ΔE^*	11.92	12.15

and –b is blue. The attributes of chroma are C*—saturation or color purity, and hue h—color wheel. The global color variation (ΔE^*) was evaluated using the formula $\Delta E^* = (\Delta L^{*2} + \Delta a^{*2} + \Delta b^{*2})^{1/2}$.

Some authors state that in order to have an acceptable ΔE^*, its value has to be lower than 5; other authors consider that this threshold value should be 10 [24,25]. One must mention that even if the acceptable chromatic variation is considered to be $\Delta E^* < 10$, some monumental stones do not fit in these limits no matter the applied treatment, their colors depend on their composing minerals. This appears to be the case of Repedea, which registers total color change values higher than 10 no matter the applied water-repellent.

9.4 Artificially Accelerated Aging

During the last decades, many attempts have been made to understand and mitigate the degradation factors and their impact on monumental buildings in a more scientific manner. This part of the study deals with the evaluation of Laspra and Repedea durability, against three of the most aggressive stone decay factors: UV irradiation, salt mist action, and SO_2 action in the presence of humidity. In order to understand and evaluate what happens with the two chosen stones when exposed to the action of these factors, artificially accelerated aging tests were performed. The tests have the purpose of simulating these natural agents' actions on the untreated and treated stones, using special climatic chambers.

In order to evaluate and compare the behavior of Laspra and Repedea—untreated samples/treated with a worldwide-used water-repellent and the newly designed nanocomposite material/treated and artificially aged—several evaluation methods were applied using standardized tests [26].

9.5 Characterization of Treated Stones after the Artificially Accelerated Aging under UV Irradiation

First accelerated aging cycle (500 h of UV irradiation) of Tegosivin HL 100 cast as film determined a significant loss of the siloxane part (elimination of the low-molecular weight components through siloxane photo-induced degradation), being registered a decrease in the major signals with approximately 50% (Figure 9.9a). The main chemical groups affected by UV irradiation were Si–CH$_3$ (1270 cm^{-1}), Si–OCH$_3$ (961 cm^{-1}), Si–O–Si, Si–O–C, and Si–CH$_3$ (1000–1200 cm^{-1}). The second aging cycle had a smaller influence on the degradation of the siloxane network. The FT-IR spectrum of unaged TMSPMA film (Figure 9.9b) shows the presence of the characteristic peaks related to the chemical structure of the synthesized compound. The first 500 h of UV aging determined an important decrease in the bands attributed to the siloxane bonds. During UV accelerated aging, several processes like hydrolysis and polycondensation reaction of methoxy units (diminishing of the Si–O–C signals (1013–1180, 1298, 908, 940, and 982 cm^{-1})), as well as the polymerization of the unreacted methacrylic double bonds (disappearance of the absorption band located at around 1640 cm^{-1} corresponding to C=C groups) can be taken into account (Figure 9.9). UV crosslinking of the unreacted double bonds leads to obtaining a highly crosslinked film, very stable and capable to resist further degradation cycles. Therefore, the second aging cycle had no influence on the degradation of the siloxane network.

The contact angle values registered by the treated and aged Laspra and Repedea samples are shown in Figure 9.10.

Table 9.4 presents the contact angle values of Laspra and Repedea, treated with Tegosivin HL 100 and TMSPMA, before and after 1000 h of artificially accelerated aging under UV irradiation.

In case of Laspra, Tegosivin HL 100 shows the same behavior in terms of contact angle, the samples presenting a slight decrease in the contact angle values. However, if one considers that the stone samples were exposed to 1000 h of continuous UV irradiation, their behavior may be evaluated as being very good—their water-repellency not being affected by the artificial exposure to UV irradiation at 40°C. Repedea treated with Tegosivin HL 100 shows an increase in the contact angle values after the exposure to 1000 h of UV irradiation, probably due to the finalization of the polymerization due to either UV irradiation or to the increased temperature conditions (40°C).

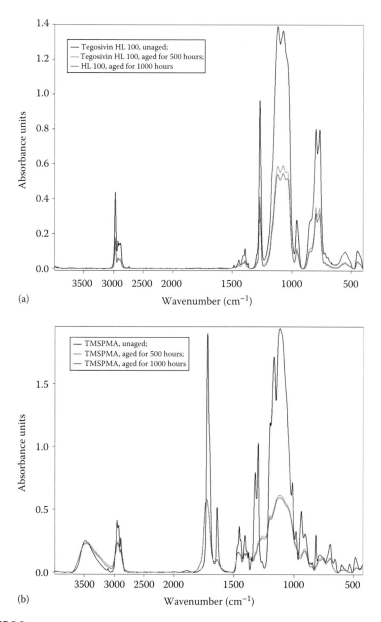

FIGURE 9.9
(a) FT-IR spectrum of Tegosivin HL 100, cast as film and (b) FT-IR spectrum of TMSPMA, cast as film.

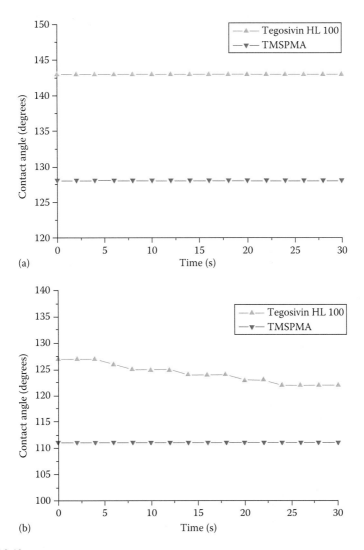

FIGURE 9.10
Contact angle values after 1000 h of artificially accelerated aging under UV irradiation: (a) Laspra and (b) Repedea.

Prior to the artificially accelerated aging under UV irradiation at 40°C, both Laspra and Repedea brushed with TMSPMA showed a slight decrease in contact angle values (more pronounced for Laspra but barely noticeable for Repedea).

A better behavior after 1000 h of UV irradiation was exhibited for Repedea treated with TMSPMA. During the artificially accelerated aging, TMSPMA polymeric layers became more hydrophobic, increasing the water-repellency. The 40°C temperature used in the experiment determines the networking of

TABLE 9.4

Contact Angle Values of Laspra and
Repedea: (a) Treated, (b) Treated/Aged
for 1000 h under UV Irradiation

(a)	Tegosivin HL 100	TMSPMA
Laspra	147	120 ↓
Repedea	100 ↓	105
(b)	Tegosivin HL 100	TMSPMA
Laspra	143	126
Repedea	127 ↓	111

the polymeric structure through the polymerization of the remaining free
methacrylate units, thus increasing the system's hydrophobicity. At the same
time, the additional formation of the cage-like fragments of silsesquioxane
type also increased the system's water-repellency.

9.6 Resistance to Salt Mist Action

Along with water, soluble salts may be the cause of building stone decay.
The growth of salt crystals within the pores of a monumental stone may
generate stresses that are able to overcome the stone's tensile strength and,
over time, transform the stone into powder. It is generally recognized that
salt crystals greatly limit the durability of porous building stones, as well.
Salt weathering is a process of rock disintegration that takes place in a variety of environments and affects many types of rocks [27]. In materials with
small pores, the crystallization pressure will be higher than in materials
having large pores [28–32], and mineral precipitation tends to occur preferably within the inner part of the stone (forming subflorescences). The
application of water-repellent treatments is expected to slow down the deterioration process and to increase the durability of the monumental stones.
The salt mist accelerated aging test was performed according to UNE-EN
14147 European Standard.

After stone cutting and treatment, the monumental stone samples were
kept in desiccators at room temperature. When they reached constant
weights, the samples treated with the commercial water-repellent and the
new designed nanocomposite material were introduced in a salt mist climatic chamber [33], which was covered to avoid the evaporation process. The
salt mist artificial aging was performed by alternating wetting by spraying
(with 5% saline solution) and drying cycles. A spraying cycle took place for
4 h and then was followed by a drying cycle of 20 h. The experiment was

(a) (b)

(c) (d)

FIGURE 9.11
(See color insert.) Laspra treated with (a) Tegosivin HL 100 and (b) TMSPMA. Repedea treated with (c) Tegosivin HL 100 and (d) TMSPMA.

performed for 30 cycles, until one of the selected limestones (Laspra) became severely deteriorated and then continued for another 30 cycles for Repedea (Figure 9.11). Figure 9.12 evidences the ESEM micrographs of stone samples coated with the studied compounds, initially and after performing the cycles of salt mist artificial accelerated aging.

These micrographs evidence that NaCl crystals tend to nucleate heterogeneously, revealing a strong interaction between the crystals in different zones. In the case of the samples treated with TMSPMA, during evaporation, NaCl crystallizes as small efflorescences on stone surfaces. As for the worldwide-used product, this process is extended to a depth of micrometers (sometimes reaching 1 mm), the outermost polymeric layer showing large fractures and being dramatically deteriorated, producing a visible granular disintegration and powdering, therefore leading to stone mass loss. In depth, NaCl has been observed to crystallize mainly as an agglomeration of crystals. This may be related to the wetting properties of the NaCl solution [34,35] and it can explain why in-depth precipitation of halite often occurs as a smooth layer instead of elongated crystals.

The contact angle values registered by the treated monumental stones after performing the resistance against salt mist action artificially accelerated aging test are given in Figure 9.13.

For Repedea, the samples are obviously less damaged, even after 60 artificially accelerated aging cycles. The samples of Laspra treated with the nanocomposite hybrid material exhibited a 96° contact angle value during the

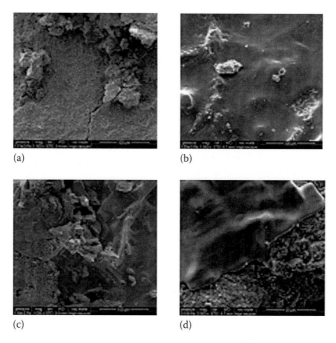

(a)   (b)

(c)   (d)

FIGURE 9.12
ESEM micrographs of Laspra coated with (a) Tegosivin HL 100 and (b) TMSPMA—after 30 cycles of salt mist artificial accelerated aging. Repedea coated with (c) Tegosivin HL 100 and (d) TMSPMA after 60 cycles of salt mist artificial accelerated aging.

first seconds of measurement, a slight decrease in the values being shown during the 30 s of measurements. This may be considered to be a very good result, especially considering that the other Laspra samples, treated with the worldwide-used water-repellent, showed no contact angle at all—the water drop being instantly sucked by Laspra's surface.

After the 60 artificially accelerated salt mist cycles, Repedea samples registered the contact angle values as shown in Figure 9.13b. More than one contact angle curve, corresponding to the same treated sample, may be observed in this figure, depending on the level of deterioration of the polymeric film on the particular spot where the water drop was placed. Figure 9.13b presents the most representative contact angle curves registered by the treated/aged Repedea samples, no generalization being possible. This demonstrates that the water repellent polymeric film is damaged in a random way, the contact angle values depending on the particular location where the measurement took place. The weight loss exhibited by a stone sample after the exposure to the aggressive action of salt mist represents an important evaluation factor, as it indicates the stone sample's decay level and its durability from a particular weathering agent's perspective. Even if Laspra samples—treated with the commercial water-repellent—were

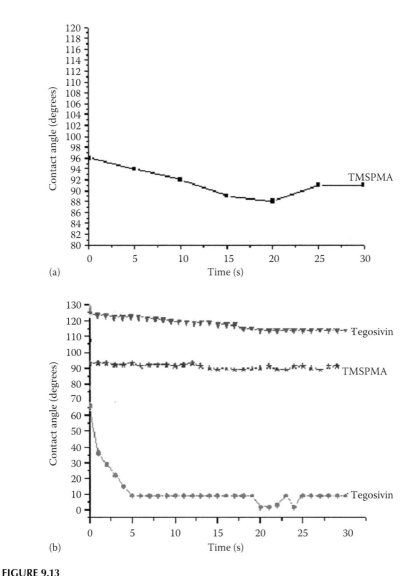

FIGURE 9.13
(a) Laspra treated with TMSPMA—contact angle values after 30 salt mist aging cycles and
(b) Repedea—contact angle values after 60 salt mist aging cycles.

severely damaged and proved to be unfit for further evaluations after their removal from the climatic chamber, an attempt to quantify their mass loss was made using the samples that were less deteriorated. The graphs in Figure 9.14 clearly show a much higher loss of mass for Laspra, due to powdering, fractures, and even the breakage of some of the aged samples, the lowest mass loss exhibited by Laspra was treated with the new nanocomposite material—1.04%. Repedea samples show a significantly reduced

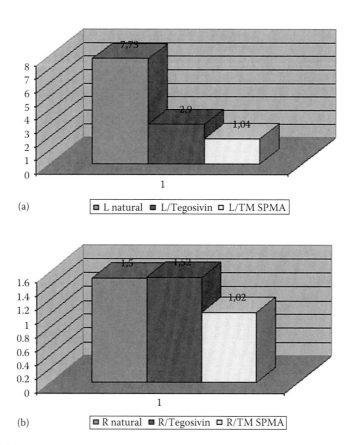

(a)

☐ L natural ☐ L/Tegosivin ☐ L/TM SPMA

(b)

☐ R natural ☐ R/Tegosivin ☐ R/TM SPMA

FIGURE 9.14
(See color insert.) (a) Laspra—weight loss after 30 salt mist aging cycles and (b) Repedea—weight loss after 60 salt mist aging cycles.

mass loss, even after 60 artificially accelerated aging cycles, with no significant differences between the samples treated with one water repellent product or another.

9.7 Resistance to SO_2 Action in the Presence of Humidity

SO_2 is considered the main atmospheric pollutant affecting the building materials, being very corrosive, reactive, and acidifying the rain. This compound is released into the environment and travels long distances in overpowering winds before coming back to earth as rain, dust, fog, or snow. The main effects of sulfur dioxide on limestones are the development of crusts and the material loss due to solubilization (30%–50% of material loss) [36].

The loss of material may also be produced where the weathering crusts reach certain thicknesses and then drop from the monumental stone surface [37].

The chemical composition and the surface morphology of the natural and treated limestones subjected to SO_2-contaminated atmosphere were evaluated by FT-IR spectroscopy and X-ray diffraction analysis [38,39]. The experiments showed differences of the gypsum crystals due to the presence of a specific water repellent product applied onto the limestones surface. Gypsum is being produced due to the reaction between sulfur-containing compounds (SO_2, SO_3, H_2SO_4), water (liquid or vapor) in the atmosphere, and calcite. The FT-IR spectra of the two natural limestones and of the stones treated with the two studied compounds and exposed to SO_2 action are exhibited in Figure 9.15. The formation of gypsum was supported by the appearance of the OH stretching at 3546 and 3405 cm^{-1}; O–H–O bending of the crystallization water at 1685 and 1621 cm^{-1}; sulfate ions at around 1141, 1115, 670, and 602 cm^{-1}; as well as sulfite ions at 983 and 944 cm^{-1} in all spectra.

The X-ray measurements were performed for the qualitative phase identification for both limestones exposed to SO_2 dry deposition. Moreover, different phases for carbonate, sulfite, and sulfate species have been identified for both limestones (Figure 9.16): Laspra: dolomite (D)–$CaMg(CO_3)_2$, calcite (C)–$CaCO_3$, ankerite (A)–$CaMg_{0.32}Fe_{0.68}(CO_3)_2$; Repedea: calcite (C)–$CaCO_3$, magnesian calcite (CM)–$Mg_{0.1}Ca_{0.9}CO_3$, quartz (Q)–SiO_2; Laspra, Repedea: hannebachite (H)–$CaSO_3$–0.5 H_2O; Laspra: gypsum (G)–$CaSO_4$ –2 H_2O, epsomite (E)–$MgSO_4$ –7 H_2O, bassanite (B)–$CaSO_4$ –0.5 H_2O; Repedea: gypsum (G)–$CaSO_4$–2 H_2O, kieserite (K)–$MgSO_4$–H_2O.

For Laspra, the presence of Tegosivin HL 100 appears to favor the formation of epsomite as one of the reaction products. Usually, this mineral is considered to be responsible for the enhancement of the sulfation process due to the formation of a thin water film that allows the formation of a thicker crust salt. The crust developed onto the surface of Laspra treated with TMSPMA is composed only of gypsum, probably due to the fact that the hybrid nanocomposite with silsesquioxane units induced a perfect balance between SO_4^{2-} and Ca^{2+} ions. A higher amount of sulfate species is observed in case of Repedea treated with Tegosivin HL 100, while a smaller amount of gypsum content is evidenced in case of Repedea treated with TMSPMA. In the latter case, the formation of only sulfite species points out the higher resistance of TMSPMA to SO_2 action. Due to the self-assembling properties, the silsesquioxane-based hybrid nanocomposite with methacrylate units (TMSPMA) bonds with both limestones and penetrates these ones deeper than Tegosivin HL 100. In this way, this nanocoating can increase the limestone resistance to SO_2 dry deposition.

The ESEM analysis was performed to evidence the limestone's surface morphology exposed to SO_2 dry deposition (Figure 9.17).

For Repedea brushed with the commercial product, the dry deposition of SO_2 leads to the formation of a considerable amount of gypsum (evidenced by a higher crust thickness). Due to the large molecular structures, Tegosivin HL 100 allows a good SO_2 penetration into stone substrates, the larger

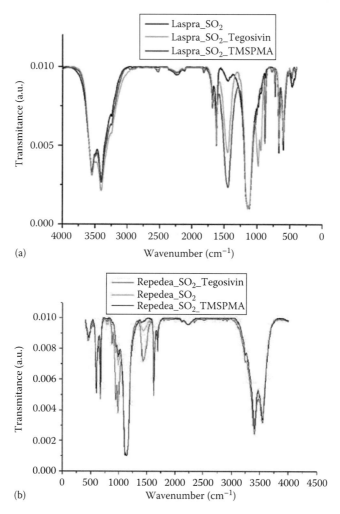

FIGURE 9.15

FT-IR spectra of (a) Laspra subjected to SO_2 action and (b) Repedea subjected to SO_2 action.

penetration, and the higher in-depth sulfation (reaction between carbonate and SO_2). The complete sulfation process in case of Laspra, as opposite to Repedea, may be correlated to a higher susceptibility of Laspra to SO_2 action. In case of TMSPMA, the nanosized porous network inhibits SO_2 penetration ($M_{SO_2} = 64$, $M_{H_2O} = 18$), allowing water vapor to enter and exit, the stones being able to "breathe" freely. Due to the homogeneous layers of TMSPMA comprising a small amount of surfactant, the sulfation proceeds only above the surface, leading to the development of a very thin gypsum layer. For Laspra treated with the silsesquioxane-based nanocomposite, the ESEM micrographs show the presence of only sheet-like crystals. Considering the fact that all the

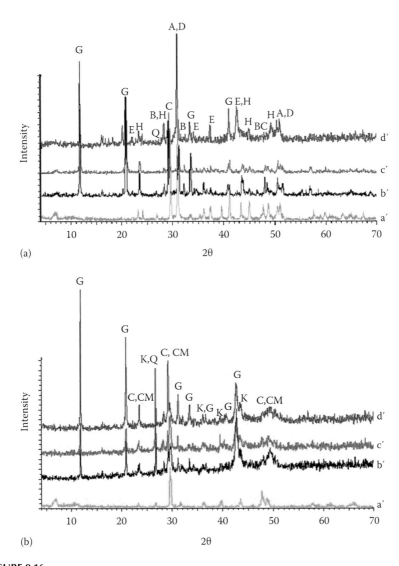

FIGURE 9.16
(a) XRD patterns of untreated Laspra (a′); Laspra + SO₂ (b′); Laspra + TMSPMA + SO₂ (c′); Laspra + Tegosivin HL 100 + SO₂ (d′); (b) XRD patterns of untreated Repedea (a′); Repedea + SO₂ (b′); Repedea + TMSPMA + SO₂ (c′); Repedea + Tegosivin HL 100 + SO₂ (d′).

samples have been exposed to the same conditions, the formation of a thin crust of rosette type (corresponding to the final stage of weathering) suggests that the sulfation process in this particular case is stopped at this level, while the appearance of needle-like crystals (initial stage) when the commercial product was used can be correlated to the extension of the sulfation process by the development of subsequent new layers of sulfite and sulfate.

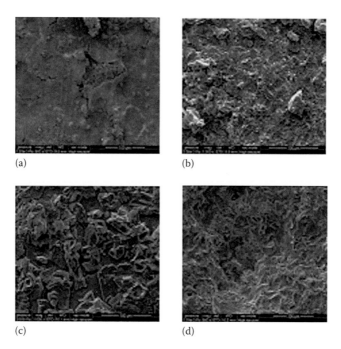

FIGURE 9.17
ESEM micrographs of Laspra coated with (a) Tegosivin HL 100 and (b) TMSPMA after exposure to SO$_2$-contaminated atmosphere; ESEM micrographs of Repedea coated with (c) Tegosivin HL 100 and (d) TMSPMA after exposure to SO$_2$-contaminated atmosphere.

The measurement of contact angle values exhibited by the treated stones before and after the exposure to SO$_2$-contaminated atmosphere is presented in Figure 9.18.

For Laspra, Tegosivin HL 100 induced either the higher contact angle value, or, on the contrary, the lowest, depending on the decay degree of the polymeric film on the spot where the water drop was deposited. The contact angle values exhibited during the first seconds of measurement have been much higher than the ones registered after 30 s of measurement. TMSPMA registered a better behavior as compared to the commercially available water-repellent: the same contact angle values have been exhibited for a stone sample, no matter where the water drop was placed.

Weight measurements have been performed for assessing the mass changes suffered by the monumental stone samples after the exposure to SO$_2$-contaminated atmosphere (Figure 9.19). As in the case of the resistance to salt mist artificially accelerated aging, Laspra showed higher mass loss values as compared to Repedea.

Looking at the two graph scales, one may state that Laspra samples have suffered a mass loss more or less twice as significant as the Repedea samples.

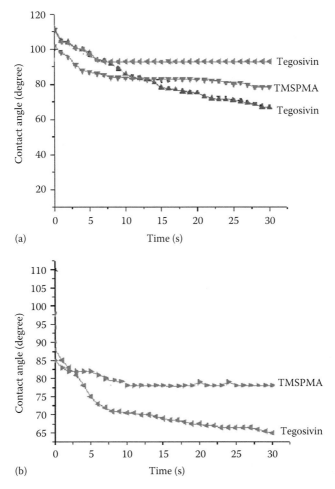

FIGURE 9.18
Contact angle values of (a) treated Laspra after exposure to SO_2 atmosphere and (b) treated Repedea after exposure to SO_2 atmosphere.

9.8 Conclusion

The present research was focused on the durability of two monumental stones treated with siloxane-based water-repellents. For this research, a worldwide-used siloxane-based water repellent product, namely, Tegosivin HL 100, and a newly designed silsesquioxane-based nanocomposite were investigated from the point of view of their protective ability and durability. The comparative assessment of the experimental data proved a significantly higher durability of the limestones treated with TMSPMA, especially

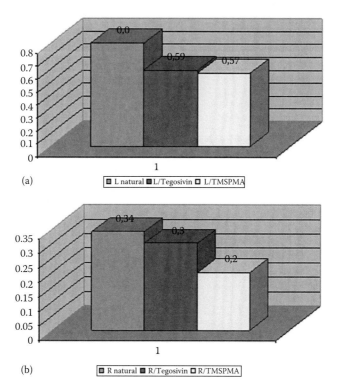

FIGURE 9.19
(See color insert.) Weight loss of (a) Laspra after exposure to SO_2 saturated atmosphere and (b) Repedea after exposure to SO_2 saturated atmosphere.

in the case of salt mist action. TMSPMA novelty consists in its high functionality—the starting monomer contains, in a single molecule, a methacrylate group, capable of thermally or photo-polymerizing, and alkoxy silane groups, which are able to react with each other or with stone functional groups. These functional groups are linked by hydrophobic short chains and they form polymeric networks *via* sol–gel reactions. On the other hand, the polymer self-assemblies in organized domains and the whole assembly have become highly crosslinked network structures.

Acknowledgment

The research leading to these results has received funding from the "Cristofor I. Simionescu" Postdoctoral Fellowship Program, Contract: POSDRU/89/1.5/S/55216.

References

1. Wheeler, G. 2005. *Alkoxysilanes and the Consolidation of Stone*. Mark Greenberg, Los Angeles, CA: Getty Publications.
2. Félix, C. and Furlan, V. 1994. Variations dimensionnelles de grès et calcaires liées à leur consolidation avec un silicate d'éthyle. In *Conservation of Monuments in the Mediterranean Basin*, eds. V. Fassina, H. Ott, and F. Zezza, pp. 855–859. Venice, Italy: Soprintendenza ai Beni Artistici e Storici di Venezia.
3. Alonso, F. J., Esbert, R. M., Alonso, J., and Ordaz, J. 1994. Saline spray action on a treated dolomitic stone. In *Conservation of Monuments in the Mediterranean Basin*, eds. V. Fassina, H. Ott, and F. Zezza, pp. 860–870. Venice, Italy: Soprintendenza ai Beni Artistici e Storici di Venezia.
4. Selwitz, C. M. 1991.The use of epoxy resins for stone conservation. In *Materials Issues in Art and Archaeology*, 2nd edn., eds. P. B. Vandiver, J. R. Druzik, and G. S. Wheeler, pp. 181–191. Pittsburgh, PA: Materials Research Society.
5. Winkler, E. M. 1975. *Stone: Properties, Durability in Man's Environment*, 2nd edn., New York: Springer-Verlag.
6. Wheeler, G. S., Schein, A., Sherrer, G., Su, S. H., and Scott Blackwell, C. 1992. Preserving our heritage stone. *Anal. Chem.* 64:347–356.
7. Cimitàn, L., Rossi, P. P., and Torraca, G. 1994. Accelerated sulphation of calcareous materials in a climatic chamber: Effect of protective coatings and inhibitors. In *Conservation of Monuments in the Mediterranean Basin*, eds. V. Fassina, H. Ott, and F. Zezza, pp. 233–241. Venice, Italy: Soprintendenza ai Beni Artistici e Storici di Venezia.
8. Ji, X., Hampsey, J. E., Hu, Q., He, J., Yang, Z., and Lu, Y. 2003. Mesoporous silica-reinforced polymer nanocomposites. *Chem. Mater.* 15:3656–3662.
9. Wight, A. P. and Davis, M. E. 2002. Design and preparation of organic–inorganic hybrid catalysts. *Chem. Rev.* 102:3589–3614.
10. Simionescu, B., Olaru, M., Aflori, M., Buruiana, E. C., and Cotofana, C. 2009. Silsesquioxane-based hybrid nanocomposites of polymethacrylate type with self-assembling properties. *Solid State Phenom.* 151:17–23.
11. Simionescu, B., Olaru, M., Aflori, M., and Cotofana, C. 2010. Silsesquioxane-based hybrid nanocomposite with self-assembling properties for porous limestones conservation. *High Perform. Polym.* 22:42–55.
12. Pescarmona, P. P. and Maschmeyer, T. 2001. Oligomeric silsesquioxanes: Synthesis, characterization and selected application. *Austr. J. Chem.* 54:583–596.
13. Eisenberg, P., Erra-Balsells, R., Ishikawa, Y., Lucas, J. C., Mauri, A. N., Nonami, H., Riccardi, C. C., and Williams, R. J. J. 2000. Cagelike precursors of high-molar-mass silsesquioxanes formed by the hydrolytic condensation of trialkoxysilanes. *Macromolecules* 33:1940–1947.
14. Feher, F. J. and Budzichowski, T. A. 1995. Silasesquioxanes as ligands in inorganic and organometallic chemistry. *Polyhedron* 14:3239–3253.
15. Du, J. and Chen, Y. 2005. Hairy nanospheres by gelation of reactive block copolymer micelles. *Macromol. Rapid Commun.* 26:491–494.
16. Du, J., Chen, Y., Zhang, Y., Han, C. C., Fischer, K., and Schmidt, M. 2003. Organic/inorganic hybrid vesicles based on a reactive block copolymer. *J. Am. Chem. Soc.* 125:14710–14711.

17. Du, J. and Chen, Y. 2004. Organic–inorganic hybrid nanoparticles with a complex hollow structure. *Angew. Chem. Int. Ed.* 43:5084–5087.
18. Galan, E. 1993. Control of physico-mechanical parameters for determining the effectiveness of stone conservation treatments. In *Stone Material in Monuments: Diagnosis and Conservation*, pp. 189–197. Crete, Greece: Comunità delle Università Mediterranee, Scuola Universiteria Conservazione dei Monumenti.
19. Sasse, H. R., Honsinger, D., and Schwamborn, B. 1993. "PINS:" New technology in porous stone conservation. In *Conservation of Stone and Other Materials*, ed. M.-J. Thiel, pp. 705–716. London, U.K.: E & F N Spon.
20. Galan, E. and Carretero, M. I. 1994. Estimation of the efficacy of conservation treatments applied to a permotriassic sandstone. In *Conservation of Monuments in the Mediterranean Basin*, eds. V. Fassina, H. Ott, and F. Zezza, pp. 947–954. Venice, Italy: Soprintendenza ai Beni Artistici e Storici di Venezia.
21. Simionescu, B., Aflori, M., and Olaru, M. 2009. Protective coatings based on silsesquioxane nanocomposite films. *Constr. Build. Mater.* 23:3426–3430.
22. Dunham, R. J. 1962. Classification of carbonate rocks according to depositional texture. In *Classification of Carbonate Rocks—A Symposium*, ed. W. E. Ham, pp. 108–121. Tulsa, OK: American Association of Petroleum Geologists Memoir.
23. Folk, R. L. 1959. Practical petrographic classification of limestones. *Am. Assoc. Petrol. Geol. Bull.* 43:1–38.
24. Folk, R. L. 1962. Spectral subdivision of limestone types. In: *Classification of Carbonate Rocks—A Symposium*, ed. W. E. Ham, pp. 62–84. Tulsa, OK: American Association of Petroleum Geologists Memoir.
25. Simionescu, B. and Olaru, M. 2009. Assessment of siloxane-based polymeric matrices as water repellent coatings for stone monuments. *Eur. J. Sci.Theol.* 5:59–67.
26. Simionescu, B. 2009. Thesis: Durability of monumental stones treated with siloxane-based water repellents, Bologna University, Bologna, Italy.
27. Lewin, S. Z. 1981. *The Mechanism of Masonry Decay through Crystallization*, Washington, DC: National Academy of Sciences.
28. Flatt, R. J. 2002. Salt damage in porous materials: How high supersaturations are generated. *J. Cryst. Growth* 242:435–454.
29. Scherer, G. W. 1999. Crystallization in pores. *Cem. Concr. Res.* 29:1347–1358.
30. Wellman, H. W. and Wilson, A. T. 1965. Salt weathering, a neglected geological erosive agent in coastal and arid environments. *Nature* 205:1097–1098.
31. Gauri, K. L., Chowdhury, A. N., Kulshreshtha, N. P., and Punuru, A. R. 1988. *Engineering Geology of Ancient Works, Monuments and Historical Sites*, p. 723. Rotterdam, the Netherlands: Balkema.
32. Scherer, G. W. 2000. *Ninth International Congress on Deterioration and Conservation of Stone*, Venice, Italy, p. 187.
33. Simionescu, B., Doroftei, F., Olaru, M., Aflori, M., and Cotofana, C. 2009. Salt-induced decay in limestones treated with siloxane-based water repellents, *J. Optoel. Adv. Mater.–Symposia* 1:1077–1082.
34. Pel, L., Huinink, H., Kopinga, K., Van Hees, R. P. J., and Zezza, F. 2004. *Proceedings of 13th International Brick and Block Masonry Conference*, Amsterdam, the Netherlands.
35. Lubelli, B. and De Rooij, M. R., 2009. NaCl crystallization in restoration plasters. *Constr. Build. Mat.* 23:1736–1742.

36. Baedeker, P. A., Reddy, M. M., Reimann, K. J., and Sciammarella, C. A. 1992. Effects of acidic deposition on the erosion of carbonate stone—Experimental results from the U.S. national acid precipitation assessment program (NAPAP). *Atmos. Environ.* 26B:147–158.
37. Camuffo, D., del Monte, M., and Sabbioni, C. 1983. Origin and growth mechanisms of the sulfated crusts on urban limestone. *Water, Air, Soil Poll.* 19:351–360.
38. Olaru, M., Aflori, M., Simionescu, B., Doroftei, F., and Stratulat, L. 2010. Effect of SO_2 dry deposition on porous dolomitic limestones. *Materials* 3:216–231.
39. Simionescu, B., Olaru, M., Aflori, M., and Doroftei, F. 2011. Siloxane-based polymers as protective coatings against SO_2 dry deposition, *High Performance Polymers* 23:326–334.

10

POSS-Containing Nanocomposite Polymer Coatings

Yasmin Farhatnia, Aaron Tan, and Alexander M. Seifalian

CONTENTS

10.1 Introduction

Medical implants are ubiquitous in clinical settings, and their rate of success is determined by their biocompatibility, which is in turn affected by its surface properties [1]. One way of ensuring the implant is not rejected by the body is to coat it with a layer of material that is highly biocompatible.

In addition to being biocompatible, the coating technique and its resultant surface topography are also important considerations [2].

Different materials have been used for biomedical applications, with an increasing focus on composite materials utilizing nanotechnology [3]. Composite materials comprise two or more elementary materials that are combined for the purpose of significantly improving their overall performance in the final product. Matrices and reinforcements are the two main constituents in a composite material. Reinforcements come in a plethora of shapes including spherical (e.g., metal nanoparticles), layered (e.g., clay), and fibrous (e.g., carbon nanotubes [CNTs]) [4]. Apart from the constituents, fabrication techniques can also affect the characteristics of the composite material. A small change in the fabrication technique can result in a direct impact on the behavior of the final product. Composite materials can be termed "nanocomposites" when one or more of the elementary constituents are less than 100 nm. Matrices in the nanocomposite can be viewed as "hosts" and the reinforcements as "guests" [5]. These guest molecules can assemble themselves in three ways: phase separated, intercalated, and exfoliated. The large surface area-to-volume ratio of nanocomposites enables various chemical reactions to occur on the surface. The higher propensity for chemical reactions will thus facilitate the reinforcements to serve as a link between the polymer molecules [6]. This will enhance its optical, electrical, thermal, and mechanical properties. For example, polyhedral oligomeric silsesquioxane (POSS) is a reinforcement that can improve the mechanical properties of the polymer and it has been used in various biomedical applications.

10.2 Polyhedral Oligomeric Silsesquioxane: A Caged Nanostructure

Silsesquioxane is defined by the empirical chemical formula $R_nSi_nO_{1.5n}$, where R is an organic group (e.g., alkene) or hydrogen. Silsesquioxane structures can be generally classified as being either caged or noncaged. Noncaged structures can be further subdivided into random, ladder, and partial cage [6,7] (Figure 10.1). Caged structures include POSS [8]. POSS has a 3-D shape (polyhedral = many-sided) formed by a few (oligomeric) units of silsesquioxanes. The most common stoichiometric formula for POSS is $R_8Si_8O_{12}$. Each POSS molecule measures 1.5 nm in diameter (including the –R groups) and it can be considered as the smallest achievable silica particle [9].

Being constituents of nanocomposite polymers [10], the organic–inorganic nature of POSS molecules confers thermal resistance and improves the overall mechanical property of the polymer [11,12]. This is in part due to the decrease in relative permittivity, increase in glass transition temperature

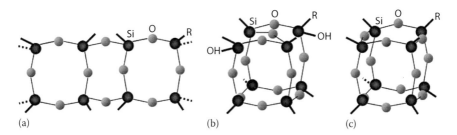

FIGURE 10.1
Structure of silsesquioxanes: (a) ladder-like, (b) partial cage, and (c) cage. Figures (Reproduced with permission from Springer Science+Business Media: *Applications of Polyhedral Oligomeric Silsesquioxanes*, Biomedical application of polyhedral oligomeric silsesquioxane nanoparticles, 2011, 363, Ghanbari, H., Marashi, S.M., Rafiei, Y., Chaloupka, K., and Seifalian, A.M., Copyright 2011.)

(Tg) [13–15], and decrease coefficient of thermal expansion [9,16]. Mechanical characteristics like tensile strength [17,18], viscoelasticity [19,20], and oxidative resistance are significantly improved with the addition of POSS into polymers. These attributes are particularly useful in a range of disciplines from aerospace technology to the biomedical field.

In addition to improving the bulk properties and its stability in various conditions, the inclusion of POSS into polymers alters the surface properties of the material. These include hydrophobicity, surface energy, and morphology (which affects biological response), thus making them viable candidates for coating medical devices [21].

10.2.1 Mechanical Characteristics of POSS-Based Nanocomposites

The mechanical properties of coatings for implants play a vital role in its viability and functionality. POSS-incorporated nanocomposites have been extensively studied. It has been revealed that POSS incorporated into poly(carbonate-urea) urethane (POSS–PCU) matrix demonstrated increased mechanical strength, while maintaining radial elasticity, tensile strength, tear strength, and hardness (Figure 10.2). A higher tensile strength was observed in POSS–PCU (53.6 ± 3.4 and 55.9 ± 3.9 N mm^{-2} at 25°C and 37°C, respectively) compared to PCU (33.8 ± 2.1 and 28.8 ± 3.4 N mm^{-2}). Young's modulus of POSS–PCU (25.9 ± 1.9 and 26.2 ± 2.0 N mm^{-2} at 25°C and 37°C respectively) was also significantly greater than PCU (9.1 ± 0.9 and 8.4 ± 0.5 N mm^{-2} at 25°C and 37°C respectively) [22].

Mechanical characterization is an important aspect in ascertaining the properties of coatings for medical devices. Hence, the superior mechanical aspects of POSS–PCU make it an ideal candidate as a coating for small diameter stents in coronary and peripheral arterial applications. Furthermore, a POSS–PCU manufactured graft was placed in a flow circuit mimicking the circulatory system. Data extracted from wall tracking ultrasound showed

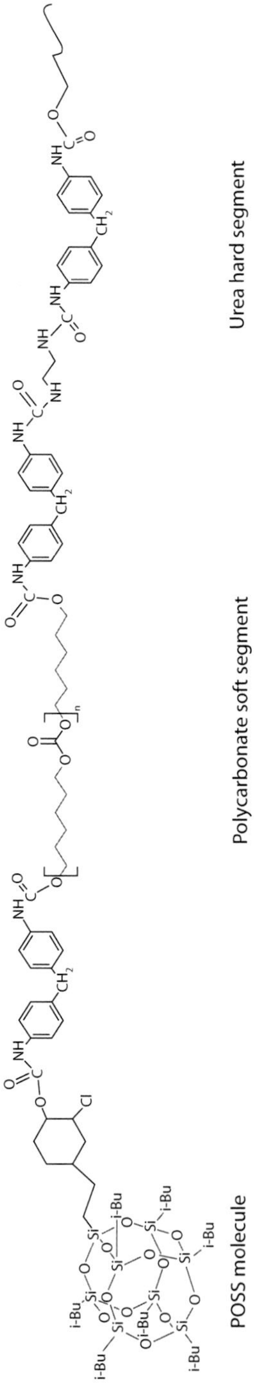

POSS molecule   Polycarbonate soft segment   Urea hard segment

FIGURE 10.2

Chemical structure of POSS–PCU. POSS–PCU comprises three segments: POSS molecule, polycarbonate soft segment, and urea hard segment.

excellent elastic and viscous behavior, suggesting a close approximation of mechanical properties between the graft and native artery. Since the mismatch between graft and native artery is greatly reduced, this further supports the hypothesis that POSS–PCU graft would exhibit long-term patency and reduced neointimal hyperplasia [23,24].

10.2.2 Degradative Resistance

It was initially postulated that incorporating POSS into PCU would strengthen the constituent components, thereby increasing the degradative resistance of the nanocomposite [25]. There are three main types of degradation: hydrolytic, oxidative, and physiological. When exposed to various plasma protein fractions (PF I–IV), FTIR analysis revealed no difference in Si–O, C–O–C, NH–CO, and (NHCO)O intensities for each bond wavelength compared to control. This suggests that biological proteins do not have a significant effect on POSS–PCU. However, hydrolytic enzymes, particularly phospholipase, might have an effect on Si–O bonds as there is a decrease in its wavelength intensity. Nevertheless, there is no observable change in color or any visible signs of degradation. A reduction in the intensity of Si–O bonds was observed when POSS–PCU was exposed to oxidative agents such as $H_2O_2/CoCl_2$ and glutathione/t-butyl $H_2O_2/CoCl_2$ [26]. A leftward shift in the peak of C–O–C bonds also indicates a decrease in hydrogen bonding. SEM and FTIR reveal that surface degradation is increased with the weakening of amide cross-linkages with a loss of crystalline peaks. Stress–strain analysis did not show any significant difference between hydrolytic and oxidative degraded POSS–PCU and control. This suggests that the soft segment of the polyurethane (PU) remains intact. However, t-butyl peroxide-degraded POSS–PCU ruptures at pressures above 220 mmHg. The Si–O bonds hold the nanocomposite together via zone intercalation between the hard and soft segments of PU. Thermal analysis shows that there is no significant difference in the glass transition temperature (Tg) of degraded and nondegraded samples of POSS–PCU.

10.2.3 Hemocompatibility

Since the clotting time is high for surfaces with low wettability, thrombosis is less likely to occur on nonwettable materials [27–29]. Surfaces with anionic charges have a reduced propensity for thrombosis. This can be alluded to the electrostatic repulsion between the polymer surface and blood [30–33].

Studies show that upon the incorporation of POSS molecule into PCU structure, POSS moiety migrate to the surface of the polymer matrix resulting in an amphiphilic lipid-like behavior in this nanocomposite with a low surface energy. This in turns reduces both platelet and protein adsorption, repelling platelet surface adsorption and lowering their binding affinity to the nanocomposite polymer. This corresponds to the poor adsorption characteristics exhibited toward fibrinogen leading to antithrombogenic properties of the surface [34,35].

Thromboelastography has been used to assess the antithrombogenic characteristics of POSS–PCU. A significantly lower maximum amplitude was observed, indicating decreased platelet bonding strength. Unstable clots were also detected on the surface of POSS–PCU, but were dissolved within 60 min [35]. Direct ELISA fibrinogen adsorption analysis showed significantly lower fibrinogen adsorption rate of POSS–PCU and PCU compared to control (PTFE). Platelet adsorption assays also showed lower platelet adsorption on POSS–PCU and PCU compared to PTFE. Hence, the antithrombogenic characteristic of POSS–PCU renders it an ideal candidate for bypass grafts, heart valves, and stent coatings.

10.2.4 Enhancing Endothelialization

In vivo, the inner layer of blood vessels is populated by a single layer of endothelial inactivated platelets. In a circulatory loop system, this stops the adherence or activation, indicating an intrinsic nonthrombogenic property and long-term hemocompatibility. Therefore, surface nanoengineering of cardiovascular devices suggests that a uniform endothelial cell (EC) layer to the endothelialization of the material before/or after implantation enhances their biocompatibility and hemocompatibility, and improves their clinical outcome of these devices [36].

Various studies showed that the presence of an EC lining on the luminal surface of bypass grafts with ECs, particularly in lower limb arterial bypass grafts [37], improves the patency and clinical outcome of these devices [38].

EC-seeding has shown to reduce the incidence of restenosis and neointimal hyperplasia (NIH) [39], particularly in the elderly. The general consensus of EC-seeded stents does not seem popular, as there are only a small number of studies for EC-seeded bare-metal stent (BMS). EC-seeded covered stents have been developed for microporous PU membranes, as well as a collagen-hybrid tissue membrane [40]. Incorporation of POSS nanocage structures modifies the surface morphology of the PCU polymer and provides a surface roughness at the nanometer scale, which is more favorable for EC interactions and cellular behavior through adhesion, growth, and proliferation than a smoother surface profile [41] (Figure 10.3). EC adhesion and proliferation are highly dependent on factors such as wettability, chemistry, surface charge, and topography of the polymer [42,43]. Charged surfaces result in a higher degree of adsorption of proteins, while the reverse is true [44–46]. In addition, the functional groups of the POSS nanocage structure within POSS–PCU have the potential to be further modified by the incorporation of bioactive peptides, growth factors, receptor ligands, and/or antibodies within the polymeric matrix [6].

Modification of the polymer surface by peptides and functional part of extracellular matrix components can enhance endothelialization potential of the surface, as demonstrated by various studies, particularly for

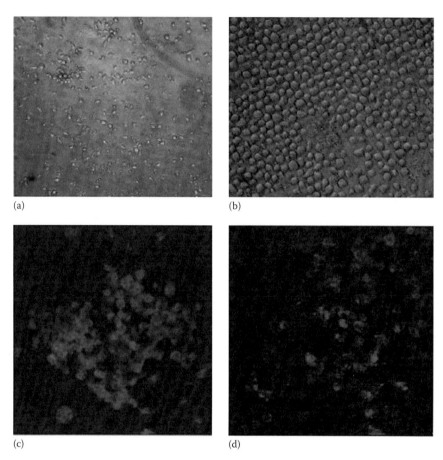

FIGURE 10.3
(See color insert.) Proliferation of endothelial progenitor cells on POSS–PCU: (a) spindle-shaped at day 7, (b) cobblestone-shaped at day 21, (c) cells stained with von Willebrand factor, and (d) cells stained with vascular endothelial growth factor receptor-2 (VEGFR2). (Reproduced from Ghanbari, H., de Mel, A., and Seifalian, A.M., *Int. J. Nanomed.*, 6, 775, 2011. With permission from Dove Press. Copyright 2011.)

cardiovascular device applications. The functionalization potential of POSS–PCU surface was observed by the conjugation of biofunctional peptide (i.e., RGD) onto the surface of nanocomposite. The results present evidence for an excellent cell adhesion and proliferation for both human umbilical vein endothelial cells (HUVECs) and stem cells [38]. SEM revealed several flattened ECs at the surface of polymer, suggesting that these ECs are capable of morphogenesis and have the ability to proliferate well [47].

As discussed earlier, migration of POSS onto the surface creates a topography that can potentially enhance attachment and proliferation of cells (Figure 10.4). This has been assessed by investigating the effect of different

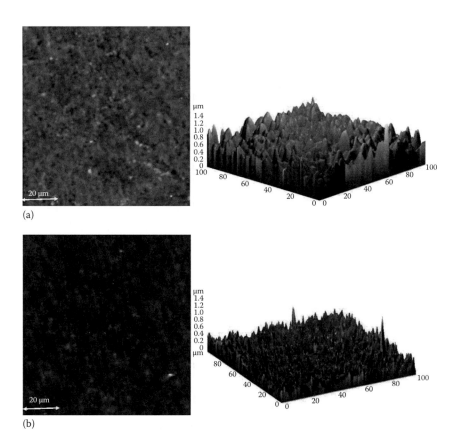

FIGURE 10.4
(See color insert.) Atomic force microscopy of (a) PCU and (b) POSS–PCU. Addition of POSS into PCU (making POSS–PCU) results in nanotopography patterns on the surface of the nanocomposite. (Reproduced with permission from Springer+Business Media: *Applications of Polyhedral Oligomeric Silsesquioxanes*, Biomedical application of polyhedral oligomeric silsesquioxane nanoparticles, 2011, 363, Ghanbari, H., Marashi, S.M., Rafiei, Y., Chaloupka, K., and Seifalian, A.M., Copyright 2011.)

percentages of POSS in the polymer on the growth of fibrinogen molecule adsorption, which can potentially enhance the attachment and proliferation of cells. Adsorption protein network is higher on the soft polymeric component of this nanocomposite, compared to the reduced fibrinogen adsorption on the hard nanocage component. A reduction in fibrinogen accumulation on the hard POSS nanocage therefore would reduce the possibility of clot formation [48].

The safety and compatibility of POSS–PCU have been assessed *in vitro* using HUVECs and were compared to conventional silicone copolymers [37,49,50], and the results showed that the POSS-based nanocomposites enhance the degree of adherence and proliferation of the ECs on the surface.

These studies reveal that when they adhere to the surface, ECs grow and proliferate to form a confluent layer of cells. While PicoGreen assay gives a quantitative indication of cell proliferation, light microscopy gives a qualitative assessment. Before growing into confluent layer, ECs appear in a reticular fashion. The gaps are then filled to form a confluent layer [25].

10.2.5 Preventing the Occurrence of Calcification

Calcification is a complex phenomenon formed as a result of various biochemical and mechanical factors. This is a limiting factor and a main cause of failure for polymeric heart valve [51]. To investigate the fatigue and calcification, features of the POSS–PCU polymer sheets were compared with glutaraldehyde-fixed bovine pericardium (BP) and PU in accelerated physiological pulsatile pressure system for a period of 31 days. Calcium solution was pumped through the circuit, operating at a frequency of 50 Hz (400 million cycles) to simulate the physiological pulsatile pressure system. The surface and mechanical characteristics of the samples were investigated, and the results confirmed a significantly lower level of calcification for POSS–PCU sheets compared to fixed BP and PU samples. However, the mechanical properties of nanocomposite samples remained unchanged while the PU samples were significantly degraded. Although the wettability of both POSS–PCU and PU samples were changed toward lower hydrophobicity, the POSS–PCU samples were more hydrophobic ($p < 0.0001$) with less platelet attachment than PU samples, all indicating the potential advantage of these nanocomposites for heart valve leaflet application [52].

10.2.6 Circumventing Biological Inflammation

Inflammatory response is a complex immunological reaction involving various types of immune cells and immunomodulators such as cytokines. This is a major factor in determining the long-term durability and biocompatibility of the medical implants.

Although the specific physiochemical properties of the biomaterials which have a direct impact on the biological response are not fully understood, it is well known that increased inflammation can result in the failure of the implant.

The potential inflammatory reaction to POSS–PCU was investigated *in vitro* by exposing the peripheral blood mononuclear cells and evaluating the cells surface markers and cytokine release. The results confirmed a significantly lower level of proinflammatory cytokines (i.e., interleukin-1 (IL-1β/IL-1F2), tumor necrosis factor (TNFα/TNFSF1A), along with the reduction of leukocyte activation marker such as CD 86 and CD 69 [3].

The biocompatibility and inflammatory reactions of POSS–PCU were also investigated *in vitro* for a period of 36 months in a healthy adult sheep model. The POSS–PCU sheets were implanted subcutaneously into the back of the animal under general anesthesia. The histopathological examination of the surrounding tissue revealed no evidence of inflammatory reaction or capsule formation [53].

10.3 Imperative for Coating Medical Devices

Coating augments the functionality of medical devices and is a major area of research and development in medical applications. It is used to improve the biocompatibility of materials, which are mechanically robust, but otherwise unsuitable for prolonged implantation in biological systems. Coatings are also used to improve the physical characteristics of the implant (i.e., reduce friction). They can reduce irritation and inflammation and the risk of infection related to the implanted device. They decrease the potential of scar tissue formation surrounding the implanted devices and encourage the growth of tissues to support the healing process. Studies have shown that the characteristics of the coating *in vivo* are directly influenced by the combination of materials, coating technique, and coating device adhesion strength [54]. The applied surface coatings can range from metal to polymer and drug-incorporated polymer with precise pores to allow a controlled drug release profile.

Medical device coatings can be accomplished through various coating procedures including Langmuir–Blodgett thin films [55,56], airbrushing [57], and chemical vapor deposition techniques (for a surface reaction to create film) [58,59], physical vapor deposition (transfer a solid source to a surface film), freeze condensation of vapor (create a thin film of frozen liquid) electrospinning [60], layer-by-layer (LBL) assembly [61], and ink jet placement (impingement of small droplets on to the surface) [62].

Prior to embarking on selecting a coating technique for a medical device, several engineering design considerations have to be taken into account such as complexity of the substrate geometry, uniform and homogeneous coating [63], consistency in coatings from device-to-device, avoiding bridging and coating buildup across web structures [62], and facilitating controlled drug release platform. Factors increasing production costs such as chemical waste, complex equipment, and the requirement of a coating environment under clean room conditions need to be avoided [64].

Ultrasonic atomization spray has been performed for various applications involving POSS–PCU (Figure 10.5). This includes coating of stents, drug tablets, and tubular structures for tissue engineering purposes. The main rationale of coating stents via ultrasonic atomization with POSS–PCU is to produce a uniform layer of biocompatible and nonthrombogenic coating, thereby reducing the possibility of restenosis and thrombosis. We have also experimented with spraying POSS–PCU onto drug tablets for controlled and

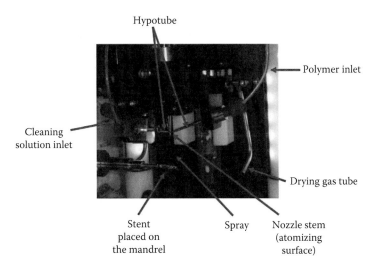

FIGURE 10.5
Digital image of ultrasonic atomization spray system.

sustained release. In addition, we have also manufactured tubular micro-structures, which can be used as artificial nerve conduits [65].

Ultrasonic atomization spray coating is a viable alternative approach for coating stents compared to traditional coating techniques. This approach involves producing an electric charge by the ultrasonic atomization genera-tor. The generator vibrates a ceramic piezoelectric transducer located inside the spray nozzle. This in turn causes the tip to vibrate and atomize any liquid that flows out of it [66]. The polymer flows to the tip of the nozzle (atomizing surface) by a minitube while compressed gas is delivered through the orifice, to render atomized droplets into a precise, targeted spray. This method is useful for devices requiring a precisely generated thin and uniform film of polymer. This method allows the stents to be fully coated with a high level of precision and uniformity, without any webbing, or polymer buildup in between the stent struts.

10.3.1 Coating Coronary Stents

Coronary stents can be classified as either BMS or drug-eluting stent (DES). A major problem seen in BMS is vessel restenosis due to neointimal hyper-plasia. This is the renarrowing of the luminal area of the blood vessel due to proliferation of vascular smooth muscle cells (VSMCs) on the metal struts, as the body perceives the metal as "foreign" and mounts an immune response [67]. DES was developed in response to this problem, as it had the added advantage of having antiproliferative drugs (e.g., paclitaxel or sirolimus) on the metal struts. These antiproliferative drugs prevent the growth and prolif-eration of VSMC, thereby circumventing restenosis. However, a serious and

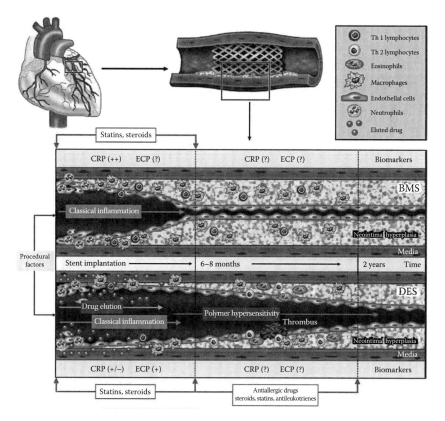

FIGURE 10.6

(See color insert.) Problems associated with bare-metal stents (BMSs) and drug-eluting stents (DESs). In-stent restenosis (ISR) is observed with BMS, necessitating repair procedures. Although ISR was largely circumvented in DES, late stent thrombosis (ST) was seen in DES, which is a potentially fatal complication. (Reproduced from Niccoli, G., Montone, R.A., Ferrante, G., and Crea, F., The evolving role of inflammatory biomarkers in risk assessment after stent implantation, *J. Am. Coll. Cardiol.*, 56, 1783–1793, 2010, Copyright 2010. With permission from American College of Cardiology Foundation.)

highly lethal side effect was observed in the use of DES: late stent thrombosis (ST). Although its pathogenesis has not been fully elucidated, it is postulated that this is due to drug/polymer coating hypersensitivity and lack of endothelialization [68] (Figure 10.6).

There are currently 11 FDA-approved DES for clinical use. Among them, they share a number of similar nonbiodegradable polymer coating platforms. These proprietary polymers act as solvents in which antiproliferative drugs are dissolved.

Polymer coatings on stents can be generally classified into two categories: nonbiodegradable and biodegradable. Nonbiodegradable polymer coatings remain on the metal stent surface permanently, and they prevent blood from coming into contact with the metal surface. The first FDA-approved DES (CYPHER), which was approved in 2003, utilized a nonbiodegradable polymer. Subsequently, all other FDA-approved DESs also utilize nonbiodegradable polymer platforms.

Biodegradable coatings for stents were also explored as potential candidates. Due to the observation of late ST seen in DES with nonbiodegradable coatings, coatings that will degrade over time were seen as an alternative, in essence transforming the polymer-coated stent to a BMS over time. However, the actual benefits of having a biodegradable coating remain to be seen, as early reports indicate possible inflammation due to the degradation of products [69]. Furthermore, FDA has not approved biodegradable coatings on stents.

Apart from functioning as a barrier between the blood–metal interface, polymer coatings also serve as a platform on which drugs can be eluted. The drug–polymer matrix can also be specifically fine-tuned for the drug to be released in a sustained and controlled manner. Sustained and controlled release of antiproliferative drugs is an important aspect in preventing in-stent restenosis (ISR).

The most important characteristic of the polymer coating is that it must be biocompatible. For instance, the endeavor stent uses a phosphorylcholine polymer, which is associated with biological membranes. The polymer coating also aids in controlled release of drugs. Polyvinyl pyrrolidone in the endeavor resolute increases the initial drug burst due to its hydrophilicity, thus enhancing the elution rate. Layers of different polymers can also be employed, as seen in the CYPHER stent. It uses a parylene C basecoat, with the next coating made up of polyethylene-co-vinyl acetate, and poly n-butyl methacrylate (PBMA) mixed with sirolimus. A topcoat of drug-free PBMA is then layered to aid in the controlled release of sirolimus.

10.3.1.1 POSS: An Ideal Stent Coating?

In order to prevent both restenosis and late ST, the coating material must not elicit an immunological response and must also be a suitable platform for reendothelialization to occur. Stents can be spray-coated with POSS–PCU to provide a protective barrier, preventing metal-to-blood contact. This type of stent can be seen as a "middle ground" between BMS and DES. Interestingly, it has also been demonstrated that POSS coatings significantly reduced corrosion in metal alloys [70].

It is important to note that antiproliferative drugs on DES are nonselective as it inhibits the formation of both VSMCs and ECs. The lack of endothelialization on DES contributes to late ST. Selective targeting can be achieved by encapsulating antiproliferative drug molecules, or even genes in nanoparticles (e.g., liposomes or exosomes), and attaching a targeting moiety (e.g., antibody) specific for VSMC [71]. This would ensure that only VSMCs are destroyed while ECs are left intact. The POSS–PCU coating on coronary stents would not only serve as a protective barrier against the blood–metal

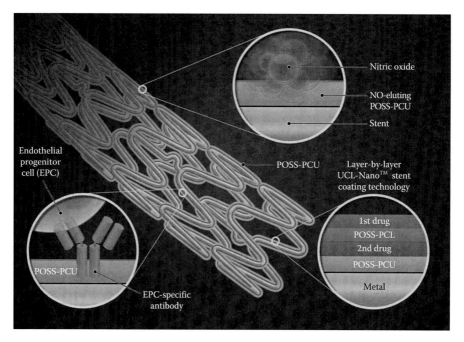

FIGURE 10.7
(See color insert.) Convergence of biotechnology and nanotechnology. We envisage that next-generation coronary stent coatings would incorporate nanotechnology, like endothelial progenitor cell capture, biofunctionalization via nitric oxide (NO), and layer-by-layer self-assembly for sustained and controlled drug release. (Reproduced from *Trends Biotechnol.*, 30, Tan, A., Alavijeh, M.S., Seifalian, A.M., Next generation stent coatings: Convergence of biotechnology and nanotechnology, 406, Copyright 2012, with permission from Elsevier.)

interface, but also function as an excellent platform to attach these targeted nanoparticles. Furthermore, POSS–PCU can be biofunctionalized with bio-active molecules like nitric oxide (NO), which is an important molecule in maintaining vessel homeostasis. POSS–PCL is a poly(ε-caprolactone)-based nanocomposite, which is a biodegradable nanocomposite and thus could be used as a substrate for sustained and controlled release of drugs. Recent evidence suggests that tailored drug release can be achieved on DES with POSS coatings [72]. Indeed, we advocate a three-pronged approach in the development of next-generation stents incorporating nanotechnology: bioactive nanocomposite polymers, endothelial progenitor cell (EPC) capture platforms, and LBL controlled drug release [73] (Figure 10.7).

10.3.1.2 Stent Coatings Utilizing Nanotechnology

As mentioned earlier, multiple layers of polymer coatings can be used on stents. The LBL self-assembly technique can be employed for tunable and multiple drug release. This confers the ability to release different drug types in a controlled manner to suit different phases of the healing process. This opens up the potential of incorporating both antiproliferative and anti-thrombogenic agents. The degradation kinetics of the multiple layers can be fine-tuned to match healing times, and they can also be functionalized with bioactive molecules such as NO. Stent coatings can also be biofunc-tionalized with NO precursors and donors (e.g., sodium nitroprusside and *S*-nitrosoglutathione). NO would subsequently be released upon contact with blood, conferring antithrombogenic and antiproliferative effects [74].

Novel stent coatings incorporating nanotechnology have also been con-ceived. Endothelialization is an important prerequisite for the regenera-tion of a healthy endothelium in the luminal area of the stented vessel. This would prevent ISR and also late ST. EPC capture technology is exemplified by the Genous stent, which has antiCD34 antibodies attached to its stent sur-face coating [75]. Other EPC capture technologies using various antibodies like antiCD133 and VEGFR-2 are also currently being explored [76,77]. Apart from antibodies, functional peptides can also be conjugated onto the coat-ings of stents to promote endothelialization, with examples such as RGD and peptide amphiphile nanofibers [78]. Furthermore, glycoprotein IIb/IIIa inte-grin complex can prevent platelet aggregation and thrombosis.

Nanoparticles can also be incorporated onto stent coatings [79] (Figure 10.8). Drugs encapsulated in nanoparticles can significantly increase the therapeu-tic index, resulting in a decreased dosage needed for its effect. This would also lower the risk of side effects. Encapsulation of pharmacologic agents into nanoparticles like liposomes would enable sustained and controlled release. There is also experimental evidence indicating that magnetic nanoparticles can be guided onto the stent area, augmenting drug localization and delivery [80].

The concept of gene therapy using gene-eluting stents, although in its infancy, has also been extensively explored [71] (Figure 10.9). It has been shown that plasmid DNA expressing endothelial nitric oxide synthase and inducible nitric oxide synthase can inhibit the smooth muscle cell prolifera-tion and enhance reendothelialization [81]. Genetic material can be delivered via adenoviruses or liposomes attached on the polymer coating on stent sur-face, which would prevent restenosis and thrombosis.

10.3.2 Orthopedic Implants

With the increasingly ageing population, hip replacements are becoming ubiquitous. Recent studies have indicated unusually high failure rates of metal-on-metal (MOM) hip implants, requiring revision surgery [82,83]. Furthermore, experimental data have indicated that metal debris resulting from wear could contribute to immunological responses in the body [84]. There is also evidence showing an increased risk of cancer in patients using MOM hip replacement [85]. Although the exact mechanism of the action of implant failure and toxicity remains contentious, there is a widening consensus that protective coatings should be incorporated onto metal implants to prevent the leaching

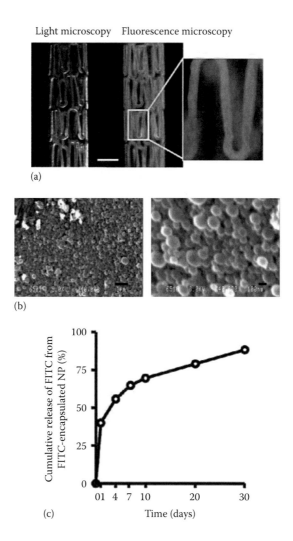

FIGURE 10.8
Nanoparticle-eluting stents: (a) conjugation of nanoparticles via cationic electrodeposition technology, (b) scanning electron microscopy of nanoparticles on stent, and (c) drug release kinetics. (Reproduced from Nakano, K., Egashira, K., Masuda, S., Funakoshi, K., Zhao, G., Kimura, S., Matoba, T., Sueishi, K., Endo, Y., and Kawashima, Y., Formulation of nanoparticle-eluting stents by a cationic electrodeposition coating technology: Efficient nano-drug delivery via bioabsorbable polymeric nanoparticle-eluting stents in porcine coronary arteries, *JACC: Cardiovasc. Interv.*, 2, 277, 2009, Copyright 2011. With permission of American College of Cardiology Foundation.)

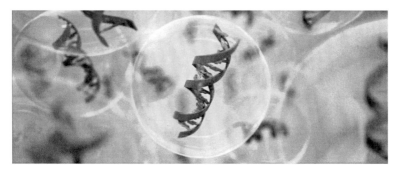

FIGURE 10.9
Genetic material can be encapsulated in POSS–PCU nanoparticles. The concept of combining nanotechnology and gene therapy is a rapidly advancing field.

of metal ions into the body [86]. Our lab has conducted preliminary studies on using an ultrasonic atomization spray system for coating metal implants with POSS–PCU to confer biocompatibility. Furthermore, we also employ a dip-coating technique for an enhanced protection during wear and tear.

10.3.3 Breast Implants

Recent reports indicated that breast implants manufactured by Poly Implant Prothèse (PIP) were prone to rupture, creating a healthcare scandal in Europe, due to its use of nonmedical grade silicone [87,88]. This increased the risk of filler materials entering the systemic circulation, although its toxic/carcinogenic effects remain contentious [89]. We have been actively pursuing this line of research, developing a new generation of POSS–PCU-based breast implants. Preliminary testing has revealed superior mechanical properties of these breast implants. Furthermore, the robust POSS–PCU layer of external coating concomitantly prevents any potential leakage of the filler material.

10.4 Coating Considerations for Theranostic Nanoparticles

10.4.1 POSS-Based Nanoscale Drug Delivery

The main rationale for utilizing nanotechnology in drug delivery is to improve its therapeutic index. Therapeutic index is defined as the ratio of the amount of drug needed for the intended (therapeutic) effect to the amount that causes toxic effect. Therefore, if the drug can be concentrated in minute quantities, its reduced dosage would translate to reduced toxicity. For example, drugs can be encapsulated in liposomes or conjugated to albumin. This increases the circulating time for the drug, while preventing premature clearance and degradation. This combined effect of utilizing nanotechnological methods allows for a lower dosage to be used, thereby reducing systemic toxicity.

Due to its nanoscale dimensions and high charge density that enhances cellular uptake, POSS nanocomposites have recently been explored as a potential candidate for drug delivery applications. A study assessed the cellular transfection efficiency of octaAmmonium-POSS by labeling it with a fluorescent dye, BODIPY [90]. Results indicated that POSS nanoparticles were dispersed in the cytosol and not the nucleus. Cell viability assays also revealed that POSS was not toxic to cells. It was demonstrated that POSS-core dendrimers can increase the amount of drug payload (Figure 10.10). Furthermore, fluorophores carried in POSS-dendrimers are protected from photobleaching. Poly-L-glutamic acid dendrimer conjugated to POSS with targeting moieties was developed, displaying pH-dependent drug release. Doxorubicin was the drug payload, and drug release at different pH was achieved [35]. The strong resistance to degradation also makes POSS nano-composites suitable as drug delivery shells [91]. A nanocomposite containing POSS as hydrophobic core and polyvinyl alcohol (PVA) as a hydrophilic

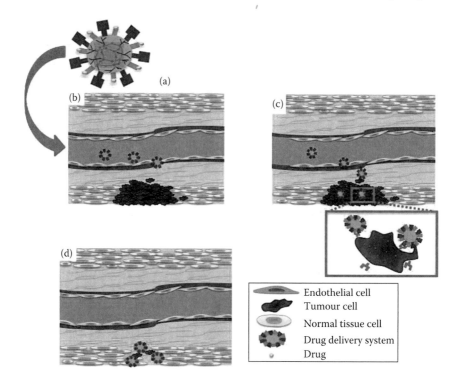

FIGURE 10.10
(See color insert.) Therapeutic drug delivery system using POSS: (a) POSS-based nanocarrier, (b) penetration of nanocarriers through capillary walls toward cancer cells, (c) targeted drug delivery using specific receptors, and (d) destruction of cancer cells via targeted POSS-based nano-drug delivery.

outer shell demonstrated a controlled release rate of drug elution. POSS–PVA has improved thermal stability and hydrophilicity, making it ideal for carrying drugs, DNA, peptides, and proteins [92,93].

10.4.2 Carbon-Based Nanomaterials

With the discovery of graphene being awarded the Nobel Prize in 2010, there has been much interest in the use of carbon-based nanomaterials for biological and medical applications. These include CNTs, graphene, and nanodiamonds (NDs).

CNTs are cylindrical nanostructures made entirely out of carbon, with applications ranging from optoelectronics to high-performance materials. More recently, CNTs have been proposed as promising candidates for therapy and diagnostic (theranostic) applications [94] (Figure 10.11). CNTs absorb strongly in the near-infrared (NIR) spectrum and dissipate high levels of

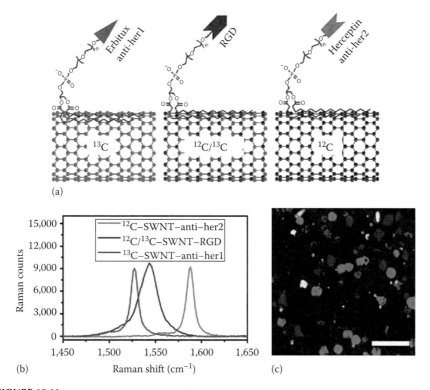

FIGURE 10.11
(See color insert.) Carbon nanotubes (CNTs) for theranostic applications: (a) CNTs can be conjugated with antibodies and peptide sequences, (b) Raman spectroscopy of different antibody-conjugated CNTs, and (c) deconvoluted Raman image of cancer cells, which internalized different antibody-conjugated CNTs. (Reproduced with permission from Z. Liu, S. Tabakman, K. Welsher, H. Dai, Carbon nanotubes in biology and medicine: in vitro and in vivo detection, imaging and drug delivery, *Nano Research*, 2 (2009) 85–120. Copyright Springer.)

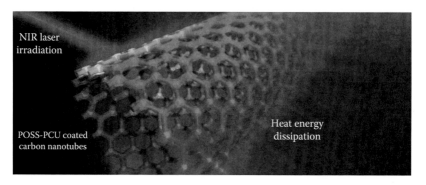

FIGURE 10.12

Thermal ablation of cancer cells using POSS–PCU-coated CNT. Increased heat dissipation is observed in POSS–PCU-coated CNTs, resulting in a higher degree of cancer cell elimination. (Reproduced with permission from Springer+Business Media: from *J. Mater. Sci.: Mater. Med.*, Microspheres leaching for scaffold porosity control, 16, 2005, 1093, Draghi, L., Resta, S., Pirozzolo, M., and Tanzi, M., Copyright 2005.)

heat energy. This can be exploited in a biomedical setting for photothermal ablation of cancer cells [95]. Alluding to their nanoscale properties, CNTs can also act as efficient transport vehicles for delivering chemotherapeutic drugs directly into cancer cells [96]. Their photoacoustic and NIR-absorbance attributes also allow CNTs to be used as *in vitro* and *in vivo* imaging agents [97]. Pure unfunctionalized CNTs are insoluble in water and toxic to biological systems. However, with proper functionalization, solubility and biocompatibility can be conferred. This can be achieved via π–π interactions between the aromatic rings of CNTs and functionalizing agents, for instance, DNA, phospholipid polyethylene glycol (PL-PEG) or POSS–PCU [98]. We have established functionalization schemes of CNTs with POSS–PCU and successfully conferred biocompatibility. In addition, a greater heating effect was observed when POSS–PCU-functionalized CNTs were exposed to NIR laser [99] (Figure 10.12). The electrical conductivity of CNTs can also be exploited in a biomedical setting, and we have developed a novel conductive polymer with POSS–PCU–CNT complexes [100,101]. These nanocomposite coatings have the potential for being used as brain implants and neural probes.

Graphene is a 2-D single layer of sp^2 carbon atoms [102]. The high surface area to the volume ratio of graphene has made it an attractive platform on which to deliver small molecules (e.g., chemotherapeutic drugs) into cells [103]. Current research in graphene for biological applications involves functionalization with PEG [104] (Figure 10.13). We envisage that it would be possible to also confer biocompatibility to graphene by functionalizing it with POSS–PCU. Indeed, it has been shown that POSS-coated graphene achieved superhydrophobicity, with possible applications in self-cleaning and antistatic products [105].

NDs have also been postulated as promising candidates for drug delivery [106] (Figure 10.14). Their high biocompatibility and versatility allow them

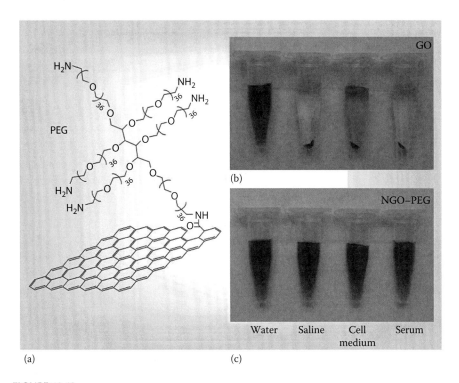

FIGURE 10.13
Graphene: A Nobel-prize-winning material for biomedical applications. (a) biofunctionalization of graphene can be achieved via conjugation to polyethylene glycol (PEG) and POSS–PCU
via π–π stacking, (b) unfunctionalized graphene is insoluble in water and biological systems,
and (c) functionalized graphene is soluble in water and can therefore be used in biomedical
applications. (Reproduced from Feng, L. and Liu, Z., *Nanomedicine*, 6, 317, 2011. With permission from Future Medicine. Copyright 2011.)

to be interfaced with various small molecules [107]. It has been shown that
chemotherapeutic drugs can be attached onto NDs for enhanced therapeutic index, with the potential for destroying cancer cells [108]. Therefore, it is
theoretically possible to coat NDs with POSS–PCU to increase its efficacy in
terms of overcoming efflux-based chemoresistance.

10.4.3 Other Functional Nanomaterials

Quantum dots (QDs) are semiconductor nanocrystals measuring 15–20 nm.
QDs are often made up of heavy metal core (for instance, CdTe) and inorganic shells (for instance, ZnS) [109]. In terms of *in vitro* and *in vivo* imaging
applications, QDs are superior to conventional fluorophores, as they have
a broad excitation spectrum, narrow and symmetrical emission spectra, a
long half-life, high quantum yield, high absorbance cross section, high saturation intensity, and resistance to photobleaching [110]. However, one main

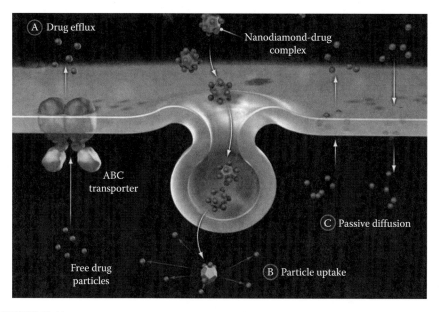

FIGURE 10.14
(See color insert.) Nanodiamond drug delivery. Enhancement of drug delivery via nanodiamonds can be achieved with coating and functionalization of POSS–PCU. (From Merkel, T.J. and DeSimone, J.M., Dodging drug-resistant cancer with diamonds, *Sci. Trans. Med.*, 3, 73–78, 2011, Copyright 2011. Reproduced with permissions of AAAS.)

disadvantage of QDs is their toxicity to biological systems, as there is a tendency for the leaching of the heavy metal core out of the protective shell [111]. This can be overcome by coating QDs with biocompatible molecules like PEG and POSS–PCU. The resultant biocompatible QD can be further functionalized by attaching bioactive molecules and targeting moieties like antibodies to increase specificity [112]. Indeed, our lab has successfully coated QDs with POSS–PCU for sentinel node imaging in breast cancer detection. Furthermore, biocompatibility can be conferred to other theranostic nanoparticles like superparamagnetic iron oxide and gold nanoparticles by coating them with POSS–PCU.

10.5 Beyond Coating: Tissue Engineering Applications

Tissue engineering revolves around the premise of creating or regenerating cells, tissues, and even organs to replace damaged or defective ones. Materials used in engineering tissues can be broadly classified as natural or synthetic. The advantage of using synthetic materials over natural ones is that its production process can be tightly regulated and controlled, ensuring batch-to-batch consistency. Characteristics like biocompatibility, robust mechanical properties that can withstand stresses and strains, and a favorable platform on which cells can adhere to and proliferate on must also be preserved. The porosity of a biomaterial is also one crucial aspect, necessary for supporting the growth of cells. Highly porous materials offer large surface area to volume ratio for increased cell density, as well as provide a sustainable environment where the delivery of nutrients and removal of waste can occur [113,114].

We have developed a novel proprietary nanocomposite polymer, POSS–PCU, which has been used in various first-in-man studies, including lacrimal duct, bypass graft, and the world's first bioartificial trachea, with excellent clinical results. It is currently undergoing preclinical trials as a transcatheter heart valve for pediatric and adult applications [115]. POSS–PCU is nonbiodegradable, making it suitable for use as scaffolds for implants as well as coatings for medical devices. The structure of POSS–PCU essentially comprises three parts: the POSS nanocage, the urethane hard section, and the carbonate soft section [116]. The soft segment has a bearing on various characteristics, like viscosity and electrical conductivity by interacting with the solvent.

A study by Gupta et al. assessed the potential POSS–PCU and POSS–PCL of being scaffolds for cartilage, liver, and small intestine, using electrohydrodynamic printing technique. Cell viability assays via Alamar Blue showed that cell growth on POSS–PCU was comparable to controls. Indirect assessment indicated some decrease in cell viability in high concentrations of POSS–PCU, but no significant increase in cell death. Although cell viability was similar in both electro-sprayed and electro-spun POSS–PCU scaffolds, cell infiltration was more pronounced in electro-spun scaffolds [19].

The degradable version of POSS–PCU and POSS–PCL was used as a model for intestinal tissue engineering. Rat intestinal epithelial cells were seeded on the polymer, and it was shown to support epithelial cell proliferation. In this case, the polymer was manufactured via solvent casting and particulate leaching to be a macroporous scaffold, with pore sizes in the range of 150–250 μm, with 40%–80% porosity. The cell-seeded scaffold displayed physiochemical properties resembling endogenous intestinal cells [117].

10.6 Future Directions and Conclusion

Medical devices implanted in the human body come into contact with biological tissues like blood. This in turn has an immunological effect, as the body tends to see implants as "foreign" and would subsequently mount an immune response. Hence, if the device itself is made of a material that would elicit an immune response, it is imperative to have a protective layer of coating that would act as a buffer between the device and the biological tissue. The recent health scare about devices like the rupturing of PIP breast implants and the failure of MOM hip replacement have once again underscored the need for having effective coatings. The superior biocompatibility, robust mechanical properties, and surface characteristics of POSS–PCU open the possibility of it being used to coat a multitude of medical devices.

With the increasing popularity of combining nanotechnology with medicine, it is also foreseeable that nanotechnology-based techniques of manufacturing and coating would feature prominently in design considerations. In view of the current research and information presented, we believe that POSS-based nanocomposites have a significant role to play in biomedical applications. Therefore, more lab-based experiments and stringent industrial testing would not only be necessary, but integral for the advancement of POSS-based nanocomposites for medical and clinical use.

References

1. N. Roohpour, A. Moshaverinia, J.M. Wasikiewicz, D. Paul, M. Wilks, M. Millar, P. Vadgama, Development of bacterially resistant polyurethane for coating medical devices, *Biomedical Materials*, 7 (2012) 015007.
2. A. Zareidoost, M. Yousefpour, B. Ghaseme, A. Amanzadeh, The relationship of surface roughness and cell response of chemical surface modification of titanium, *Journal of Materials Science: Materials in Medicine*, 23 (2012) 1479–1488.
3. R.Y. Kannan, H.J. Salacinski, P.E. Butler, A.M. Seifalian, Polyhedral oligomeric silsesquioxane nanocomposites: The next generation material for biomedical applications, *Accounts of Chemical Research*, 38 (2005) 879–884.
4. A.C.O.N. Dufresne, *Natural Polymers: Volume 2: Nanocomposites*, Royal Society of Chemistry, 2012.
5. H. Ghanbari, B.G. Cousins, A.M. Seifalian, A nanocage for nanomedicine: Polyhedral oligomeric silsesquioxane (POSS), *Macromolecular Rapid Communications*, 32 (2011) 1032–1046.
6. H. Ghanbari, S.M. Marashi, Y. Rafiei, K. Chaloupka, A.M. Seifalian, Biomedical application of polyhedral oligomeric silsesquioxane nanoparticles, *Applications of Polyhedral Oligomeric Silsesquioxanes*, London, U.K.: (2011) 363–399.
7. R.Y. Kannan, H.J. Salacinski, M.J. Edirisinghe, G. Hamilton, A.M. Seifalian, Polyhedral oligomeric silsesquioxane-polyurethane nanocomposite microvessels for an artificial capillary bed, *Biomaterials*, 27 (2006) 4618–4626.
8. G.Z. Li, T. Yamamoto, K. Nozaki, M. Hikosaka, Crystallization of ladderlike polyphenylsilsesquioxane (PPSQ)/isotactic polystyrene (i-PS) blends, *Polymer*, 42 (2001) 8435–8441.
9. G.Z. Li, L. Wang, H. Toghiani, T.L. Daulton, K. Koyama, C.U. Pittman, Viscoelastic and mechanical properties of epoxy/multifunctional polyhedral oligomeric silsesquioxane nanocomposites and epoxy/ladderlike polyphenyl-silsesquioxane blends, *Macromolecules*, 34 (2001) 8686–8693.
10. L. Draghi, S. Resta, M. Pirozzolo, M. Tanzi, Microspheres leaching for scaffold porosity control, *Journal of Materials Science: Materials in Medicine*, 16 (2005) 1093–1097.

11. C. Sanchez, G.J.D.A.A. Soler-Illia, F. Ribot, T. Lalot, C.R. Mayer, V. Cabuil, Designed hybrid organic–inorganic nanocomposites from functional nano-building blocks, *Chemistry of Materials*, 13 (2001) 3061–3083.

12. H. Hosseinkhani, M. Hosseinkhani, F. Tian, H. Kobayashi, Y. Tabata, Osteogenic differentiation of mesenchymal stem cells in self-assembled peptide-amphiphile nanofibers, *Biomaterials*, 27 (2006) 4079–4086.

13. C.M. Leu, Y-T. Chang, K.H. Wei, Polyimide-side-chain tethered polyhedral oligomeric silsesquioxane nanocomposites for low-dielectric film applications, *Chemistry of Materials*, 15 (2003) 3721–3727.

14. B.X. Fu, M.Y. Gelfer, B.S. Hsiao, S. Phillips, B. Viers, R. Blanski, P. Ruth, Physical gelation in ethylene–propylene copolymer melts induced by polyhedral oligomeric silsesquioxane (POSS) molecules, *Polymer*, 44 (2003) 1499–1506.

15. T.S. Haddad, J.D. Lichtenhan, Hybrid organic-inorganic thermoplastics: Styryl-based polyhedral oligomeric silsesquioxane polymers, *Macromolecules*, 29 (1996) 7302–7304.

16. J. Huang, C. He, Y. Xiao, K.Y. Mya, J. Dai, Y.P. Siow, Polyimide/POSS nanocom-posites: Interfacial interaction, thermal properties and mechanical properties, *Polymer*, 44 (2003) 4491–4499.

17. S. Pellice, D. Fasce, R. Williams, Properties of epoxy networks derived from the reaction of diglycidyl ether of bisphenol A with polyhedral oligomeric silses-quioxanes bearing OH-functionalized organic substituents, *Journal of Polymer Science Part B: Polymer Physics*, 41 (2003) 1451–1461.

18. M. Oaten, N.R. Choudhury, Silsesquioxane-urethane hybrid for thin film appli-cations, *Macromolecules*, 38 (2005) 6392–6401.

19. A. Gupta, A.M. Seifalian, Z. Ahmad, M.J. Edirisinghe, M.C. Winslet, Novel elec-trohydrodynamic printing of nanocomposite biopolymer scaffolds, *Journal of Bioactive and Compatible Polymers*, 22 (2007) 265–280.

20. R. Langer, J.P. Vacanti, Tissue engineering, *Science*, 260 (1993) 920–926.

21. E. Jeoung, J.B. Carroll, V.M. Rotello, Surface modification via 'lock and key' specific self-assembly of polyhedral oligomeric silsesquioxane (POSS) deriva-tives to modified gold surfaces, *Chemical Communications*, (2002) 1510–1511.

22. A.G. Kidane, G. Burriesci, M. Edirisinghe, H. Ghanbari, P. Bonhoeffer, A.M. Seifalian, A novel nanocomposite polymer for development of synthetic heart valve leaflets, *Acta Biomaterialia*, 5 (2009) 2409–2417.

23. S. Sarkar, H. Salacinski, G. Hamilton, A. Seifalian, The mechanical properties of infrainguinal vascular bypass grafts: Their role in influencing patency, *European Journal of Vascular and Endovascular Surgery*, 31 (2006) 627–636.

24. M. Ahmed, G. Hamilton, A.M. Seifalian, Viscoelastic behaviour of a small cali-bre vascular graft made from a POSS-nanocomposite, in: *Engineering in Medicine and Biology Society (EMBC), 2010 Annual International Conference of the IEEE*, IEEE, London, U.K.: 2010, pp. 251–254.

25. R.Y. Kannan, H.J. Salacinski, M. Odlyha, P.E. Butler, A.M. Seifalian, The degra-dative resistance of polyhedral oligomeric silsesquioxane nanocore integrated polyurethanes: An in vitro study, *Biomaterials*, 27 (2006) 1971–1979.

26. Z. Zhang, A. Gu, G. Liang, P. Ren, J. Xie, X. Wang, Thermo-oxygen degradation mechanisms of POSS/epoxy nanocomposites, *Polymer Degradation and Stability*, 92 (2007) 1986–1993.

27. S. Gogolewski, Selected topics in biomedical polyurethanes. A review, *Colloid and Polymer Science*, 267 (1989) 757–785.

28. J. Chen, Y. Leng, X. Tian, L. Wang, N. Huang, P. Chu, P. Yang, Antithrombogenic investigation of surface energy and optical bandgap and hemocompatibility mechanism of Ti (Ta^{+5}) O$_2$ thin films, *Biomaterials*, 23 (2002) 2545–2552.

29. W. Sharp, B. Taylor, J. Wright, A. Finelli, Experience with negatively charged polyurethane-backed velours, *Journal of Biomedical Materials Research*, 5 (1971) 75–81.

30. D. Kleem, B. Severich, H. Höcker, Correlation between chemical and physical surface properties and blood compatibility of PPE/EVA-blends, in: *Macromolecular Symposia*, Wiley Online Library, Weinheim, Germany: 1996, pp. 19–29.

31. Z. Aiping, C. Tian, Blood compatibility of surface-engineered poly (ethylene terephthalate) via o-carboxymethylchitosan, *Colloids and Surfaces B: Biointerfaces*, 50 (2006) 120–125.

32. P.V. Murphy, A. La Croix, S. Merchant, W. Bernhard, Development of blood compatible polymers using the electret effect, *Journal of Biomedical Materials Research*, 5 (1971) 59–74.

33. A. Bantjes, Clotting phenomena at the blood-polymer interface and development of blood compatible polymeric surfaces, *British Polymer Journal*, 10 (1978) 267–274.

34. L. Zheng, R.J. Farris, E.B. Coughlin, Novel polyolefin nanocomposites: Synthesis and characterizations of metallocene-catalyzed polyolefin polyhedral oligomeric silsesquioxane copolymers, *Macromolecules*, 34 (2001) 8034–8039.

35. H. Yuan, K. Luo, Y. Lai, Y. Pu, B. He, G. Wang, Y. Wu, Z. Gu, A novel poly (l-glutamic acid) dendrimer based drug delivery system with both pH-sensitive and targeting functions, *Molecular Pharmaceutics*, 7 (2010) 953–962.

36. W. Wang, Y. Liu, J. Wang, X. Jia, L. Wang, Z. Yuan, S. Tang, M. Liu, H. Tang, Y. Yu, A novel copolymer poly (lactide-co-β-malic acid) with extended carboxyl arms offering better cell affinity and hemocompatibility for blood vessel engineering, *Tissue Engineering Part A*, 15 (2008) 65–73.

37. P. Zilla, R. Fasol, M. Deutsch, T. Fischlein, E. Minar, A. Hammerle, O. Krupicka, M. Kadletz, Endothelial cell seeding of polytetrafluoroethylene vascular grafts in humans: A preliminary report, *Journal of Vascular Surgery*, 6 (1987) 535–541.

38. N. Alobaid, H. Salacinski, K. Sales, B. Ramesh, R. Kannan, G. Hamilton, A. Seifalian, Nanocomposite containing bioactive peptides promote endothelialisation by circulating progenitor cells: An in vitro evaluation, *European Journal of Vascular and Endovascular Surgery*, 32 (2006) 76–83.

39. N. Kipshidze, J.J. Ferguson, M.H. Keelan, H. Sahota, R. Komorowski, L.R. Shankar, P.S. Chawla, C.C. Haudenschild, V. Nikolaychik, J.W. Moses, Endoluminal reconstruction of the arterial wall with endothelial cell/glue matrix reduces restenosis in an atherosclerotic rabbit, *Journal of the American College of Cardiology*, 36 (2000) 1396–1403.

40. T. Shirota, H. Yasui, H. Shimokawa, T. Matsuda, Fabrication of endothelial progenitor cell (EPC)-seeded intravascular stent devices and in vitro endothelialization on hybrid vascular tissue, *Biomaterials*, 24 (2003) 2295–2302.

41. H. Ghanbari, A. de Mel, A.M. Seifalian, Cardiovascular application of polyhedral oligomeric silsesquioxane nanomaterials: A glimpse into prospective horizons, *International Journal of Nanomedicine*, 6 (2011) 775.

42. M. Khorasani, H. Mirzadeh, Effect of oxygen plasma treatment on surface charge and wettability of PVC blood bag—In vitro assay, *Radiation Physics and Chemistry*, 76 (2007) 1011–1016.

43. J.L. Dewez, A. Doren, Y.J. Schneider, P.G. Rouxhet, Competitive adsorption of proteins: Key of the relationship between substratum surface properties and adhesion of epithelial cells, *Biomaterials*, 20 (1999) 547–559.

44. T. Kumar, L. Krishnan, Fibrin-mediated endothelial cell adhesion to vascular biomaterials resists shear stress due to flow, *Journal of Materials Science: Materials in Medicine*, 13 (2002) 751–755.

45. G. Altankov, K. Richau, T. Groth, The role of surface zeta potential and substratum chemistry for regulation of dermal fibroblasts interaction, *Materialwissenschaft und Werkstofftechnik*, 34 (2004) 1120–1128.

46. M. Khorasani, S. MoemenBellah, H. Mirzadeh, B. Sadatnia, Effect of surface charge and hydrophobicity of polyurethanes and silicone rubbers on L929 cells response, *Colloids and Surfaces B: Biointerfaces*, 51 (2006) 112–119.

47. A. de Mel, G. Punshon, B. Ramesh, S. Sarkar, A. Darbyshire, G. Hamilton, A.M. Seifalian, in situ endothelialisation potential of a biofunctionalised nanocomposite biomaterial-based small diameter bypass graft, *Bio-Medical Materials and Engineering*, 19 (2009) 317–331.

48. M. Yaseen, X. Zhao, A. Freund, A.M. Seifalian, J.R. Lu, Surface structural conformations of fibrinogen polypeptides for improved biocompatibility, *Biomaterials*, 31 (2010) 3781–3792.

49. H. Ai, Y.M. Lvov, D.K. Mills, M. Jennings, J.S. Alexander, S.A. Jones, Coating and selective deposition of nanofilm on silicone rubber for cell adhesion and growth, *Cell Biochemistry and Biophysics*, 38 (2003) 103–114.

50. Y. Hesse, J. Kampmeier, G.K. Lang, A. Baldysiak-Figiel, G.E. Lang, Adherence and viability of porcine lens epithelial cells on three different IOL materials in vitro, *Graefe's Archive for Clinical and Experimental Ophthalmology*, 241 (2003) 823–826.

51. H. Ghanbari, H. Viatge, A.G. Kidane, G. Burriesci, M. Tavakoli, A.M. Seifalian, Polymeric heart valves: New materials, emerging hopes, *Trends in Biotechnology*, 27 (2009) 359–367.

52. H. Ghanbari, A.G. Kidane, G. Burriesci, B. Ramesh, A. Darbyshire, A.M. Seifalian, The anti-calcification potential of a silsesquioxane nanocomposite polymer under in vitro conditions: Potential material for synthetic leaflet heart valve, *Acta Biomaterialia*, 6 (2010) 4249.

53. R.Y. Kannan, H.J. Salacinski, J. Ghanavi, A. Narula, M. Odlyha, H. Peirovi, P.E. Butler, A.M. Seifalian, Silsesquioxane nanocomposites as tissue implants, *Plastic and Reconstructive Surgery*, 119 (2007) 1653–1662.

54. L. Indolfi, F. Causa, P.A. Netti, Coating process and early stage adhesion evaluation of poly (2-hydroxy-ethyl-methacrylate) hydrogel coating of 316L steel surface for stent applications, *Journal of Materials Science: Materials in Medicine*, 20 (2009) 1541–1551.

55. F. McGillicuddy, I. Lynch, Y. Rochev, M. Burke, K. Dawson, W. Gallagher, A. Keenan, Novel "plum pudding" gels as potential drug-eluting stent coatings: Controlled release of fluvastatin, *Journal of Biomedical Materials Research Part A*, 79 (2006) 923–933.

56. T. Sharkawi, D. Leyni-Barbaz, N. Chikh, J.N. Mcmullen, Evaluation of the in vitro drug release from resorbable biocompatible coatings for vascular stents, *Journal of Bioactive and Compatible Polymers*, 20 (2005) 153–168.

57. M.C. Chen, H.F. Liang, Y.L. Chiu, Y. Chang, H.J. Wei, H.W. Sung, A novel drug-eluting stent spray-coated with multi-layers of collagen and sirolimus, *Journal of Controlled Release*, 108 (2005) 178–189.

58. P. Hanefeld, U. Westedt, R. Wombacher, T. Kissel, A. Schaper, J.H. Wendorff, A. Greiner, Coating of poly (p-xylylene) by PLA-PEO-PLA triblock copolymers with excellent polymer-polymer adhesion for stent applications, *Biomacromolecules*, 7 (2006) 2086–2090.

59. J. Lahann, D. Klee, H. Thelen, H. Bienert, D. Vorwerk, H. Höcker, Improvement of haemocompatibility of metallic stents by polymer coating, *Journal of Materials Science: Materials in Medicine*, 10 (1999) 443–448.

60. S. Shanmugasundaram, K.A. Griswold, C.J. Prestigiacomo, T. Arinzeh, M. Jaffe, Applications of electrospinning: Tissue engineering scaffolds and drug delivery system, in: *Bioengineering Conference, 2004. Proceedings of the IEEE 30th Annual Northeast*, IEEE, Springfield, MA, 2004, pp. 140–141.

61. P. DeMuth, J.J. Moon, H. Suh, P.T. Hammond, D.J. Irvine, Releasable layer-by-layer assembly of stabilized lipid nanocapsules on microneedles for enhanced transcutaneous vaccine delivery, *ACS Nano*, 6 (2012) 8041–8051.

62. P.J. Tarcha, D. Verlee, H.W. Hui, J. Setesak, B. Antohe, D. Radulescu, D. Wallace, The application of ink-jet technology for the coating and loading of drug-eluting stents, *Annals of Biomedical Engineering*, 35 (2007) 1791–1799.

63. R. Bakhshi, M.J. Edirisinghe, A. Darbyshire, Z. Ahmad, A.M. Seifalian, Electrohydrodynamic jetting behaviour of polyhedral oligomeric silsesquioxane nanocomposite, *Journal of Biomaterials Applications*, 23 (2009) 293–309.

64. R. Okner, M. Oron, N. Tal, A. Nyska, N. Kumar, D. Mandler, A. Domb, Electrocoating of stainless steel coronary stents for extended release of paclitaxel, *Journal of Biomedical Materials Research Part A*, 88 (2009) 427–436.

65. A. Tan, J. Rajadas, A.M. Seifalian, Biochemical engineering nerve conduits using peptide amphiphiles, *Journal of Controlled Release*, 163 (2012) 342–352.

66. H. Berger, Using ultrasonic spray nozzles to coat drug-eluting stents, *Medical Device Technology*, 17 (2006) 44.

67. S. Garg, P.W. Serruys, Coronary stents: Current status, *Journal of the American College of Cardiology*, 56 (2010) S1–S42.

68. G. Niccoli, R.A. Montone, G. Ferrante, F. Crea, The evolving role of inflammatory biomarkers in risk assessment after stent implantation, *Journal of the American College of Cardiology*, 56 (2010) 1783–1793.

69. W.J. Van Der Giessen, A.M. Lincoff, R.S. Schwartz, H.M.M. van Beusekom, P.W. Serruys, D.R. Holmes, S.G. Ellis, E.J. Topol, Marked inflammatory sequelae to implantation of biodegradable and nonbiodegradable polymers in porcine coronary arteries, *Circulation*, 94 (1996) 1690–1697.

70. I. Jerman, A.Š. Vuk, M. Kozelj, B. Orel, J. Kovač, A structural and corrosion study of triethoxysilyl functionalized POSS coatings on AA 2024 alloy, *Langmuir*, 24 (2008) 5029–5037.

71. A. Tan, J. Rajadas, A.M. Seifalian, Exosomes as nano-theranostic delivery platforms for gene therapy, *Advanced Drug Delivery Reviews*, 65 (2012) 357–367.

72. Q. Guo, P.T. Knight, P.T. Mather, Tailored drug release from biodegradable stent coatings based on hybrid polyurethanes, *Journal of Controlled Release*, 137 (2009) 224–233.

73. A. Tan, M.S. Alavijeh, A.M. Seifalian, Next generation stent coatings: Convergence of biotechnology and nanotechnology, *Trends Biotechnology*, 30 (2012) 406–409.

74. L.K. Keefer, Biomaterials: Thwarting thrombus, *Nature Materials*, 2 (2003) 357–358.

75. M.A.M. Beijk, M. Klomp, N.J.W. Verouden, N. Van Geloven, K.T. Koch, J.P.S. Henriques, J. Baan, M.M. Vis, E. Scheunhage, J.J. Piek, Genous™ endothelial progenitor cell capturing stent vs. the Taxus Liberté stent in patients with de novo coronary lesions with a high-risk of coronary restenosis: A randomized, single-centre, pilot study, *European Heart Journal*, 31 (2010) 1055–1064.

76. H.F. Langer, J.W. von der Ruhr, K. Daub, T. Schoenberger, K. Stellos, A.E. May, H. Schnell, A. Gauß, R. Hafner, P. Lang, Capture of endothelial progenitor cells by a bispecific protein/monoclonal antibody molecule induces reendothelialization of vascular lesions, *Journal of Molecular Medicine*, 88 (2010) 687–699.

77. C. Stefanadis, K. Toutouzas, E. Stefanadi, A. Lazaris, E. Patsouris, N. Kipshidze, Inhibition of plaque neovascularization and intimal hyperplasia by specific targeting vascular endothelial growth factor with bevacizumab-eluting stent: An experimental study, *Atherosclerosis*, 195 (2007) 269–276.

78. R. Blindt, F. Vogt, I. Astafieva, C. Fach, M. Hristov, N. Krott, B. Seitz, A. Kapurniotu, C. Kwok, M. Dewor, A novel drug-eluting stent coated with an integrin-binding cyclic Arg-Gly-Asp peptide inhibits neointimal hyperplasia by recruiting endothelial progenitor cells, *Journal of the American College of Cardiology*, 47 (2006) 1786–1795.

79. K. Nakano, K. Egashira, S. Masuda, K. Funakoshi, G. Zhao, S. Kimura, T. Matoba, K. Sueishi, Y. Endo, Y. Kawashima, Formulation of nanoparticle-eluting stents by a cationic electrodeposition coating technology: Efficient nano-drug delivery via bioabsorbable polymeric nanoparticle-eluting stents in porcine coronary arteries, *JACC: Cardiovascular Interventions*, 2 (2009) 277–283.

80. M. Chorny, I. Fishbein, B.B. Yellen, I.S. Alferiev, M. Bakay, S. Ganta, R. Adamo, M. Amiji, G. Friedman, R.J. Levy, Targeting stents with local delivery of paclitaxel-loaded magnetic nanoparticles using uniform fields, *Proceedings of the National Academy of Sciences*, 107 (2010) 8346–8351.

81. I. Fishbein, I.S. Alferiev, O. Nyanguile, R. Gaster, J.M. Vohs, G.S. Wong, H. Felderman, I.W. Chen, H. Choi, R.L. Wilensky, Bisphosphonate-mediated gene vector delivery from the metal surfaces of stents, *Proceedings of the National Academy of Sciences of the United States of America*, 103 (2006) 159–164.

82. S.E. Graves, A. Rothwell, K. Tucker, J.J. Jacobs, A. Sedrakyan, A multinational assessment of metal-on-metal bearings in hip replacement, *The Journal of Bone and Joint Surgery*, 93 (2011) 43–47.

83. C.C.P.M. Verheyen, J.A.N. Verhaar, Failure rates of stemmed metal-on-metal hip replacements, *Journal of Bone Joint Surgery British*, 93 (2011) 298–306.

84. I. Polyzois, D. Nikolopoulos, I. Michos, E. Patsouris, S. Theocharis, Local and systemic toxicity of nanoscale debris particles in total hip arthroplasty, *Journal of Applied Toxicology*, 32 (2012) 255–269.

85. A.J. Smith, P. Dieppe, M. Porter, A.W. Blom, Risk of cancer in first seven years after metal-on-metal hip replacement compared with other bearings and general population: Linkage study between the National Joint Registry of England and Wales and hospital episode statistics, *British Medical Journal*, 344 (2012) e2383.

86. C. Myant, R. Underwood, J. Fan, P. Cann, Lubrication of metal-on-metal hip joints: The effect of protein content and load on film formation and wear, *Journal of the Mechanical Behavior of Biomedical Materials*, 6 (2011) 30–40.

87. M. Berry, J.J. Stanek, The PIP mammary prosthesis: A product recall study, *Journal of Plastic, Reconstructive and Aesthetic Surgery*, 65(6) (2012) 697–704.

88. A. O'Dowd, UK launches inquiry into safety of PIP breast implants, *British Medical Journal*, 344 (2012) e11.

89. Z. Kmietowicz, PIP implants don't pose risk to health, expert group concludes, *British Medical Journal*, 344 (2012) e4234.

90. P.A. Wheeler, B.X. Fu, J.D. Lichtenhan, J. Weitao, L.J. Mathias, Incorporation of metallic POSS, POSS copolymers, and new functionalized POSS compounds into commercial dental resins, *Journal of Applied Polymer Science*, 102 (2006) 2856–2862.

91. F.A. Sheikh, N.A.M. Barakat, M.A. Kanjwal, S. Aryal, M.S. Khil, H.Y. Kim, Novel self-assembled amphiphilic poly (ε-caprolactone)-grafted-poly (vinyl alcohol) nanoparticles: Hydrophobic and hydrophilic drugs carrier nanoparticles, *Journal of Materials Science: Materials in Medicine*, 20 (2009) 821–831.

92. L.A. Dailey, M. Wittmar, T. Kissel, The role of branched polyesters and their modifications in the development of modern drug delivery vehicles, *Journal of Controlled Release*, 101 (2005) 137–149.

93. J.B. Carroll, A.J. Waddon, H. Nakade, V.M. Rotello, "Plug and play" polymers. Thermal and X-ray characterizations of noncovalently grafted polyhedral oligomeric silsesquioxane (POSS)-polystyrene nanocomposites, *Macromolecules*, 36 (2003) 6289–6291.

94. Z. Liu, S. Tabakman, K. Welsher, H. Dai, Carbon nanotubes in biology and medicine: In vitro and in vivo detection, imaging and drug delivery, *Nano Research*, 2 (2009) 85–120.

95. N.W.S. Kam, M. O'Connell, J.A. Wisdom, H. Dai, Carbon nanotubes as multifunctional biological transporters and near-infrared agents for selective cancer cell destruction, *Proceedings of the National Academy of Sciences of the United States of America*, 102 (2005) 11600–11605.

96. Z. Liu, A.C. Fan, K. Rakhra, S. Sherlock, A. Goodwin, X. Chen, Q. Yang, D.W. Felsher, H. Dai, Supramolecular stacking of doxorubicin on carbon nanotubes for in vivo cancer therapy, *Angewandte Chemie International Edition*, 48 (2009) 7668–7672.

97. A. Zerda, Z. Liu, S. Bodapati, R. Teed, S. Vaithilingam, B.T. Khuri-Yakub, X. Chen, H. Dai, S.S. Gambhir, Ultrahigh sensitivity carbon nanotube agents for photoacoustic molecular imaging in living mice, *Nano Letters*, 10 (2010) 2168–2172.

98. Z. Liu, S.M. Tabakman, Z. Chen, H. Dai, Preparation of carbon nanotube bioconjugates for biomedical applications, *Nature Protocols*, 4 (2009) 1372–1381.

99. A. Tan, S.Y. Madani, J. Rajadas, G. Pastorin, A.M. Seifalian, Synergistic photothermal ablative effects of functionalizing carbon nanotubes with a POSS-PCU nanocomposite polymer, *Journal of Nanobiotechnology*, 10 (2012) 34.

100. E. Antoniadou, R.K. Ahmad, R.B. Jackman, A. Seifalian, Next generation brain implant coatings and nerve regeneration via novel conductive nanocomposite development, in: *Engineering in Medicine and Biology Society, 2011 Annual International Conference of the IEEE*, IEEE, Boston, MA, 2011, pp. 3253–3257.

101. E.V. Antoniadou, B.G. Cousins, A.M. Seifalian, Development of conductive polymer with carbon nanotubes for regenerative medicine applications, in: *Engineering in Medicine and Biology Society , 2010 Annual International Conference of the IEEE*, IEEE, London, U.K., 2010, pp. 815–818.

102. Z. Liu, J.T. Robinson, X. Sun, H. Dai, PEGylated nanographene oxide for delivery of water-insoluble cancer drugs, *Journal of the American Chemical Society*, 130 (2008) 10876–10877.

103. H. Hong, K. Yang, Y. Zhang, J.W. Engle, L. Feng, Y. Yang, T.R. Nayak, S. Goel, J. Bean, C.P. Theuer, in vivo targeting and imaging of tumor vasculature with radiolabeled, antibody-conjugated nanographene, *ACS Nano*, 6 (2012) 2361–2370.

104. L. Feng, Z. Liu, Graphene in biomedicine: Opportunities and challenges, *Nanomedicine*, 6 (2011) 317–324.

105. J. Jin, X. Wang, M. Song, Graphene-based nanostructured hybrid materials for conductive and superhydrophobic functional coatings, *Journal of Nanoscience and Nanotechnology*, 11 (2011) 7715–7722.

106. T.J. Merkel, J.M. DeSimone, Dodging drug-resistant cancer with diamonds, *Science Translational Medicine*, 3 (2011) 73ps78.

107. V.N. Mochalin, O. Shenderova, D. Ho, Y. Gogotsi, The properties and applications of nanodiamonds, *Nature Nanotechnology*, 7 (2011) 11–23.

108. A. Adnan, R. Lam, H. Chen, J. Lee, D.J. Schaffer, A.S. Barnard, G.C. Schatz, D. Ho, W.K. Liu, Atomistic simulation and measurement of pH dependent cancer therapeutic interactions with nanodiamond carrier, *Molecular Pharmaceutics*, 8 (2011) 368.

109. X. Gao, L. Yang, J.A. Petros, F.F. Marshall, J.W. Simons, S. Nie, in vivo molecular and cellular imaging with quantum dots, *Current Opinion in Biotechnology*, 16 (2005) 63–72.

110. A. Tan, L. Yildirimer, J. Rajadas, H. De La Peña, G. Pastorin, A. Seifalian, Quantum dots and carbon nanotubes in oncology: A review on emerging theranostic applications in nanomedicine, *Nanomedicine*, 6 (2011) 1101–1114.

111. S. Ghaderi, B. Ramesh, A.M. Seifalian, Fluorescence nanoparticles "quantum dots" as drug delivery system and their toxicity: A review, *Journal of Drug Targeting*, 19 (2011) 475–486.

112. A. de Mel, J.T. Oh, B. Ramesh, A.M. Seifalian, Biofunctionalized quantum dots for live monitoring of stem cells: Applications in regenerative medicine, *Regenerative Medicine*, 7 (2012) 335–347.

113. D.J. Mooney, D.F. Baldwin, N.P. Suh, J.P. Vacanti, R. Langer, Novel approach to fabricate porous sponges of poly (D, L-lactic-co-glycolic acid) without the use of organic solvents, *Biomaterials*, 17 (1996) 1417–1422.

114. S. Yang, K.F. Leong, Z. Du, C.K. Chua, The design of scaffolds for use in tissue engineering. Part I. Traditional factors, *Tissue Engineering*, 7 (2001) 679–689.

115. B. Rahmani, S. Tzamtzis, H. Ghanbari, G. Burriesci, A.M. Seifalian, Manufacturing and hydrodynamic assessment of a novel aortic valve made of a new nanocomposite polymer, *Journal of Biomechanics*, 45 (2012) 1205–1211.

116. M.S. Motwani, Y. Rafiei, A. Tzifa, A.M. Seifalian, in situ endothelialization of intravascular stents from progenitor stem cells coated with nanocomposite and functionalized biomolecules, *Biotechnology and Applied Biochemistry*, 58 (2011) 2–13.

117. A. Gupta, D.S. Vara, G. Punshon, K.M. Sales, M.C. Winslet, A.M. Seifalian, in vitro small intestinal epithelial cell growth on a nanocomposite polycaprolactone scaffold, *Biotechnology and Applied Biochemistry*, 54 (2009) 221–229.

11

Nanocomposite PPy Coatings for Al Alloys Corrosion Protection

Kirill L. Levine

CONTENTS

11.1 Introduction

Aluminum (Al) alloys, such as Al 2024-T3, are extensively used in aircraft and construction industries because of their low density and excellent mechanical properties. Chemically pure Al is naturally protected from corrosion by an oxide layer. Al alloys are vulnerable to corrosion because they contain intermetallic impurities, such as copper, zinc, and molybdenum, whose contents can be considerably high. For example, Cu content in Al2024-T3 is close to 5%.* Impurities are distributed in alloys in a nonuniform way, resulting in the presence of clusters which become the centers for galvanic corrosion. A galvanic corrosion pair is formed between the impurity and the surrounding metal. The material that is consumed in this corrosion is usually Al as it is more reactive.

Corrosion of Al alloys results in approximately 3% annual loss of Al constructions worldwide; therefore, their corrosion protection is a serious industrial concern. This problem was recently solved by using hexavalent chromium (Cr^{6+}) coatings. Although Cr^{6+} provides satisfactory corrosion protection to Al alloys, Cr^{6+} coatings are carcinogenic. Technologies utilizing this sort of protection are therefore hazardous for human health and the environment, and currently their replacement with chromium-free alternatives has been approved in the United States and almost all other developed countries.

One of the alternatives to Cr^{6+} coatings is Mg-rich coatings [1]. The ideology of these coatings is that Mg is sacrificed for the sake of Al, thus protecting Al. These coatings have demonstrated their efficiency at a short-term run. However, their long-lasting efficiency is under question. It is not clear how these coatings will perform after all the Mg is consumed.

Among the approaches replacing Cr^{6+}, utilizing intrinsically conducting polymers (ICPs) for corrosion protection is one of the most promising.

An example of ICPs is shown in Figure 11.1. ICPs contain a double bond in the backbone of their macromolecule which is responsible for their specific properties. The property which is important for their corrosion protection performance is electroactivity. Electroactivity is the ability of a polymer film coated on an electrode to change its reduction or oxidation state under the influence of an applied potential, when the electrode is immersed in the

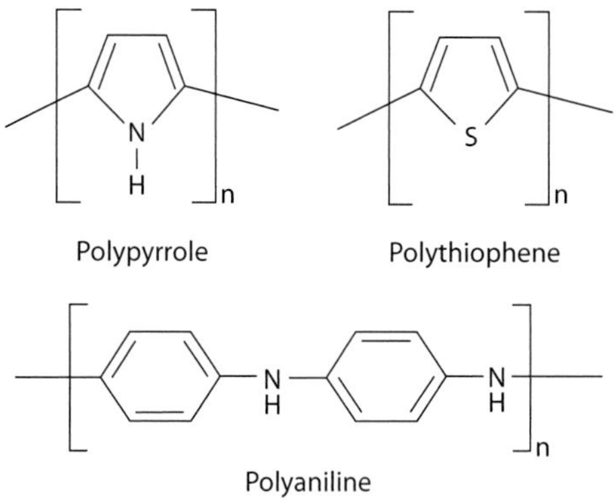

FIGURE 11.1
Example of conducting polymers.

---

* Composition of Al 2024-T3 (w/w %) is 3.8–4.9 Cu, 1.2–1.8 Mg, 0.50 Fe, 0.3–0.9 Mn, 0.50 Si, 0.25 Zn, 0.15 Ti, 0.1 Cr, 0.15 other unspecified elements, and Al to 100%.

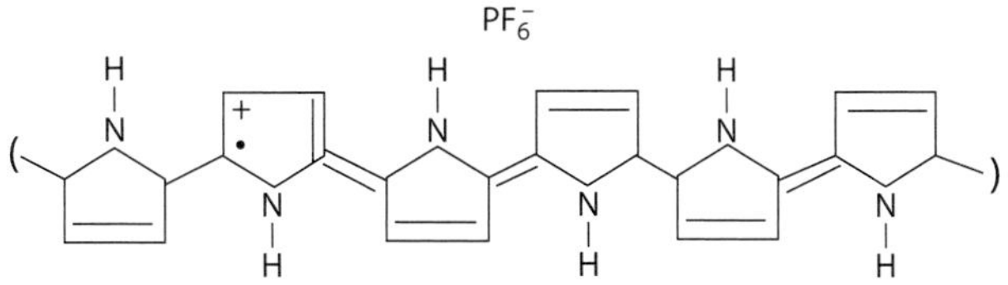

FIGURE 11.2
Example of PPy doped by a hexafluorophosphate ion.

solution of an electrolyte. Changing the reduction or oxidation state is followed by the movement of an electrolyte ion in or out of the film. This phenomenon is called "doping" and ions participating in this are referred to as "dopants." An example of a PPy film doped with PF_6^- is shown in Figure 11.2. This chapter will later discuss how ICP doping can be utilized in corrosion protection. It should be pointed out that the majority of ICPs—and PPy is not an exclusion—possess more positive oxidation potential than Al, and therefore its coating on Al surfaces possesses ennobling properties.

11.2 ICPs and Their Anticorrosion Performance

The method suggested by Kendig et al. [2] utilizes oxygen reduction inhibitors (ORIs) as dopants. ORIs, such as 2,5-dimercapto-1,3,4-thiadiazole (DMCT) (Figure 11.3), were found to significantly slow down corrosion. The mechanism of their action is shown in Figure 11.4, which can be explained by means of the following sequence:

1. As a result of corrosion, an electron was released (corrosive pit at the left).
2. An electron was absorbed by a conjugated double-bond electronic structure.
3. From the electroneutrality considerations, an ORI$^-$ ion was released at the location of a corrosive pit (shown at the right-hand side of the figure).
4. Further corrosion was slowed down due to an ORI action.

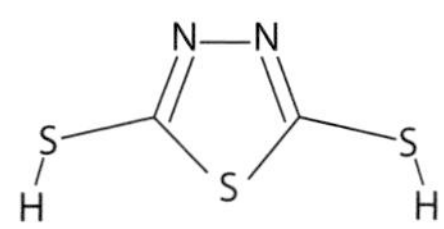

FIGURE 11.3
2,5-dimercapto 1,3,4-thiadiazole (acid form).

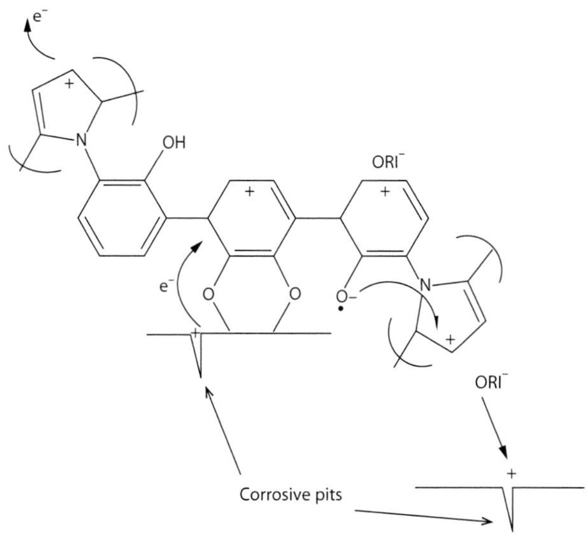

FIGURE 11.4
Schematic representation of ORI release.

Opponents of this theory have justified that this mechanism needs the corrosion to be started in order to be in force [2]. Other methods utilizing "smart" materials, such as described in the literature review [3], were suggested.

When conducting polymers are obtained on an active metal surface, they are capable of filling pores caused by imperfections in metal oxide. ICPs can form a barrier to aggressive ions which penetrate through those pores to a metal surface, therefore slowing down the corrosion [4,5].

11.3 Problems and Approaches

11.3.1 Electron Transfer Mediators and Their Application for the Deposition of Conducting Polymers

When a monomer of pyrrole or another ICP is dissolved in the solution, it can be polymerized electrochemically by a mechanism first described by Diaz [6]. But the surface of the active metal is covered with an oxide that prevents the polymer from forming an adherent uniform layer. The electrochemical method allows overcoming this difficulty with the help of so-called electron transfer mediators (ETMs). Examples of different ETMs are shown in Table 11.1. By applying positive potential to the surface of an active metal in a solution of an electrically conducting liquid containing hydroxyl ions, the surface starts to oxidize. In the presence of a monomer of a conducting

TABLE 11.1

Different Compounds with Electron Transfer Mediating Properties

| 4,5-dihydroxy-1,3-benzenedisulfonate (DHBDS or Tiron) | 3,6-dihydroxy-benzenesulfonate (DHBS) | 1,3 dibenzenesulfonate (DBS) |
| 1,2 dihydroxybenzene (Catechol) | 1,4 dihydroxybenzene (Hydroquinone) | 1,3 dihydroxybenzene (Resorcinol) |

polymer, such as pyrrole, oxidation goes more readily than polymerization and a uniform ICP layer does not form. An approach that uses ETMs allows the reduction of PPy deposition potential on Al and its alloys by nearly 500 mV, permitting ICP film deposition from an aqueous solution with high current efficiency [7–10]. A mediator whose high efficiency was shown first was disodium salt of 4,5-dihydroxy-1,3-benzenedisulfonate, (DHBDS, also known as Tiron). The mechanism of this mediator's performance is shown in Figure 11.5. When DHBDS is dissolved in an aqueous solution, it dissociates, forming anions. At the anode, their hydroxyl groups release protons and acquire positive charge, which is immediately transferred to a monomer with the proton returned back. Therefore, ETM is returned to its anionic state while the monomer is oxidized. This process repeats continuously, resulting in a collection of oxidized monomers in a pre-anodic space and their polymerization and film growth. Anions of DHBDS can also serve as dopants to ICP.

The presence of hydroxyl substitutions in a phenyl ETM phenyl ring prompted the evaluation of different compounds with hydroxyl in different positions as shown in Table 11.1.

Among those compounds, three in the top row can dissociate and serve as dopants; the three in the bottom row cannot dissociate and need additional electrolyte to make their solution electrically conductive. DBS does not have hydroxyl substitutions and was examined for comparison. In agreement with the theory of electron transfer mediation, compounds with hydroxyl substitutions in 1,2 and 1,3 positions decrease pyrrole oxidation potential most effectively. Resorcinol does not perform any mediation effect. DBS, which

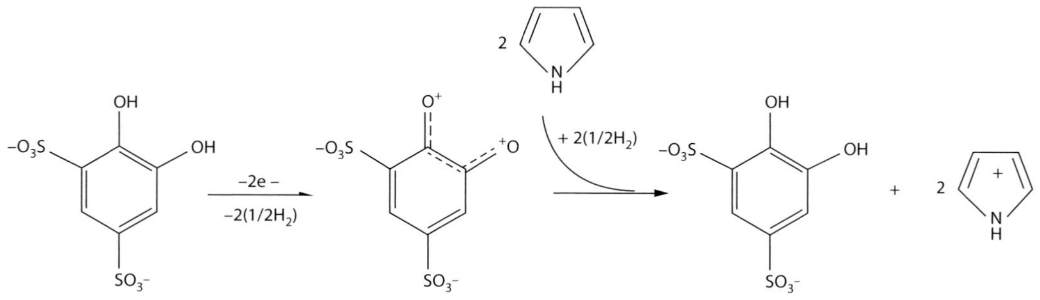

FIGURE 11.5
Schematic representation of Tiron's action on pyrrole oxidation.

did not have hydroxyl substitutions, however, was also observed to decrease polymerization potential. In conclusion, in substituted benzenes, hydroxyl groups are responsible for the electron exchange between the monomer and the mediator, while sulfonate groups facilitate monomer penetration into Al oxide pores [11]. In the case where a combination of both sulfonate and hydroxyl substitutions is within the same mediator, both of the aforementioned mechanisms act simultaneously.

While designing a nanocomposite system which would slow down corrosion at the surface of active metals, issues of electron exchange have to be considered when the surface is coated by an oxide layer. The necessity to do so occurs, for example, when a "smart" response of a corroding surface is expected which results in ORI release by the schematic suggested by Kendig et al. [2], and shown in a modified way in Figure 11.4. In this figure, the reaction that provides the electron is

$$Al \rightarrow Al^{3+} + 3e^-$$

and the reaction that consumes the electron is

$$\frac{1}{2}\, O_2 \rightarrow O^{2-} - 2e^-$$

The electric driving force generated by this reaction is 1.7 V, which far exceeds PPy dedoping potential. In order to examine the process of electron transfer through an oxide layer formed in the presence of different mediators, we are interested in studying the dielectric properties of an oxide, such as the concentration of oxygen vacancies, and electrochemical potential. According to Levine et al. [12], the surface of an active metal covered by an ultrathin oxide layer was analyzed by Mott–Schottky spectroscopy.

11.3.2 Studying an Oxide Layer Formed in the Presence of Different Mediators

The fundamental property of any dielectric material is the size and structure of its bandgap, the location of energy levels introduced by donor or acceptor impurities, and Fermi level, which in the case of electrochemical equilibrium is called "electrochemical potential" [13]. The most precise electrochemical method that allows looking at the electrochemical potential is the Mott–Schottky analysis [14], in which the "electrochemical potential" for historical reasons is called the "flat band potential," and has the same meaning as Fermi level.

In order to conduct the Mott–Schottky experiment, an oxide layer was obtained in the presence of ETM by cyclic potential scanning using the procedure of the cyclic voltammetry (CV) experiment. The potential scanned from negative 1 to positive E_{max} equal to 900 mV versus Ag/AgCl, as in Ref. [10]. After that the electrolyte was replaced by one without a mediator and the Mott–Schottky analysis was performed. The studied compounds were the benzenes with either hydroxyl, or sulfonate substitutions, or a combination of both.

The physical nature of the type of contact between an oxide film on the top of an Al and metal surface is responsible for the mechanism of electron transfer. From the microfabrication it is known that there exist three types of junction: metal-to-metal, metal-to-semiconductor (Shottky barrier), and semiconductor-to-semiconductor. As there are no specific differences between the intrinsic semiconductor and dielectric, except the size of a bandgap, it can be assumed that the oxide–Al interface possesses Shottky junction properties affected by the level of oxygen vacancies in Al oxide.

Al oxide can be described as a dielectric with an impurity comprised of oxygen vacancies (Figure 11.6).

The Mott–Schottky method is based on the determination of space charge layer capacitance C_{sc}^{-2} measured at a fixed frequency as a function of potential around an open-circuit potential (OCP)* (Figure 11.4). The Mott–Schottky relationship for a p-type semiconductor can be written as

$$C_{sc}^{-2} = -\frac{2}{\varepsilon\varepsilon_0 e N_A A^2}(E - E_{FB} - kT) \tag{11.1}$$

where
　　ε is the dielectric constant for Al oxide
　　ε_0 is the dielectric constant
　　e the electron charge
　　E is the applied potential
　　E_{FB} is the flat band potential
　　k is the Boltzmann constant

FIGURE 11.6
Zone diagram of Al oxide with the acceptor type of conductivity.

---

* Physical meaning of ICP is the potential of the free surface, that is, the potential of the surface which is in equilibrium with an electrolyte solution. It is characteristic to the energy which has to be spent to remove an electron from the surface, which is in some sense characteristic to corrosion.

T is the temperature

A is the sample area

N_A is the acceptor concentration

E_{FB} was found to depend on the position of hydroxyl substitutions in the benzene ring (Table 11.2). The oxide formed in the presence of the mediator with closer positions of hydroxyls possessed less electron affinity than the one formed in the presence of the mediator with larger displacement of hydroxyl groups. The oxide formed at larger E_{max} did not show this correlation (Figure 11.7). E_{max} governs the thickness of the oxide layer, and possibly the E_{FB}: barrier that electrons have to overcome to be removed from an oxide. This principle was confirmed for computerized axial tomography (CAT) and somewhat for RES; however, it did not hold for HQ, showing a correlation with the proximity of hydroxyls in the benzene ring on $E_{FB} - E_{max}$ dependence. References for electrochemical oxidative polymerization of HQ [10,15,16] and CAT [17] can be found in the literature. Summarizing, it can be said that below positive 0.6 V (below CAT and HQ polymerization potential at Al), a correlation between the position of the hydroxyl group and E_{FB} was observed.

TABLE 11.2

Flat Band Potential of Different Compounds with Only Hydroxyl Substitutions Obtained at Different E_{max}.

Compound	Position of Hydroxyl Substitutions	E_{FB}, V, versus Ag/AgCl		
		0.6	1.2	1.8
CAT	1, 2	−4.53	−4.60	−4.63
RES	1, 3	−4.60	−4.64	−4.64
HQ	1, 4	−4.63	−4.62	−4.61

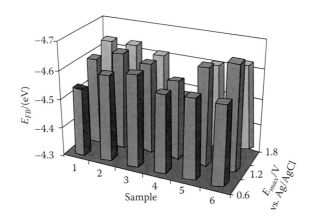

FIGURE 11.7

(See color insert.) E_{FB} of different samples as a function of potential at which oxide was obtained. 1-CAT, 2-RES, 3-HQ, 4-HQS, 5-BDS, and 6-DHBDS.

TABLE 11.3

Relative Acceptor Concentration ($N_A/N_{A\,max}$) in Oxide
Obtained at 0.6 V in the Presence of Different Mediators
(Normalized to CAT)

Compound	CAT	RES	HQ	HQS	BDS	DHBDS
N/N_{ACC}	1	0.602	0.679	0.270	0.292	0.138

The largest E_{FB} values were observed for DHBDS. The best corrosion protection performance of samples prepared with DHBDS was supported by monitoring experiments.

Due to the complicated structure of the Al alloy, Al oxide formed on its surface contains impurities and deviations from stoichiometry. Si^{4+} substitutions create vacancies, while Al^{2+} or Mg^{2+} generate interlattice electrons. The concentration of these charge carriers defines the type and magnitude of electrical conductivity of an oxide. Electron transfer mediators applied during the oxide deposition affect the balance between p and n defects in the Al oxide that is a p type due to oxygen vacancies. E_{FB} and impurities concentration were found to correlate with the structure of the mediator applied during the oxide deposition (Table 11.3). These data are believed to be important to understand the mechanism of electron transfer from an oxide surface to a conducting polymer which is part of a triggering device responsible for "smart" behavior.

11.3.3 PPy Deposition on Anodic Nanoporous Alumina and Anticorrosion Properties of this Coating

The rationale for depositing PPy on the oxide layer is to combine the barrier properties of Al oxide with corrosion-inhibiting properties of PPy. Chemically pure Al oxide (Al_2O_3, sometimes referred as alumina) is a dielectric and in order to deposit ICP onto an oxide of a few microns thickness, it has to be porous. In Ref. [18] Al oxide was obtained electrochemically by a method that is a simplified version of oxidation used to prepare anodic porous alumina (APA), such as described in Ref. [19]. Electrochemically, it is possible to obtain oxide in a wide range of thicknesses. The mechanism of oxide growth is described in Ref. [20]. Due to variations in the composition of the alloy, it is impossible to obtain a regular pore structure. Surface pore distribution was not regular but uniform as shown by scanning electron microscopy (SEM) (Figure 11.8). Electrochemical constant current deposition of PPy on a porous oxide (PO) surface required potentials that were sufficiently higher than the potentials necessary for PPy deposition on an Al alloy that is not oxidized (which is in fact covered by a very thin, natural oxide layer of atmospheric origination). The curve at the bottom of Figure 11.9 shows deposition of PPy onto a bare Al alloy surface in the presence of DHBDS.

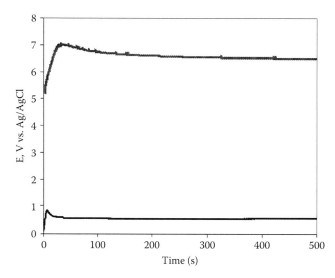

FIGURE 11.8
Galvanostatic deposition of PPy on the surface of aluminum alloy with oxide (top) and without oxide layer (bottom).

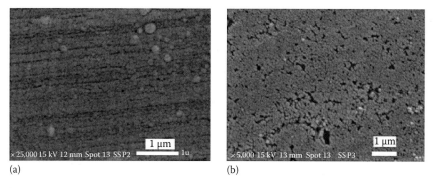

FIGURE 11.9
SEM images of (a) untreated aluminum and (b) porous aluminum surface.

In this case, the deposition potential was 0.6–0.65 V*, which is a typical value for the deposition in the presence of this mediator [10]. When the Al alloy was preliminary coated with PO, the deposition started at 5.2 V, reaching approximately 7 V after the first 50 s of the experiment, followed by a slow decrease to 6.5 V. Without DHBDS, deposition on the oxide-coated

---

* Unless specified or table values, potentials are given versus silver/silver chloride (Ag/AgCl) reference electrode.

TABLE 11.4

OCP of Different Surfaces
Determined by Potentiodynamic
Experiments

Material	OCP, V vs. SCE
Bare Al control	−0.95
PPy on bare Al	−0.55
Al oxide on Al	−0.4
PPy on oxide on Al	0.0

surface was not possible (the deposition potential was above 8 V, and no PPy formation was observed). The chemical structure was identified by Fourier transform infrared spectroscopy (FTIR) as PPy in both cases with the presence of all major picks characteristic to PPy [18], suggesting that in the presence of ETM the overpotential of PPy deposition was negligible and did not result in overoxidation of a conducting polymer, which is usually observed if the potential of PPy deposition is too high [21].

Corrosion protection properties of different types of coatings were assessed by potentiodynamic (Tafel) experiments. Tafel plots allow the determination of OCP for different types of surfaces. Experiments were conducted in contact with a corrosive electrolyte, diluted Harrison solution (DHS). The results are shown in Table 11.4.

PPy on Al surface shows relatively high OCP that was shown previously by a scanning vibrating electrode technique (SVET) and reported in Ref. [22]. High thickness of Al oxide in PPy/oxide/Al structure brings OCP to the level of mercury, which is sometimes referred to as a noble metal because it can be found in some geological deposits in metallic form. Therefore, the PPy/oxide/Al structure can be considered to possess properties of noble metals.

11.3.4 Porous Structure of Oxide Film and Its Characterization by Impedance Methods

Potential drop at metal/oxide/PPy/solution interface can be formulated as

$$E_{appl} = E_{Sh} + E_{Ox} + E_{PPy\,dep} \tag{11.2}$$

where
  E_{appl} is the applied potential
  E_{Sh} is the potential drop due to Shottky junction
  E_{ox} is the potential drop at oxide
  $E_{PPy\,dep}$ is the potential drop required to for PPy deposition

The Mott–Shottky junction possesses rectifying properties; therefore, the potential drop in straight direction in a rough estimate can be neglected. Assuming that $E_{PPy\,dep}$ is the same in the presence and absence of an oxide (600 mV) and subtracting it from a steady-state potential of deposition on an oxide (6500 mV), pore resistance can be determined as 0.17 Ω from Ohms law. Potential drop at an oxide surface required for PPy deposition when the oxide was previously developed is created by the resistance of the solution inside the pores. However, taken separately, the improved mechanical contact cannot be responsible for an increased OCP; therefore, E_{Sh}, which was neglected during the calculation of the pore resistance, must be taken into consideration. The existence of this potential drop in straight direction can be seen at the volt–ampere diode curves depending on the type of electrical contact. The junction properties of the ICP–metallic interface have to be considered when surface-ennobling properties are discussed [12,23].

By calculating the area occupied by pore per unit area (approximately 3.5%) and dividing by integral pore resistance from Equation 11.1, the resistance of a single pore, assuming their connection in parallel, can be roughly estimated as 4.8 MΩ. From ultraviolet–visible (UV–VIS) interference measurements, oxide thickness was determined as 2.52 μm, assuming 5.9 V potential drop on an oxide, and field gradient along a pore was on the order of 10^6 V/m, which is enough to form PPy nucleation sites inside an oxide pore.

11.3.5 Characterization of Porous Layers by Impedance Methods

Electrical conductivity of the solution σ is contributed by the conductivity of several types of ions i:

$$\sigma = \sum n_i \mu_i |q_i| \tag{11.3}$$

where
    n_i is the concentration
    q_i the charge
    μ_i the mobility of ith ion

An oxide layer can be taken as a permeable membrane which is placed between a metal and a solution. If the membrane resistance r_{pore} is much higher than the solution resistance r_{sol} ($r_{pore} \gg r_{sol}$), the measurable quantity characterizes the resistance of pores (pore resistance) [24]. The resistance of a cylindrical pore (Figure 11.10a), with index j ($r_{pore\,j}$), can be written as

$$r_{porej} = \rho \frac{d}{A_p} \tag{11.4}$$

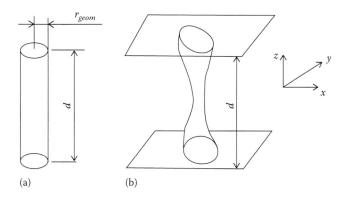

FIGURE 11.10
(a) Cylindrical and (b) noncylindrical pore.

and

$$A_p = \pi r_{geom}^2 \qquad (11.5)$$

where
  d is the pore length
  A_p is the cross-sectional area
  r_{geom} is the pore radius
  ρ is the unit resistance of the solution in the pore equal to

$$\rho = \sigma^{-1} \qquad (11.6)$$

If the pore is not cylindrical (Figure 11.10b), its resistance can be determined by integrating

$$r_{pore\,j} = \int_V \frac{dV}{A(x,y,z)} \qquad (11.7)$$

By taking the sum of separate parallel pores by index j, the entire film resistance can be determined as

$$r_{pore}^{-1} = \sum_{j=1}^{N} r_{pore(j)}^{-1} \qquad (11.8)$$

which gives a mathematical representation of pore resistance.

Measured at small frequency ($1/2\pi$ Hz), the cell resistance r_{cell}, determined from Bode plot, can be written as

$$r_{cell} = r_{sol} + r_{pore} + r_{pol} \qquad (11.9)$$

In this equation, r_{sol} is the solution resistance, which can be neglected, and r_{pol} is the polarization resistance, which is on the same order as r_{pore} [25]. Therefore:

$$r_{cell} \approx r_{pol} + r_{pore} \tag{11.10}$$

where r_{pol} is no more of interest because it can be eliminated by selecting zero potential, which equilibrates the film with the solution, giving

$$r_{cell} \approx r_{pore} \tag{11.11}$$

and in this case, it can be determined from the low-frequency part of the Bode plot.

In principle, the problem of determining pore size distribution from electrochemical impedance spectroscopy (EIS) data can be possibly solved. It opens possibilities of monitoring corrosion, progressive characterization of porous materials, such as fuel cell catalytic membranes, and other applications. This can be a subject for future investigation.

11.3.6 Visualizing Pore Structure by SEM Imaging

A SEM image of PPy obtained on PO (Figure 11.11a and b) shows a dense and uniform surface that is possibly a PPy-coated oxide with PPy globules on the top of the pores.

The visual appearance of the samples with PPy on an oxide surface was the same after 7 days of corrosion test as it was before the test: no delamination spots were observed, while PPy obtained by normal procedure was roughly half delaminated. Such an effect can be explained by better

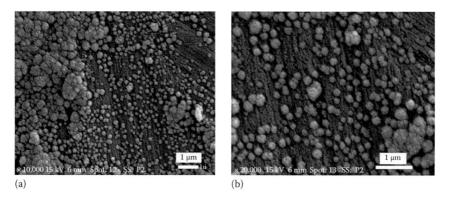

(a)   (b)

FIGURE 11.11
SEM images of PPy on PO aluminum at (a) ×10,000 and (b) ×20,000 magnification. (PPy deposition: 500 s, 1 mA/cm², 0.1 M Py, 1 M Na₂SO₄, pH3.)

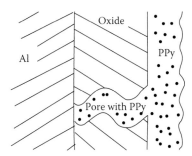

FIGURE 11.12
Schematic representation of PPy fixation on Al surface with developed Al oxide.

PPy adherence on the oxide surface because of the penetration and locking mechanism inside the pores (Figure 11.12). Adherence was responsible for the prolonged lifetime of a coating in a corroding solution. Nucleation sites for PPy deposition were formed inside the pores, while the formed PPy globules on the surface of an oxide possessed typical PPy "mushroom" morphology.

11.3.7 Smart Coatings for Corrosion Protection

In this chapter, electron kinetics has not been included as there are many publications and reviews regarding this topic in the literature [3]. However, issues related to varied porosity have not yet been discussed. If ICP concentration is high enough, ICP particles in mechanical contact with each other form the electrical percolation cluster. It was shown that there is an additional mechanism of corrosion prevention by ICP-modified nanoparticles: the nanoparticles in mechanical contact with each other provide passes for aggressive ions to penetrate into the coating [26], thus enhancing corrosion. Consequently, approaches where ICP concentration is below the mechanical percolation level are now of interest. When ICP concentration is decreased, it is still possible for electrons to move within the coating by a hopping or tunneling mechanism (such as variable range hopping [VRH] between clusters of variable size [27] or fluctuation-induced tunneling [28]). Ionic movement is not strongly affected by ICP.

While electrons hop or tunnel between nanoparticles, ions penetrate through the coating by diffusion. Without any fillers or inhomogeneities, there is only uniform diffusion [29]: penetration of ions and moisture into the coating through intermolecular spaces. In the presence of additives, such as nanoparticles or nanoflakes (Figure 11.13), uniform diffusion (Figure 11.14a) is interrupted (Figure 11.14b). However, nanoadditives introduce the mechanism of nonuniform diffusion.

ICP undoping caused changing volume, which was experimentally confirmed [30–33]. A volume increase of 8% during undoping was shown for

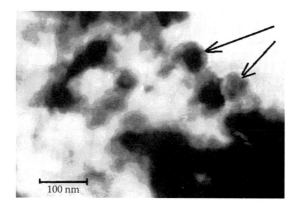

FIGURE 11.13
TEM image of nanoparticle modified with PPy (shown by arrows). (From Tallman, D.E. et al., *Appl. Surf. Sci.*, 254, 5452, 2008.)

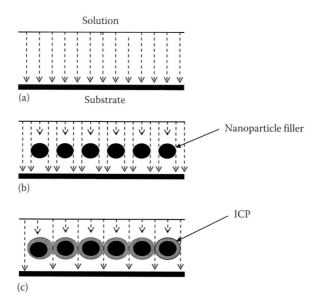

FIGURE 11.14
Schematic representation of diffusion into the coating: (a) Coating without filler. (b) Diffusion through coating with PPy-modified nanoparticle filler. (c) PPy-modified nanoparticles swelling under the influence of an electrolyte.

poly(3-octyl-thiophene) gel [31], and 12.5% volume increase for PPy doped with polystyrene sulphonate [32]. A similar effect was discovered for PPy doped with perchlorate ions [33]. The application of this effect in corrosion protection is shown schematically in Figure 11.14c. During swelling, PPy absorbed additional moisture that penetrated inside the coating through the voids [34]. Levine et al. [34] suggested that swelling of PPy-filled voids blocks passes for ion permeation inside the coating (Figure 11.14c). Experimental confirmation of this finding was provided by EIS.

EIS is a technique in which alternating frequency potential is applied to study the system and current response is recorded [35]. EIS data are displayed with the help of the Nyquist plot, which is the imaginary part of impedance versus the real part, and the Bode plot, which is the real part of the impedance versus frequency. EIS data revealed two phases of a coating containing a nanoparticle filler (Figure 11.15). Phase I, associated with high capacitance, was related to the nanoparticle filler, while phase II, associated with low capacitance, was due to the continuous phase of the coating [4]. While phase I changed its shape and size due to swelling, phase II remained relatively intact.

From the Bode plot it was established that phase I changed its charge transfer resistance from 53 to 22.5 KΩ when nanoparticle contents decreased from 2.0% to 0.5%. The high-frequency semicircle resembles the same pattern regardless of the amount of nanoparticle filler with charge transfer resistance within the range of 9.5–12.5 KΩ. After immersion, samples with

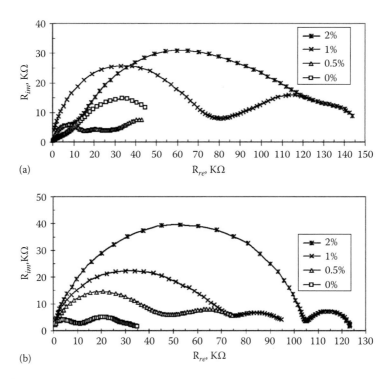

FIGURE 11.15
Nyquist plots for samples after (a) 24 and (b) 168 h of corrosion test. "Large" semicircles starting from the origin (high-frequency phase), and "small" semicircles on the right (low-frequency phase) responsible for diffusion. Percentages show nanoparticle contents.

the nanoparticle filler increased their impedance, which is usually related to the barrier property, which was attributed to the swelling of PPy-modified nanoparticles that blocked the passes to aggressive ions through the coating.

11.4 Conclusions

In this study, some aspects of corrosion protection methods and approaches related to intrinsically conducting polymers, such as polypyrrole, and active metals, such as aluminum were summarized. This study, however, did not intend to cover the entire problem in its fundamental meaning; it was focused on aspects of diffusion-linked smart behavior with relation to oxide properties that seemed important to us. Two models were discussed and the correlation between them was shown: surface ennobling by ICP, slowing down corrosion, and clogging pores by swelled ICP, thus blocking passes to aggressive ions. In practically applied anticorrosion coatings, these processes are carried out simultaneously. EIS was reviewed as a suitable method for taking a close look at diffusion through pores of variable size. The creation of new anticorrosion smart materials was also discussed.

Acknowledgments

The author acknowledges Dr. Dennis E. Tallman from North Dakota State University for his guidance throughout a significant part of this project and Dr. Valery V. Malev from St. Petersburg State University for valuable discussions.

References

1. D. Battocchi, A. M. Simões, D. E. Tallman, and G. P. Bierwagen, *Corros. Sci.*, 2006, 48(5), 1292–1306.
2. M. Kendig, M. Hon, and L. Warren, *Prog. Organ. Coat.*, 2003, 47, 183–189.
3. K. B. Kashi and V. J. Gelling, *Smart Nanocompos.*, 2(1), 55–83.
4. D. E. Tallman, K. L. Levine, C. Siripirom, V. J. Gelling, J. P. Bierwagen, and S.G. Croll, *Appl. Surf. Sci.*, 2008, 254, 5452–5459.
5. K. L. Levine, V. Fen, N. N. Nikonorova, V. M. Svetlichnyi, V. E. Yudin, L. A. Myagkova, and N. S. Pshchelko, *Smart Nanocompos.*, 3(1), 49–58.

6. T. A. Scotheim, (ed.), *Handbook of Conducting Polymers*, 1st edn., 1983. Marcel Dekker, New York.

7. D. E. Tallman, C. Vang, G. G. Wallace, and G. P. Bienragen, *J. Electrochem. Soc.*, 2002, 149, 173.

8. D. E. Tallman, M. P. Dewald, C. K. Vang, G. G. Wallace, and G. P. Bierwagen. *Curr. Appl. Phys.*, 2004, 4, 137.

9. D. E. Tallman, C. K. Vang, M. P. Dewald, G. G. Wallace, and G. P. Bierwagen, *Synth. Met.*, 2003, 135–136, 33.

10. K. L. Levine, D. E. Tallman, and G. P. Bierwagen, *Aust. J. Chem.*, 2005, 58(4), 294–301.

11. K. Naoi, M. Takeda, H. Kanno, M. Sakakura, and A. Shimada, *Electrochim. Acta*, 2000, 45(20), 3413–3421.

12. K. L. Levine, D. E. Tallman, and G. P. Bierwagen, *J. Mater. Process. Technol.*, 2008, 199, 321–326.

13. L. I. Antropov, *Theoretical Electrochemistry*, Mir Publishers, Moscow, Russia, (1978).

14. N. F. Mott, *Proc. Roy. Soc.*, 1939, A 171, 27.

15. J. S. Foos and S. M. Erker, *J. Electrochem. Soc.*, 1986, 133, 836.

16. K. Yamamoto, T. Asada, and K. Nishide, *Bull. Chem. Soc. Jpn.*, 1990, 63, 1211.

17. J. Davis, D. H. Vaughan, and M. F. Cardosi, *Electrochim. Acta*, 1997, 43(3–4), 291–300.

18. K. L. Levine, D. E. Tallman, and G. P. Bierwagen, *ECS Transactions*, 2005, 1(4), 81–91.

19. Y. Zhao, M. Chen, Y. Zhang, T. Xu, and W. Liu, *Mater. Lett.*, 2005, 59(1), 40–43.

20. M. M. Lorengel, *Mater. Sci. Eng.*, 1993, R11, 243–294.

21. I. Rodriguez, B. R. Scharifker, and J. Mostany, *J. Electroanal. Chem.* 2000, 491(1–2), 117–125.

22. J. He, D. E. Tallman, and G. P. Bierwagen, *J. Electrochem. Soc.*, 2004, 151(12), B644–B651.

23. B. Wessling, *Mater. Corros.*, 1996, 47, 439.

24. K. L. Levine and J. O. Iroh, *J. Porous Mater.*, 2004, 11, 87–95.

25. K. L. Levine and N. S. Pshchelko, *Polym. Sci. Series A (Polymer Physics)*, 2011, 53(6), 510–520.

26. M. Rohwerder and A. Michalik, *Electrochim. Acta*, 2007, 53, 1300–1313.

27. N. F. Mott and E. A. Davis, *Electronic Properties in Non-Crystalline Materials*, 1971. Clarendon press, Oxford, New York.

28. P. Sheng, *Phys. Rev. B*, 1980, 21, 2180.

29. E. L. Cussler, *J. Membr. Sci.*, 1990, 52(3), 275–288.

30. T. Okamoto, Y. Kato, K. Tada, and M. Onoda, *Thin Solid Films*, 2001, 393(1–2), 383–387.

31. X. Chen and O. Inganas, *Synth. Met.*, 1995, 74, 159–164.

32. M. F. Suarez and R. G. Compton, *J. Electroanal. Chem.*, 1999, 462(2), 211–221.

33. M. Pyo and C.-H. Kwak, *Synth. Met.*, 2005, 150, 133–137.

34. K. L. Levine, D. E. Tallman, and G. P. Bierwagen, *5th International Symposium "Molecular order and Mobility,"* Publication in meeting proceedings, St. Petersburg, Russia, June 20, 2005.

35. A. J. Bard and L. R. Faulkner, *Electrochemical Methods; Fundamentals and Applications*, 2000. Wiley Interscience Publications, New York.

12

Ultrasound-Assisted Synthesis and Its Effect on the Properties of CaCO₃–Polymer Nanocomposites

Bharat A. Bhanvase and Shirish H. Sonawane

CONTENTS

12.1 Introduction

Calcium carbonate ($CaCO_3$) is an abundant mineral comprising approximately 4% of the earth's crust. Nanoparticles of $CaCO_3$ are gaining very much importance due to its versatile applications in various industries. They are widely used in a variety of industries; for example, in paper industry, $CaCO_3$ nanoparticles are used to make whiter and brighter sheets. It also reduces the cost by replacing highly expensive fillers/pigments. Nano-$CaCO_3$ is also used to produce nonacidic papers. The plastic industry is the largest consumer of $CaCO_3$. $CaCO_3$ nanoparticles can be used as the major engineered filler in plastics such as poly(vinyl chloride) (PVC). It gives controlled whiteness and improved impact strength. It is used as an aid in easy processing and also it acts as a heat sink for exothermic reactions. $CaCO_3$ is used as a pigment and extender in paints to control color and gloss because of its low cost (Chen et al. 2000; Tsuzuki et al. 2000; Dagaonkar et al. 2004; Gupta 2004; Hu et al. 2004; He et al. 2005; Huber et al. 2005).

There are different methods available for the synthesis of $CaCO_3$ nanoparticles such as carbonation of lime solutions in reverse micellar systems (Dagaonkar et al. 2004), flame synthesis (Huber et al. 2005), using modified emulsion membranes (Gupta 2004), high-gravity reactive precipitation (Chen et al. 2000), mechanochemical processing (Tsuzuki et al. 2000), two-membrane system (Hu et al. 2004), and ultrasound-assisted synthesis (He et al. 2005; Sonawane et al. 2008). It has been reported that ultrasound-assisted synthesis method was proved to be useful to obtain narrow particle size distribution, low crystallite size, and high specific surface area of $CaCO_3$ particles.

The ultrasound-assisted method has been proved to be useful to obtain inorganic nanoparticles with narrow distribution (Cole and Cole 1942; Barsoukov and Macdonald 2005; Prasad et al. 2010a,b; Pinjari and Pandit 2011). Ultrasound is the cyclic sound wave whose frequency is above the limit of human hearing, which is usually taken to be above 20 kHz. The chemical effects of ultrasound do not come from a direct interaction with molecular species. When ultrasonic waves pass through a liquid medium, large numbers of microbubbles form, grow, and collapse in very short time leading to the cavitational effects of intense turbulence, liquid circulation currents, and also formation of the free radicals (Gogate 2008). The collapse of cavitation bubbles near the interface of immiscible liquids will cause disruption of the phases due to the generated microjets. Extreme pressure (>500 atm) and temperature (>10,000 K) with a cooling rate of $>10^{10}$ K/s conditions caused due to ultrasonic irradiation leading to intense micromixing improve solute transfer and nucleation rate in aqueous suspension, which lead to the formation of nanometer-sized particles (Pinjari and Pandit 2011). The physical and chemical effects of ultrasound irradiation occurred because acoustic cavitations are useful for the synthesis of smaller-size nanoparticles. Ultrasonic irradiations also prevent the agglomeration of the particles leading to the formation of nanoparticles.

Several reports are available that report the sonochemical synthesis of nanoparticles are presented in the following paragraphs. Arami et al. (2007) have prepared the titania nanoparticles by dissolving the TiO_2 particles into NaOH solution. They have found that in NaOH solution, the structures of TiO_2 get broken and new structures are formed in the presence of ultrasound. The sonochemical preparation of amorphous silver nanoparticles is discussed by Salkar et al. (1999). Amorphous silver nanoparticles of ca. 20 nm size have been prepared by the sonochemical reduction of an aqueous silver nitrate solution in an atmosphere of argon–hydrogen. In their work, they have attempted the synthesis of silver nanoparticle using sonochemical technique by the reduction reaction. Also the role of cavitation in the radical generation and mechanism of the formation of silver nanoparticles of smaller and uniform size has been discussed. Jeevanandam et al. (2000) have synthesized a cobalt hydroxide colloid nanoparticle using ultrasound irradiation. Shirsath et al. (2013) study the influence of ultrasound on the phase composition, structure, and performance of pure and doped TiO_2 nano-catalysts. Ce- and Fe-doped TiO_2 nano-catalysts with different amounts of doping elements were prepared by a single-step sonochemical method. Shai et al. (1998) have also used a sonochemical preparation technique for the Co–Ni alloy powder synthesis.

Extensive research efforts have been directed toward the preparation of polymer/inorganic nanocomposite (Siegel 1994). It is challenging to combine properties of organic and inorganic components to form unique composite material. Several methods have been used to produce polymer nanocomposite, such as miniemulsion polymerization (Tiarks et al. 2001), intercalative polymerization (Kong et al. 1999), solution casting method (Liu et al. 2004), hybrid latex polymerization (Tissot et al. 2001), and so on. Further, strong interfacial adhesion and dispersion of nanoparticles are two factors that decide the property profile of the polymer nanocomposite matrix; hence functionalization of the inorganic fillers is a key factor during *in situ* emulsion-polymerized nanocomposite synthesis. Surface modification of an inorganic particle with an organic molecule is one of the key methods to reduce its surface energy and increase its compatibility with the polymer matrix (Mishra et al. 2005; Sh et al. 2006; Wang et al. 2006a and 2007a, 2007; Sonawane et al. 2009). It has been reported that the polymer dispersed with nano-$CaCO_3$ particles provides excellent mechanical properties due to their high surface area-to-volume ratio (Avella et al. 2001; Chan et al. 2002; Xie et al. 2004; Mishra et al. 2005). It is well known that the driving force for uniform dispersion of nano-$CaCO_3$ lies in the synergetic effect of chemical or mechanical interactions. There have been some problems regarding the uniform dispersion of nanofiller in polymer matrix, stability of colloidal suspension, and improvement in the properties of nanocomposite using water-based suspensions in conventional *in situ* emulsion polymerization (Sahoo and Mohapatra 2003; Zhang et al. 2006). These problems are expected to be resolved using the cavitational effects induced by ultrasonic irradiations in the conventional emulsion polymerization approach operated in a semibatch mode. This leads to an improvement in the dispersion of nanofillers, which controls the effectiveness of the overall encapsulation procedure and the surface properties of the final synthesized product (Qi et al. 2001; Ryu et al. 2004).

12.2 Literature Review on Polymer/CaCO₃ Nanocomposites

$CaCO_3$ is a commercially available inorganic material and has been extensively used as the particulate filler in the manufacture of paint, paper, and plastic and is also used as particulate filler in the rubber industries. Also as an important biomineral and an industrially useful material, hydrophobic $CaCO_3$ with excellent properties is desirable in paints, inks, papers, plasticizers, and so on (Wang et al. 2006b, 2007b; Zhang et al. 2008). However, the incompatibility of its highly energetic hydrophilic surface with the low-energy surface of hydrophobic polymers is a problem that needs to be solved before it can be used as a functional filler. For this and other reasons, the surface of calcite is often rendered organophilic by a variety of surface modifiers such as silane coupling agents (Demjén et al. 1997), titanate coupling agents (Monte and Sugerman 1976), phosphate coupling agents (Nakatsuka et al. 1982), or stearic acid (Papirer et al. 1984). Further, it has been already reported that polymers uniformly dispersed with nano-$CaCO_3$ particles provide excellent mechanical properties due to their high surface area-to-volume ratio (Chan 2002; Mishra et al. 2005). Bhanvase and Sonawane (2010) have attempted an encapsulation of $CaCO_3$ into polyaniline using miniemulsion polymerization. It has been reported that the dispersion of nano-$CaCO_3$ showed an improvement in the performance of composite for anticorrosive and mechanical properties when polymer nanocomposite was dispersed in alkyd resin for coating applications. When $CaCO_3$ is used as the filler during polymerization, it is required to exhibit certain properties, such as excellent dispersion ability in the polymer, and should stabilize the viscosity of the adhesive during storage. $CaCO_3$ coated with myristic acid (MA) introduces functional groups on the particles, and it exhibits hydrophobic nature (Sonawane et al. 2008). Hence, during the polymerization, it remains embedded in the polymer matrix forming a well-dispersed composite (Bhanvase and Sonawane 2010). Aggregation of inorganic nanoparticles leads to a poor performance of polymer nanocomposite, which could be resolved by the application of cavitation technique for the effective micromixing. The collapse of cavitation bubbles at or near the interface of immiscible liquids will cause disruption due to microjets, resulting in the formation of very fine emulsions. Due to ultrasonication, the development of microdroplets takes place in the emulsion, and hydrophobic nanoparticles get dispersed in fine emulsified organic droplets, which lead to a uniform dispersion of the nanoparticles in the resultant polymer matrix (Ryu et al. 2004).

Sheng et al. (2006) have carried out the synthesis of polystyrene (PS)/$CaCO_3$ composite nanoparticles (80 nm) with a core/shell structure by *in situ* emulsion polymerization of styrene on the surface of modified $CaCO_3$ nanoparticles. Modification of $CaCO_3$ nanoparticles has been carried out via carbonation of a $Ca(OH)_2$ slurry in the presence of sodium oleate at room temperature in order to introduce functional groups onto its surface by Sheng et al. (2006). Transmission electron microscope (TEM) images showed the irregular cubic shape of untreated $CaCO_3$ particles and the modified $CaCO_3$ particles with a particle size of about 40–70 nm. Further, it has been reported that PS/$CaCO_3$ nanocomposite particles exhibit core–shell morphology with a particle size of 80 nm. The possible reasons for core–shell morphology are (1) surface modification of $CaCO_3$ nanoparticles and (2) *in situ* preparation method.

Wu et al. (2006) have prepared polymethyl methacrylate (PMMA)/CaCO$_3$ composite particles by soapless emulsion polymerization method in the aqueous suspension of nano-CaCO$_3$. They have studied the effects of temperature, agitation rate, dosages of the intiator, and CaCO$_3$ concentrations on monomer conversion. The reported results indicated that nano-CaCO$_3$ that is present in the suspension can boost the monomer conversion (Wu et al. 2006). Appropriate stirring rate can enhance the probabilities of the polymer to coat on the surface of nano-CaCO$_3$. The basic technology of the preparation of composite particles is soapless emulsion polymerization of methyl methacrylate (MMA) in nano-CaCO$_3$ aqueous suspension. With an increase in the temperature, the decomposing rate of initiating agent is raised; hence the collision probability among monomers, polymers, and particles is increased; and thus the conversion is fast. At higher temperature, it has no obvious effects on the decomposition of initiating agent, so the conversion basically remains constant. The rate of polymerization reaction increases apparently in the presence of nano-CaCO$_3$. It has been reported that the stable emulsions could be formed when there were CaCO$_3$ particles on the interfaces between water and oil. It can be predicted that CaCO$_3$ nanoparticles can be located on the interfaces of monometer–water to stabilize the formed micelles that provide more polymerization sites than without nano-CaCO$_3$. Bhanvase et al. (2009) have successfully synthesized water-based PMMA/CaCO$_3$ nanocomposite using *in situ* emulsion polymerization of MMA. Formation of PMMA/CaCO$_3$ nanocomposite has been confirmed from x-ray diffraction and TEM analysis. It has been also reported that the presence of nanosized CaCO$_3$ in the PMMA matrix significantly improves the mechanical properties of PMMA/CaCO$_3$ nanocomposite. Ma et al. (2008) have used soapless emulsion polymerization method to synthesize PMMA/CaCO$_3$ spherical composite with different loading of oleic acid-treated CaCO$_3$. The results of photon correlation spectroscopy reported by them showed that the particle size of composites is more than the pure PMMA, and it increases with an increase in the content of CaCO$_3$ nanoparticles. The reported morphological analysis showed uniform and well-encapsulated CaCO$_3$ composite microspheres. Further, it has been predicted that loading of oleic acid-treated CaCO$_3$ nanoparticles significantly improves the thermal stability of PMMA and acid-resistance of CaCO$_3$ nanofillers.

Jian-ming et al. (2004) have attempted the synthesis of PMMA/CaCO$_3$ composite by the emulsion polymerization of MMA in the presence of nano-CaCO$_3$ surface modified with γ-methacryloxypropyltrimethoxysilane. Chen et al. (2006) have carried out the synthesis of PMMA-coated CaCO$_3$ for the application of interfacial adhesion and mechanical properties of PMMA-coated CaCO$_3$ nanoparticle-reinforced PVC composites. Results indicate that the interfacial adhesion between CaCO$_3$ nanoparticles and PVC matrix was significantly improved when the CaCO$_3$ nanoparticles were coated with PMMA, which led to increased Young's moduli and tensile strengths of the PMMA-coated CaCO$_3$/PVC composites. The izod impact strengths of the composites have been strongly affected by the PMMA coating thickness and are increased significantly by increasing the volume fraction of CaCO$_3$ filler in the composites.

Xie et al. (2004) have synthesized PVC/CaCO$_3$ nanocomposite by *in situ* polymerization of vinyl chloride in the presence of CaCO$_3$ nanoparticles. Morphological analysis of PVC/CaCO$_3$ nanocomposites shows that CaCO$_3$ is uniformly distributed as nanosized particles in the PVC matrix. It has been found that the glass transition of PVC phase in PVC/CaCO$_3$ nanocomposites is shifted toward higher temperatures by the restriction of CaCO$_3$ nanoparticles on the segmental and chain mobility of the PVC phase. The nanocomposites showed shear thinning behaviors. Further, it has been reported that the "ball bearing" effect of the spherical nanoparticles decreased the apparent viscosity of the PVC/CaCO$_3$ nanocomposite melts, and the viscosity sensitivity on the shear rate of the PVC/CaCO$_3$ nanocomposite is higher than that of pristine PVC. Moreover, CaCO$_3$ nanoparticles stiffen and toughen PVC simultaneously.

The ultrasonic irradiations for improving the dispersion process, which controls the effectiveness of the overall encapsulation procedure and the surface properties of the final synthesized product, have been used by Bhanvase et al. (2010, 2011). The use of ultrasonic irradiation can create very fine emulsions (Qi et al. 2001). Also it is expected that the problem of agglomeration of nanoparticles during its loading in polymer nanocomposite might be resolved by the use of cavitation during the emulsion polymerization process (Ryu et al. 2004). The cavitational effects produced due to the use of ultrasonic irradiations have been shown to enhance the dispersion of functional nano-inorganic particles into the monomer during polymerization process. Further, the generation of free radical due to chemical effects of acoustic cavitation leads to an improvement in the polymerization rate. Bhanvase and Sonawane (2010) have used ultrasound-assisted *in situ* semibatch emulsion polymerization method for an encapsulation of CaCO$_3$ into polyaniline. It has been reported that the dispersion of nano-CaCO$_3$ is significantly improved with the use of ultrasonic irradiations, which leads to the fine dispersion of CaCO$_3$ nanoparticle in the polyaniline matrix. This fine dispersion leads to an improvement in the performance of PANI/CACO$_3$ nanocomposite for anticorrosive and mechanical properties when it is being dispersed in alkyd resin for coating applications. Further, Bhanvase et al. (2011) have used intensified ultrasound-assisted *in situ* semibatch emulsion polymerization for the improvement of the process of encapsulation of inorganic nanoparticles into polymer during nanocomposite synthesis process. With the use of this method, Bhanvase et al. (2011) have successfully synthesized PMMA/CaCO$_3$ nanocomposite by ultrasound-assisted semibatch emulsion polymerization in the presence of MA-treated CaCO$_3$ nanoparticles. It has been reported that the use of cavitation generated due to the ultrasonic irradiations during *in situ* emulsion polymerization improves the dispersion of modified CaCO$_3$ into PMMA latex compared to conventional *in situ* emulsion polymerization and also significantly intensifies the process with overall reduction in the energy requirements. Further, functionalized $CaCO_3$ nanoparticles were also encapsulated in the PMMA due to the good compatibility between the nanofillers and the polymer matrix. It has been concluded that the ultrasound-based method is effective to increase the loadings of $CaCO_3$ in the composites, which can improve the thermal stability of PMMA/$CaCO_3$ nanocomposites.

12.3 Properties of Functionalized CaCO₃ Nanoparticle-Based Polymer/CaCO₃ Nanocomposite

12.3.1 Comparative Study on Conventional and Ultrasound-Assisted In Situ Emulsion Polymerization for Preparation of PMMA/CaCO₃ Nanocomposite

As reported in the earlier section, polymer dispersed with nano-$CaCO_3$ particles offers outstanding mechanical properties. Surface modification plays an important role in the uniform dispersion of nano-$CaCO_3$ in PMMA matrix. However, there are some problems related to the uniform dispersion of nanoparticles in polymer matrix, stability of colloidal suspension, and enhancement of the properties of nanocomposite in conventional *in situ* emulsion polymerization. These problems are effectively addressed with the use of cavitational effects induced by ultrasonic irradiations.

Bhanvase et al. (2011) have used ultrasound-assisted semibatch emulsion polymerization for the synthesis of PMMA/$CaCO_3$ nanocomposite in the presence of MA-treated $CaCO_3$ nanoparticles. $CaCO_3$ nanoparticles used in this study have been synthesized by the process described by Sonawane et al. (2008). Sonochemical carbonization was carried out in 1 L batch-type reactor, which consists of sonicator probe (Dakshin make, 240 W, 22 kHz) arrangement. CO_2 gas was passed through 20 mm diameter probe with hole (4 mm) drilled along the length and diameter. The probe hole passage has been used for effective micromixing of CO_2 gas in the diluted $Ca(OH)_2$ slurry. Further synthesis of functionalized $CaCO_3$ nanoparticles has been carried out by passing CO_2 gas through calcium hydroxide slurry. MA solution was prepared in methanol by keeping methanol to MA ratio as 4:1 by weight. The MA solution was added dropwise into the flask at 60°C under continuous sonication for 1 h during the carbonation reaction. The solution was centrifuged so as to remove unreacted MA present in the reaction mixture.

The sonochemical reactor has been used for the synthesis of PMMA/$CaCO_3$ nanocomposite provided with ultrasonic horn (13 mm diameter, stainless steel) and equipped with a generator (Sonics Vibra-cell, USA) operating at a frequency of 22 kHz and rated output power of 750 W. The reported actual power dissipated (measured using calorimetric method) was 45.9 W with

12.24% energy transfer efficiency. The polymerization reaction was carried out under inert atmosphere. Initially, aqueous solution of sodium lauryl sulfate (SLS) solution was prepared along with $CaCO_3$ nanoparticles. After subjecting this solution for 10 min irradiation, it was transferred to the reactor. Also an aqueous initiator solution was prepared separately and transferred to the reactor. The polymerization has been reported to be carried out by adding 1 mL of MMA monomer in the reactor initially and 9 mL in continuous manner at a constant rate of 0.3 mL/min over a time period of 30 min. The complete reaction was carried out over a period of 60 min at 65°C (±1°C). The product was separated and then dried in an oven at 120°C for 2 h.

During the preparation of PMMA/$CaCO_3$ nanocomposite, MA-treated $CaCO_3$ nanoparticles have been used in sonochemical *in situ* carbonation process. The surface modification results in reducing the surface energy of $CaCO_3$ nanoparticles. The reported elementary reactions that occur in the carbonation process to give functionalized $CaCO_3$ particles are given as follows:

$$Ca(OH)_2 \rightarrow Ca^{2+} + 2OH^- \tag{12.1}$$

$$Ca^{2+} + 2C_{13}H_{27}COOH \rightarrow Ca(C_{13}H_{27}COO)_2 + 2H^+ \tag{12.2}$$

$$CO_2 + H_2O \rightarrow H_2CO_3 \tag{12.3}$$

$$H_2CO_3 \rightarrow CO_3^{2-} + 2H^+ \tag{12.4}$$

$$CO_3^{2-} + Ca^{2+} \rightarrow CaCO_3 \tag{12.5}$$

In the surface treatment process, initially, Ca^{2+} ions react with MA to form a hydrophobic salt of $Ca(C_{13}H_{27}COO)_2$, which acts as a cosurfactant in the reaction mixture. The formed $Ca(C_{13}H_{27}COO)_2$ deposits on the surface of $CaCO_3$ lead to the formation of hydrophobic $CaCO_3$ nanoparticles. The presence of ultrasonic irradiations helps in reducing the particle size of the final product than the conventional method due to a reduction in the induction time. The schematics of the functionalization of nanosize $CaCO_3$ process have been depicted in Figure 12.1.

The hydrophobic properties of the functionalized $CaCO_3$ nanoparticles have been confirmed with the measurement of the active ratio and contact angle. The dependency of the active ratio, which gives an indication of the hydrophobicity, on the loading of MA, has been shown in Figure 12.2. It has been observed from the figure that with an increase in the dosage of MA from 0.05 to 0.3 wt%, the active ratio was improved substantially from 45.6% to 99.1%; however, a further increase to 0.4 wt% yielded only a marginal improvement in the active ratio (value of 99.5%). Thus, it is reported that for achieving the effective hydrophobicity of $CaCO_3$ nanoparticles, the recommended optimum dosage of MA is 0.3 wt%.

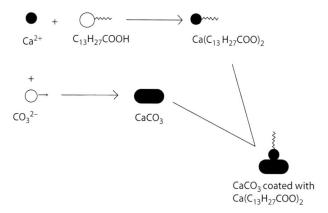

FIGURE 12.1
Schematic representation of the mechanism of the formation of functionalized CaCO₃. (Reprinted from Bhanvase, B.A. et al., *Chem. Eng. Process: Process Intensif.*, 50, 1160, 2011. With permission.)

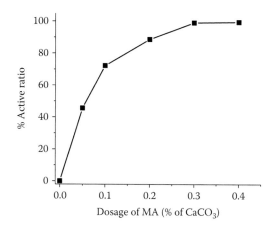

FIGURE 12.2
Influence of dosage of MA on the active ratio of modified CaCO₃. (Reprinted from Bhanvase, B.A. et al., *Chem. Eng. Process: Process Intensif.*, 50, 1160, 2011. With permission.)

The mechanism of the formation of PMMA/CaCO₃ nanocomposite by ultrasound-assisted *in situ* emulsion polymerization has been depicted schematically in Figure 12.3. As reported by Bhanvase et al. (2011), in the initiation stage, radicals are generated by dissociation of initiator and water molecules due to the cavitational effects generated by acoustic cavitation. Subsequently, in the nucleation stage, the formation of micelles around MMA droplet takes place. The size of monomer droplet is expected to reduce significantly due to the turbulence effects of the cavitation during emulsion polymerization. The free radicals enter the monomer droplet resulting into polymerization

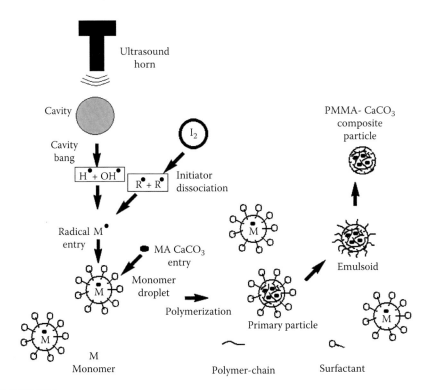

FIGURE 12.3
Schematic mechanism of the formation process of PMMA/CaCO$_3$ nanocomposite. (Reprinted from Bhanvase, B.A. et al., *Chem. Eng. Process: Process Intensif.*, 50, 1160, 2011. With permission.)

process. Thus, the cavitating effect is expected to improve the nanocomposite formation due to both the chemical effects of formation of free radicals and the physical effects of intense turbulence and liquid circulation currents. In a recent work by Dobie and Boodhoo (2010), it has been shown that the use of low-frequency ultrasound over a power dissipation range of 60–160 W (supplied electric power) does not give the desired degree of beneficial effects at an operating temperature of 45°C. The quantum of free radicals generated in the system will be dependent on the operating temperature and ultrasonic power dissipation, whereas the mixing effects (which are significant at low-frequency operation) will be mainly controlled by the ultrasonic power dissipation.

Further, Bhanvase et al. (2011) have addressed the effect of CaCO$_3$ loading on the extent of conversion of MMA during ultrasound-assisted semi-batch *in situ* emulsion polymerization (Figure 12.4). The reported percentage conversion of MMA was increased from 80.47% to 94.42% for 1%–9% CaCO$_3$ loading. It has been reported that the rate of polymerization increases with an increase in the concentration of the presence of functionalized CaCO$_3$. Further, similar observations have been reported by Wu et al. (2006).

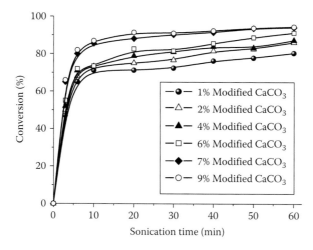

FIGURE 12.4
Effect of percentage CaCO$_3$ loading (of MMA) on conversion (quantity of MMA = 7.2 wt%, quantity of surfactant = 2% of MMA, quantity of initiator = 4% of MMA, temperature = 65°C) in ultrasound-assisted *in situ* emulsion polymerization. (Reprinted from Bhanvase, B.A. et al., *Chem. Eng. Process: Process Intensif.*, 50, 1160, 2011. With permission.)

The observed increase in the rates of polymerization can be attributed to the fact that the presence of nanosized CaCO$_3$ on the interface between monomer and water stabilizes the formed micelles and provides more polymerization sites (nucleating effect of MA-coated CaCO$_3$), which in turn enhances the rate of polymerization.

The TEM images of PMMA/CaCO$_3$ nanocomposites prepared by ultrasound-assisted and conventional *in situ* emulsion polymerization methods for 4% modified CaCO$_3$ loading have been given in Figure 12.5. The 4% modified CaCO$_3$ loading in PMMA latex has been taken as a representative loading, which has been observed to give the required alteration in the property profile of the nanocomposite. TEM image of PMMA/CaCO$_3$ nanocomposite confirms the encapsulation of CaCO$_3$ in PMMA matrix. Also it has been established that the CaCO$_3$ nanoparticles of size 50–100 nm are uniformly dispersed in polymer latex. PMMA/CaCO$_3$ nanocomposite synthesized by conventional *in situ* emulsion polymerization when with ultrasound-assisted *in situ* emulsion polymerization process, it has been clearly observed that the use of cavitation generated due to the ultrasonic irradiations during *in situ* emulsion polymerization improves the dispersion of modified CaCO$_3$ (Figure 12.5a) into PMMA latex compared to conventional *in situ* emulsion polymerization (Figure 12.5b).

The reported DSC results (Table 12.1) by Bhanvase et al. (2011) have confirmed that well-dispersed modified CaCO$_3$ nanoparticles restrict the motion of PMMA segmental chains. The glass transition temperature reported by Bhanvase et al. (2011) was found to increase from 119.5°C (for pure PMMA)

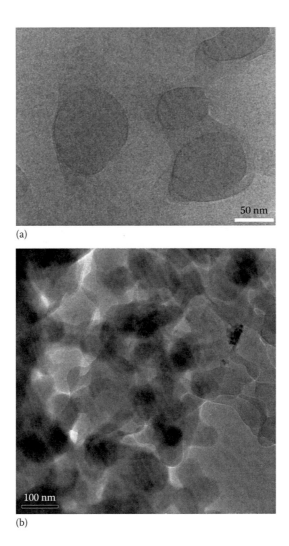

(a)

(b)

FIGURE 12.5
TEM images of PMMA/4% CaCO₃ nanocomposite synthesized by (a) ultrasound-assisted and
(b) conventional *in situ* emulsion polymerization method. (Reprinted from Bhanvase, B.A.
et al., *Polym.-Plast. Technol. Eng.*, 48, 939, 2009. With permission.)

to 210.6°C (at 4% CaCO₃ loading), but any further increase in the CaCO₃ loading resulted in decreased glass transition temperature. The observed decrease
was attributed to the agglomeration effect due to higher loading of modified
CaCO₃ (from 6% to 9%). This agglomeration weakens the interaction between
PMMA chains and modified CaCO₃ nanoparticles, which leads to decreases
in glass transition temperature to 160.5°C. Under ultrasonic conditions as used
in the present work, agglomeration of CaCO₃ in PMMA/CaCO₃ nanocomposite cannot be completely eliminated but can only be reduced to an extent.

TABLE 12.1

Effect of Modified $CaCO_3$ Loading on Glass Transition Temperature of PMMA/$CACO_3$ Nanocomposite Synthesized by Ultrasound-Assisted *In Situ* Emulsion Polymerization

Sr. No.	% $CaCO_3$ (of MMA)	Glass Transition Temperature (°C)
1	0	119.5
2	2	210.4
3	4	210.6
4	6	164.5
5	8	179.7
6	10	182.2

Source: Bhanvase, B.A. et al., *Chem. Eng. Process: Process Intensif.*, 50, 1160, 2011.

Overall, the use of cavitation generated due to the ultrasonic irradiations during *in situ* emulsion polymerization improves the dispersion of modified $CaCO_3$ into PMMA latex compared to conventional *in situ* emulsion polymerization and also significantly intensifies the process with overall reduction in the energy requirements. Further, functionalized $CaCO_3$ were also encapsulated in the PMMA due to the good compatibility between the nanofillers and the polymer matrix. It has been concluded that the ultrasound-based method is effective to increase the loadings of $CaCO_3$ in the composites, which can improve the thermal stability of PMMA/$CaCO_3$ nanocomposites.

12.3.2 PANI/$CaCO_3$ in Alkyd Coating for Mechanical and Anticorrosion Properties

Water-based polyaniline (PANI) is added during the coating formation along with alkyd resin so as to enhance the anticorrosive property of the surface coating (Jiang et al. 2006; Stejskal et al. 2006). Although a variety of conducting polymers have been synthesized and investigated, PANI is known for its good conductivity and environmental stability (Pud et al. 2003). PANI and its derivatives have been extensively used as anticorrosive coatings on metals (Jeyaprabha et al. 2006; Chang et al. 2007). However, a pure coating of polyaniline and its derivatives suffers from low mechanical properties and adhesion to the substrate (Dispenza et al. 2006). Simultaneous improvement in the mechanical, adhesive, and anticorrosive properties of PANI composite could be possible with inorganic additives/filler. $CaCO_3$ has been used as common filler in a variety of sizes and shapes. It has been reported that surface treatment of $CaCO_3$ plays an important role for the miscibility of nano-$CaCO_3$ in composite coatings. It is also known that the use of water-repellent nano-$CaCO_3$ coated with MA will show barrier to water and hence can enhance the corrosion inhibition mechanism, which in turn improves both mechanical and anticorrosive properties.

With this objective, Bhanvase and Sonawane (2010) have carried out the synthesis of PANI/$CaCO_3$ nanocomposite by semibatch *in situ* emulsion polymerization by indirect ultrasound (in ultrasound bath, Sonics and Materials, 20 kHz, 600 W) technique using aniline as a monomer, ammonium persulfate as an initiator, and SLS as a surfactant so as to enhance the reaction rate and micromixing of reaction mass. In order to study the effect of ultrasound, the synthesis of PANI/$CaCO_3$ nanocomposite has also been accomplished with the use of ultrasound-assisted as well as conventional *in situ* emulsion polymerizations. Alkyd resin (soy alkyd semidrying type) and xylene have been used during the preparation of PANI-$CaCO_3$/alkyd coatings. *In situ* emulsion polymerization reaction has been accomplished in 90 min at 4°C. MA-coated nano-$CaCO_3$ percentage was varied from 2% to 8% of monomer quantity. X-ray diffraction analysis depicts that PANI/$CaCO_3$ nanocomposite exhibits a semicrystalline nature. It has been also reported that MA-functionalized $CaCO_3$ nanoparticles are finely dispersed in the PANI matrix. The reported possible reason for the fine and uniform dispersion of PANI/$CaCO_3$ nanocomposite is the hydrophobic nature of MA-treated $CaCO_3$ nanoparticles and micromixing caused by ultrasonic irradiations.

As reported by Bhanvase and Sonawane (2010), Figure 12.6a represents the typical TEM images of PANI nanoparticles prepared by ultrasound-assisted emulsion polymerization technique. It has been observed from the reported TEM image that the PANI particles appeared in the range of 50–100 nm with uniform particle size distribution. Figure 12.6b and c shows the TEM image of PANI/$CaCO_3$ nanocomposite synthesized by ultrasound-assisted *in situ* emulsion polymerization with 4 wt% $CaCO_3$ loading. It is observed that MA-functionalized $CaCO_3$ particles (some black dots) having a diameter of 10–20 nm are embedded and finely dispersed in PANI matrix (Figure 12.6c). This fine distribution of MA-functionalized $CaCO_3$ particles in PANI matrix is due to ultrasonic irradiation. During the initial stage of emulsion polymerization, due to the hydrophobic nature, MA-coated $CaCO_3$ nanoparticles remain within aniline (monomer/organic phase), and micromixing caused by ultrasound generates fine $CaCO_3$-embedded monomer droplets. Further, ultrasound-assisted *in situ* emulsion polymerization of aniline in the presence of functionalized $CaCO_3$ leads to the formation of finely dispersed PANI/$CaCO_3$ nanocomposite. It has been also observed from Figure 12.6b that the formed PANI/$CaCO_3$ nanocomposite has a crystalline nature. It was also attributed to the homogeneous nucleation of PANI in the aqueous phase due to ultrasonic irradiation, and $CaCO_3$ nanoparticles act as templates for the nucleation of PANI. This results in the encapsulation of $CaCO_3$ nanoparticles

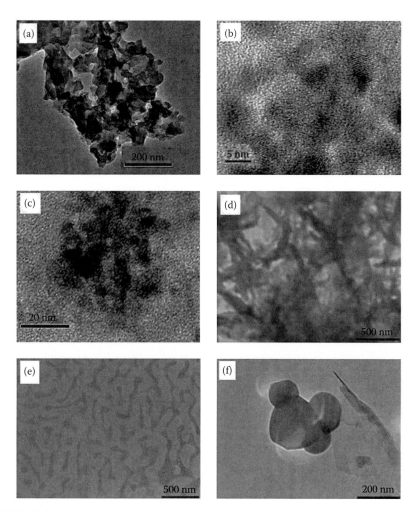

FIGURE 12.6
TEM photographs of (a) PANI particles prepared by novel ultrasound emulsion polymerization, (b) and (c) PANI/CaCO₃ nanocomposite by ultrasound-assisted *in situ* emulsion polymerization, (d) PANI by conventional emulsion polymerization, and (e) and (f) PANI/CaCO₃ nanocomposite by conventional *in situ* emulsion polymerization. (Reprinted from Bhanvase, B.A. and Sonawane, S.H., *Chem. Eng. J.*, 156, 177, 2010. With permission.)

in PANI, which leads to the formation of solid PANI/CaCO₃ nanocomposite particles. Figure 12.6d through f depicts the TEM images of pure PANI and PANI/CaCO₃ nanocomposite, synthesized by conventional emulsion polymerization technique. As reported in Figure 12.6d, with an increasing reaction time, the nanofibers into thicker structure and then agglomerate into irregular shape. The particle size of pure PANI synthesized by conventional emulsion polymerization technique was found around 500 nm. TEM images (Figure 12.6e and f) of PANI/CaCO₃ nanocomposite prepared using

conventional *in situ* emulsion polymerization technique show a phase separation between PANI and $CaCO_3$. This phase separation was observed due to
improper dispersion of nano-$CaCO_3$ in PANI matrix. Further, this phase separation in the PANI matrix is responsible for the roughness of the polymer
film. These results show that ultrasound-assisted *in situ* emulsion polymerization is more effective than conventional *in situ* emulsion polymerization
method for the preparation of fine dispersed colloidal nanocomposites.

The corrosion rates (V_C) of alkyd resin, PANI/alkyd, and PANI/$CaCO_3$/
alkyd coatings have been monitored by Bhanvase and Sonawane (2010)
for a period of 200 h. It has been reported that pure alkyd coatings applied
on MA panel disappear when placed in different corrosive media, that is,
NaOH, HCl, and NaCl, exhibiting a rapid corrosion rate of the pure alkyd
organic coating. Rate of corrosion in 5% HCl solution is reported in Figure
12.7. Corrosion rate at zero (neat coating) loading of PANI or PANI/$CaCO_3$
nanocomposites prepared by ultrasound-assisted method is 0.87 cm/year
in the case of 5% HCl solution. Corrosion rate of alkyd coating has been
reported to be decreased with an increase in the addition of PANI and
PANI/$CaCO_3$ nanocomposite in alkyd resin. For 1% loading of PANI synthesized by ultrasound-assisted emulsion polymerization, the rate of corrosion is of the order of 0.66 cm/year and reduced to 0.24 cm/year units for 5%
loading of PANI in alkyd resin. With 1% PANI/$CaCO_3$ nanocomposite loading (synthesized by ultrasound-assisted *in situ* emulsion polymerization),

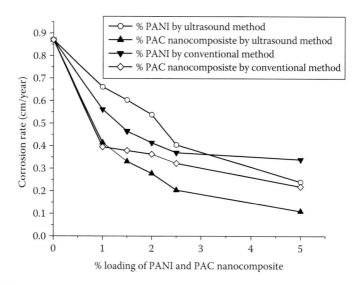

FIGURE 12.7
Corrosion rate of PANI and PANI/$CaCO_3$ nanocomposite through an ultrasonic method and
simple conventional mixing method dispersed in alkyd resin coatings recorded in 5% HCl
solution. (Reprinted from Bhanvase, B.A. and Sonawane, S.H., *Chem. Eng. J.*, 156, 177, 2010. With
permission.)

the rate of corrosion is reduced from 0.87 cm/year (for pure alkyd resin in 5% HCl solution) to 0.42 cm/year, and it reduces to 0.1 cm/year when the PANI/CaCO₃ nanocomposite loading is maintained at a constant value of 5% in alkyd resin in HCl solution.

Further, it has been reported that for 1% loading of PANI synthesized by conventional emulsion polymerization, the corrosion rate is 0.56 cm/year, and it is being reduced to 0.34 cm/year for 5% loading of PANI prepared by conventional method. Loading of PANI/CaCO₃ nanocomposite in alkyd resin also improves the corrosion resistance. The corrosion rate for 1% loading of PANI/CaCO₃ nanocomposite prepared by conventional method is 0.40 cm/year, and it has been reduced to 0.22 cm/year at 5% loading. It has been reported that PANI and PANI/CaCO₃ nanocomposite prepared by ultrasound method show better anticorrosion properties than that of conventional method. The reported possible reason is that fine dispersion of MA-coated CaCO₃ in PANI matrix improves compactness of the coating film and in turn improves the corrosion resistance.

Rate of corrosion of PANI and PANI/CaCO₃ nanocomposite coating in 5% NaCl solution has been depicted in Figure 12.8. As discussed in the earlier section, similar trends in corrosion rate of coatings were observed. The corrosion rate for 1% loading of PANI synthesized by ultrasound-assisted emulsion polymerization is of the order of 0.52 cm/year and reduced to 0.20 cm/year units for 5% loading of PANI in alkyd resin. With 1% PANI/CaCO₃ nanocomposite loading (synthesized by ultrasound-assisted *in situ* emulsion polymerization), the rate of corrosion is decreased from

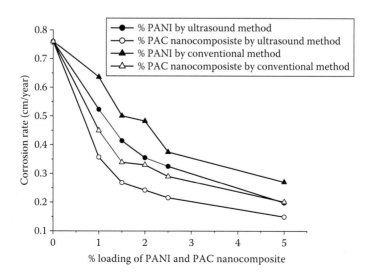

FIGURE 12.8
Corrosion rate of PANI and PANI/CaCO₃ nanocomposite through an ultrasonic method and conventional mixing method dispersed in alkyd resin coatings recorded in 5% NaCl solution. (Reprinted from Bhanvase, B.A. and Sonawane, S.H., *Chem. Eng. J.*, 156, 177, 2010. With permission.)

0.76 cm/year (for pure alkyd resin in 5% NaCl solution) to 0.36 cm/year, and it reduces to 0.15 cm/year when the PANI/CaCO$_3$ nanocomposite loading has been maintained at a constant value of 5% in alkyd resin. It has been also reported that for 1% loading of PANI synthesized by conventional emulsion polymerization, the corrosion rate is 0.63 cm/year, and it is decreased to 0.27 cm/year for 5% loading of PANI prepared by conventional method. Further, the corrosion rate for 1% loading of PANI/CaCO$_3$ nanocomposite prepared by conventional method is 0.45 cm/year, and it has been decreased to 0.20 cm/year at 5% loading. In this case also, PANI and PANI/CaCO$_3$ nanocomposite prepared by ultrasound method show better anticorrosion properties than that of conventional method with the same possible reason reported earlier. Further, similar trends of corrosion rate have been reported for the case of 5% NaOH solution, which is shown in Figure 12.9.

Overall, the value of corrosion rate is found to be very less in the case of PANI and PANI/CaCO$_3$ nanocomposite prepared by ultrasound-assisted method than that of conventional method. It has been also observed that both PANI and PANI/CaCO$_3$ nanocomposite show the best performance in the case of NaCl solution, when the loading of PANI or PANI/CaCO$_3$ nanocomposite prepared by ultrasound method is 5% in alkyd resin. It can be concluded that 5% loading of PANI/CaCO$_3$ nanocomposite synthesized by ultrasound-assisted *in situ* emulsion polymerization in alkyd resin is more effective for decreasing the rate of corrosion in acid, alkali, and salt

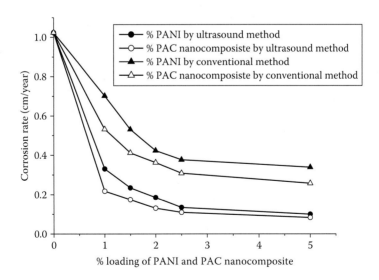

FIGURE 12.9
Corrosion rate of PANI and PANI/CaCO$_3$ nanocomposite through an ultrasonic method and simple conventional mixing method dispersed in alkyd resin coatings recorded in 5% NaOH solution. (Reprinted from Bhanvase, B.A. and Sonawane, S.H., *Chem. Eng. J.*, 156, 177, 2010. With permission.)

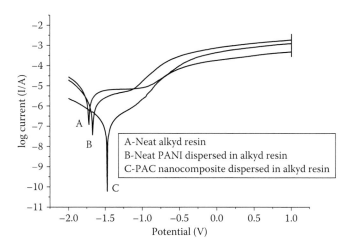

FIGURE 12.10
Tafel plots of alkyd resin, PANI, and PANI/CaCO$_3$ nanocomposite through an ultrasonic method dispersed in alkyd resin coatings recorded in 5% NaCl solution. (Reprinted from Bhanvase, B.A. and Sonawane, S.H., *Chem. Eng. J.*, 156, 177, 2010. With permission.)

solutions. From the corrosion rate data, it can be accomplished that PANI can be replaced with PANI/CaCO$_3$ nanocomposite.

Electrochemical corrosion analysis has been studied by plotting a Tafel plot (log $|I|$ vs E), which was carried out in 5% NaCl solution as electrolyte at room temperature (25°C). Figure 12.10 shows the Tafel plot for neat alkyd resin, PANI/alkyd, and PANI/CaCO$_3$ nanocomposite/alkyd coating. It is found that electrochemical current (I_{corr}) was decreased from 0.89 to 0.03 µA/cm^2, when neat alkyd resin and PANI/CaCO$_3$ nanocomposite were tested in NaCl electrolyte solution. Observed corrosion current for neat alkyd coating was 0.89 µA/cm^2 while that for PANI/alkyd and PANI/ CaCO$_3$ nanocomposite/alkyd coatings are 0.50 and 0.03 µA/cm^2 respectively. Additionally, E_{corr} value shows shifts in positive side from −1.74 to −1.47 V by an addition of PANI/CaCO$_3$ nanocomposite into alkyd coating. Overall results indicate that PANI/CaCO$_3$ nanocomposite shows improvement in anticorrosive properties, which supports the earlier corrosion rate data obtained by dip test method.

The adhesion of coating could be an important factor that controls the corrosion rate. PANI/Alkyd and PANI/CaCO$_3$/Alkyd coatings undergo surface crazing phenomenon due to the diffusion of oxygen, acidic, and alkaline ions. Due to the attack of acid, salt, and alkali, metal iron complexes (Fe^{2+}, etc.) are formed. Due to alkyd cross-linking and hydrogen bonding of alkyd with PANI/CaCO$_3$ nanocomposite, coating becomes more compact. Compact cross-linking of alkyd resin and hydrophobic nature of CaCO$_3$ inhibit the moisture infiltration and moisture contact with metal through barrier mechanism, which is inferior in the case of neat alkyd coatings. PANI has an ability to score some charges (Fe^{2+}, etc.) generated by corrosion on mild steel panel (passivation of metal). It has been reported (Bhanvase and Sonawane 2010) that small quantity of PANI acts as electrochemical inhibitor, and electrochemical properties of PANI nanocomposite film provide additional backup for protection. Both electrochemical and barrier mechanisms communally give the corrosion protection to mild steel (Dispenza et al. 2006).

12.3.3 Synthesis of CaCO$_3$/P(MMA–BA) Nanocomposite and Its Application in Water-Based Alkyd Emulsion Coating for Mechanical and Thermal Properties

Gumfekar et al. (2011) have prepared P(MMA–BA)/nano-CaCO$_3$ water-based nanocomposite by using *in situ* hybrid emulsion polymerization with acrylic monomers [MMA, n-butyl acrylate (BA)]. During the synthesis of P(MMA–BA)/nano-CaCO$_3$ nanocomposite, they have prepared aqueous solution of surfactant (based on 3 wt% of total monomer) and initiator (0.5 wt% of total monomer) separately in deionized water followed by high-speed stirring at 1000 RPM. Monomers MMA and BA were taken in a ratio of 10:90, 30:70, and 50:50 (by wt), and separate polymerization was carried out for each one of the nanocomposite in the presence of functionalized nano-CaCO$_3$ (4 wt% of total monomer) at 1000 RPM and 78°C. The reaction was continued for a period of 160 min. Further polymer–alkyd coating has been prepared by blending alkyd emulsion with polymer nanocomposite in various ratios, such as 90:10, 70:30, and 50:50 by wt%. Also the synthesis of copolymer and its nanocomposite from MMA and n-BA (50:50 wt%) by *in situ* emulsion polymerization method was reported and used during the preparation of polymer–alkyd combinations. They have also studied the effect of nano-CaCO$_3$ on coating properties.

Gumfekar et al. (2011) have reported the morphological analysis of CaCO$_3$ nanoparticles and hybrid nanocomposite (Figure 12.11). It has been reported that the structure of CaCO$_3$ nanoparticles prepared by sonochemical carbonization shows cubic shape, which has calcite phase. Also it has been reported that formation of the composite using bulk CaCO$_3$ was not uniform, and it showed no miscibility and compatibility with the MMA and BA during the polymerization process. This might be due to the hydrophilic nature of bulk CaCO$_3$ particles, which do not remain in the organic phase during the polymerization of BA and MMA to form composite. TEM image of hybrid nanocomposite clearly indicates an encapsulation of CaCO$_3$ nanoparticles into the polymer. The reported size of hybrid composite spheres was about 150 nm, and CaCO$_3$ nanoparticles were seemed to be well encapsulated in the polymeric matrix.

Further, Gumfekar et al. (2011) have used thermogravimetric analysis to investigate the thermal stability of the prepared samples (Figure 12.12). Degradation of neat polymer might occur due to relatively weak head-to-head linkage, impurities, and solvent in the lattices; however, addition of

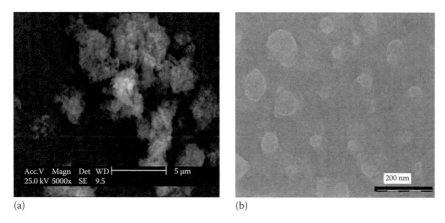

(a)   (b)

FIGURE 12.11
(a) Scanning electron microscopy image of nano-CaCO₃ particles and (b) formation of hybrid nanocomposite using functionalized CaCO₃. (Reprinted from Gumfekar, S.P. et al., *Prog. Org. Coat.*, 72, 632, 2011. With permission.)

4% nano-CaCO₃ to this neat polymer resists the thermal degradation substantially. It is possibly due to the establishment of strong interfacial bond between polymer matrix and functionalized nanoparticles. Also thermal degradation of the copolymer would be attributed to the random bond scission of PMMA–BA matrix (DTA curves shown in Figure 12.12). The nano-CaCO₃ embedded PMMA–BA matrix shows better thermal stability. It is also reported that the nano-CaCO₃ acts as thermal sink leading to a decrease in the degradation process hence shows the higher degradation temperature.

Gumfekar et al. (2011) also reported the mechanical properties of alkyd emulsion coating with varying compositions of P(MMA–BA) and its nanocomposite with alkyd emulsion. It has been reported that addition of P(MMA–BA) latex to alkyd decreases gloss of the film, while addition of nanocomposite increases the gloss. In both the cases, gloss is relatively low as compared to solvent-borne coatings. This fact may be attributed to the surface tension gradient generated during solvent evaporation, which is observed in the case of solvent-based coatings. The reduction of gloss is also due to surface roughness and incompatibility of acrylic polymer with the alkyd emulsion resin. This matting effect is predominant even at lower as well as equal loading of acrylic in alkyd resin. While in case of addition of polymer nanocomposite addition which contains nano-CaCO₃ particles shows increase in gloss value with increase in the loading of PNC. The increase in gloss is attributed to a decrement of surface roughness due to the less percentage of the inorganic particle aggregates, and there is good compatibility and dispersion of PNC in the alkyd emulsion resin. Further, pencil hardness of all coatings containing nanoparticles increases due to restriction of the motion

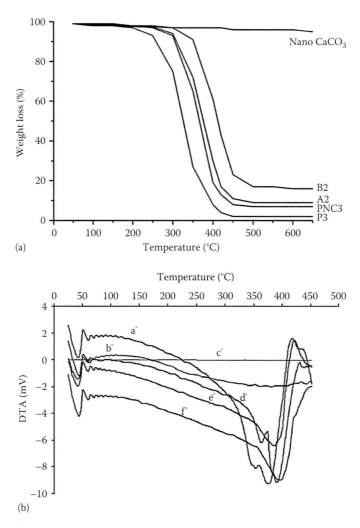

FIGURE 12.12
(a) Thermal stability analysis of composite by thermogravimetric analysis. P3:BA:MMA = 50:50, no CaCO$_3$, PNC3:BA:MMA = 50:50, 4% CaCO$_3$. A2:composite P3:alkyd = 30:70. B2:composite PNC3:alkyd = 30:70. (b) DTA curves of polymer matrix and nanocomposites. (a') P3 is BA:PMMA, 50:50 without CaCO$_3$; (b') alkyd emulsion; (c') nano-CaCO$_3$, (d') PNC3 is (BA/MMA) 50:50, and 4% CaCO$_3$; (e') A2 is P3:alkyd 30:70; (f') B2 is PNC3:alkyd, 30:70. (Reprinted from Gumfekar, S.P. et al., *Prog. Org. Coat.*, 72, 632, 2011. With permission.)

of polymeric chains filled with nano-CaCO$_3$. Overall, pencil hardness has been reported to be increased with an addition of either neat polymer or polymer nanocomposite. Overall, uniform dispersion of sonochemically prepared CaCO$_3$ nanoparticles with cubic shape in polymer matrix showed significant enhancement in mechanical and thermal properties.

12.4 Future Prospective

The synthesis of different nanocomposites has been reported by ultrasound-assisted semibatch *in situ* emulsion polymerization in the presence of initiator at high temperature. There are potentials to extend this work without the use of initiator and emulsifier at room temperature. Ultrasound-assisted *in situ* semibatch emulsion polymerization for the synthesis of polymer nanocomposite at room temperature can save energy, and without the use of initiator and emulsifier, the process becomes environmentally friendly and a pure product can be prepared. Further study of the property profile of polymer nanocomposite prepared by ultrasound-assisted *in situ* semibatch emulsion polymerization at room temperature and without the use of initiator and surfactant can be possible.

12.5 Conclusion

Surface modification of CaCO$_3$ nanoparticles has significant effect on the compatibility of polymer matrix, which leads to the fine dispersion with superior properties of polymer nanocomposite. PMMA/CaCO$_3$ nanocomposites have been successfully synthesized using ultrasound-assisted *in situ* semibatch emulsion polymerization of MMA in the presence of an initiator. Use of cavitation generated due to the ultrasonic irradiations during *in situ* emulsion polymerization improves the dispersion of modified CaCO$_3$ into PMMA latex compared to conventional *in situ* emulsion polymerization and also significantly intensifies the process with overall reduction in the energy requirements. Further, the ultrasound-based method has proved to be effective to increase the loadings of CaCO$_3$ in the composites, which can improve the thermal stability of PMMA/CaCO$_3$ nanocomposites, compared to composite prepared by conventional method. PANI/CaCO$_3$ nanocomposite has higher capacity to form more homogeneous films. Addition of PANI/CaCO$_3$ and PANI in alkyd resin improves the mechanical and anticorrosion properties. Use of ultrasound method has been found to be effective for fine dispersion of modified CaCO$_3$ in polymer nanocomposite, which leads to improvement in mechanical, anticorrosive, and rheological properties. In the synthesis of nanoparticles and nanocomposites, cavitation plays a vital role in case of the acceleration of chemical reactions with acoustic energy, which effectively shortens reaction time; improves solute transfer rate, nucleation rate, and rates of polymerization; reduces reaction condition; and achieves some reactions that cannot be accomplished by conventional methods.

References

Arami, A., M. Mazloumi, R. Khalifehzadeh, and S. K. Sadrnezhaad. 2007. Sono-chemical preparation of TiO$_2$ nanoparticles. *Materials Letters* 61:4559–4561.

Avella, M., M. Errico, and E. Martuscelli. 2001. Novel PMMA/CaCO$_3$ nanocomposites abrasion resistant prepared by an in situ polymerization process. *Nano Letters* 1:213–217.

Barsoukov, E. and J. R. Macdonald. 2005. *Impedance Spectroscopy, Theory, Experiment and Applications*, Wiley-Interscience, New York.

Bhanvase, B. A., S. P. Gumfekar, and S. H. Sonawane. 2009. Water-based PMMA-nano-CaCO$_3$ nanocomposites by in situ polymerization technique: Synthesis, characterization and mechanical properties. *Polymer-Plastics Technology and Engineering* 48:939–944.

Bhanvase, B. A., D.V. Pinjari, P. R. Gogate, S. H. Sonawane, and A. B. Pandit. 2011. Process intensification of encapsulation of functionalized CaCO$_3$ nanoparticles using ultrasound assisted emulsion polymerization. *Chemical Engineering and Processing: Process Intensification* 50:1160–1168.

Bhanvase, B. A. and S. H. Sonawane. 2010. New approach for simultaneous enhancement of anticorrosive and mechanical properties of coatings: Application of water repellent nano CaCO$_3$–PANI emulsion nanocomposite in alkyd resin. *Chemical Engineering Journal* 156:177–183.

Chan, C. M., J. Wu, J. X. Li, and Y. K. Cheung. 2002. Polypropylene/calcium carbonate nanocomposites. *Polymer* 43:2981–2992.

Chang, K. C., G. W. Jang, C. W. Peng, C. Y. Lin, J. C. Shieh, J. M. Yeh, J. C. Yang, and W. T. Li. 2007. Comparatively electrochemical studies at different operational temperatures for the effect of nanoclay platelets on the anticorrosion efficiency of DBSA-doped polyaniline/Na+-MMT clay nanocomposite coatings. *Electrochemica Acta* 52:5191–5200.

Chen, X., C. Li, S. Xu, L. Zhang, W. Shao, and H. L. Du. 2006. Interfacial adhesion and mechanical properties of PMMA-coated CaCO$_3$ nanoparticle reinforced PVC composites. *China Particuology* 4:25–30.

Chen, J., Y. Wang, F. Guo, X. Wang, and C. Zheng. 2000. Synthesis of nanoparticles with novel technology: High-gravity reactive precipitation, American Chemical Society. *Industrial Engineering Chemistry Research* 39:948–954.

Cole, K. S. and R. H. Cole. 1942. Dispersion and absorption in dielectrics, direct current characteristics. *The Journal of Chemical Physics* 10:98–105.

Dagaonkar, M. V., A. Mehra, R. Jain, and H. J. Heeres. 2004. Synthesis of CaCO$_3$ nanoparticles by carbonation of lime in reverse micellar systems. *Chemical Engineering Research and Design* 82:1438–1443.

Demjén, Z., B. Pukánszky, E. Földes, and J. Nagy. 1997. Interaction of silane coupling agents with CaCO$_3$. *Journal of Colloid and Interface Science* 190:427–436.

Dispenza, C., C. L. Presti, C. Belfiore, G. Spadaro, and S. Piazza. 2006. Electrically conductive hydrogel composites made of polyaniline nanoparticles and poly(N-vinyl-2-pyrrolidone). *Polymer* 47:961–971.

Dobie, C. G. and K. V. K. Boodhoo. 2010. Surfactant-free emulsion polymerisation of methyl methacrylate and methyl acrylate using intensified processing methods. *Chemical Engineering and Processing: Process Intensification* 49:901–911.

Gogate, P. R. 2008. Cavitational reactors for process intensification of chemical processing applications: A critical review. *Chemical Engineering and Processing: Process Intensification* 47:515–527.

Gumfekar, S. P., K. J. Kunte, L. Ramjee, K. H. Kate, and S. H. Sonawane. 2011. Synthesis of CaCO₃–P(MMA–BA) nanocomposite and its application in water based alkyd emulsion coating. *Progress in Organic Coatings* 72:632–637.

Gupta, R. 2004. *Synthesis of Precipitated Calcium Carbonate Nanoparticles Using Modified Emulsion Membranes.* Georgia Institute of Technology, Atlanta, GA, pp. 21–26.

He, M., E. Forssberg, Y. Wang, and Y. Han. 2005. Ultrasonication-assisted synthesis of calcium carbonate nanoparticles. *Chemical Engineering Communications* 192:1468–1481.

Hu, Z., Y. Deng, and Q. Sun. 2004. Synthesis of precipitated calcium carbonate nanoparticles using a two-membrane system. *Colloid Journal* 66:745–750.

Huber, M., W. J. Stark, S. Loher, M. Maciejewski, F. Krumeich, and A. Baiker. 2005. Flame synthesis of calcium carbonate nanoparticles. *Chemical Communications* 5:648–650.

Jeevanandam, P., Y. Koltypin, A. Gedanken, and Y. Mastai. 2000. Synthesis of a-cobalt (II) hydroxide using ultrasound irradiation. *Journal of Materials Chemistry* 10:511–514.

Jeyaprabha, C., S. Sathiyanarayanan, and G. Venkatachari. 2006. Effect of cerium ions on corrosion inhibition of PANI for iron in 0.5 M H₂SO₄. *Applied Surface Science* 253:432–438.

Jiang, J., L. C. Li, and F. Xu. 2006. Preparation, characterization and magnetic properties of PANI/Lasubstituted LiNi ferrite nanocomposites. *Chinese Journal of Chemical* 24:1804–1809.

Jian-ming, S., B. Yong-zhong, H. Zhi-ming, and W. Zhi-xue. 2004. Preparation of poly (methyl methacrylate)/nanometer calcium carbonate composite by in-situ emulsion polymerization. *Journal of Zhejiang University Science* 5:709–713.

Kong, Y. X., C. Kan, and C. Sun. 1999. Encapsulation of calcium carbonate by styrene polymerization. *Polymers Advanced Technologies* 10:54–59.

Liu, P., W. M. Liu, and Q. J. Xue. 2004. In situ radical transfer addition polymerization of styrene from silica nanoparticles. *European Polymer Journal* 40:267–271.

Ma, X., B. Zhou, Y. Deng, Y. Sheng, C. Wang, Y. Pan, and Z. Wang. 2008. Study on CaCO₃/PMMA nanocomposite microspheres by soapless emulsion polymerization. *Colloids and Surfaces A: Physicochemical and Engineering Aspects* 312: 190–194.

Mishra, S., S. H. Sonawane, and R. P. Singh. 2005. Studies on characterization of nano CaCO₃ prepared by in situ deposition technique and its application in PP nanocomposites. *Journal of Polymer Science Part B Polymer Physics* 43:107–113.

Monte, S. J. and G. Sugerman. 1976. Presented at the *110th Meeting of the Rubber Division American Chemical Society,* San Francisco, CA, p. 43 (October 5–8, 1976).

Nakatsuka, T., H. Kawasaki, and K. Itadani. 1982. Phosphate coupling agents for calcium carbonate filler. *Journal of Applied Polymer Science* 27:259–269.

Papirer, E., J. Schultz, and C. Turchi. 1984. Surface properties of a calcium carbonate filler treated with stearic acid. *European Polymer Journal* 20:1155–1158.

Pinjari, D. V. and A. B. Pandit. 2011. Room temperature synthesis of crystalline CeO₂ nanopowder: Advantage of sonochemical method over conventional method. *Ultrasonics and Sonochemistry* 18:1118–1123.

Prasad, K., D. V. Pinjari, A. B. Pandit, and S. T. Mhaske. 2010a. Phase transformation of nanostructured titanium dioxide from anatase-to-rutile via combined ultrasound assisted sol–gel technique. *Ultrasonics and Sonochemistry* 17:409–415.

Prasad, K., D. V. Pinjari, A. B. Pandit, and S. T. Mhaske. 2010b. Synthesis of titanium dioxide by ultrasound assisted sol–gel technique: Effect of amplitude (power density) variation. *Ultrasonics and Sonochemistry* 17:697–703.

Pud, A., N. Ogurtsov, A. Korzhenko, and G. Shapoval. 2003. Some aspects of preparation methods and properties of polyaniline blends and composites with organic polymers. *Progress in Polymer Science* 28:1701–1753.

Qi, W., X. Hesheng, and Z. Chuhong. 2001. Preparation of polymer/inorganic nanoparticles composites through ultrasonic irradiation. *Journal of Applied Polymer Science* 80:1478–1488.

Ryu, J. G., S. W. Park, H. Kim, and J. W. Lee. 2004. Power ultrasound effects for in situ compatibilization of polymer–clay nanocomposites. *Materials Science and Engineering* 24:285–288.

Sahoo, P. K. and R. Mohapatra. 2003. Synthesis and kinetic studies of PMMA nanoparticles by non-conventionally initiated emulsion polymerization. *European Polymer Journal* 39:1839–1846.

Salkar, R. A., P. Jeevanandam, S. T. Aruna, Y. Koltypin, and A. Gedanken. 1999. The sonochemical preparation of amorphous silver nanoparticles. *Journal of Materials Chemistry* 9:1333–1335.

Sh, Y., B. Zhou, C. Wang, X. Zhao, Y. Deng, and Z. Wang. 2006. In situ preparation of hydrophobic $CaCO_3$ in the presence of sodium oleate. *Applied Surface Science* 253:1983–1987.

Shai, K. V. P. M., A. Gedanken, and R. Prozorov. 1998. Sonochemical preparation and characterization of nanosized amorphous Co–Ni alloy powders. *Journal of Materials Chemistry* 8:769–773.

Sheng, Y., J. Zhao, B. Zhou, X. Ding, Y. Deng, and Z. Wang. 2006. In situ preparation of $CaCO_3$/polystyrene composite nanoparticles. *Materials Letters* 60:3248–3250.

Shirsath, S. R., D. V. Pinjari, P. R. Gogate, S. H. Sonawane, and A. B. Pandit. 2013. Ultrasound assisted synthesis of doped TiO_2 nano-particles: Characterization and comparison of effectiveness for photocatalytic oxidation of dyestuff effluent. *Ultrasonics Sonochemistry* 20:277–286.

Siegel, R. W. 1994. Nanostructured materials-mind over matter. *Nanostructured Materials* 4:121–138.

Sonawane, S. H., P. K. Khanna, S. Meshram, C. Mahajan, M. P. Deosarkar, and S. Gumfekar. 2009. Combined effect of surfactant and ultrasound on nano calcium carbonate synthesized by crystallization process. *International Journal of Chemical Reactor Engineering* 7:A47.

Sonawane, S. H., S. R. Shirsath, P. K. Khanna, S. Pawar, C. M. Mahajan, V. Paithankar, V. Shinde and C. V. Kapadnis. 2008. An innovative method for effective micromixing of CO_2 gas during synthesis of nano-calcite crystal using sonochemical carbonization. *Chemical Engineering Journal* 143:308–313.

Stejskal, J., M. Trchová, J. Brodinová, P. Kalenda, S. V. Fedorova, J. Prokeš, and J. Zemek. 2006. Coating of zinc ferrite particles with a conducting polymer Polyaniline. *Journal of Colloids Interface Science* 298:87–93.

Tiarks, F., K. Landfester, and M. Antonietti. 2001. Silica nanoparticles as surfactants and fillers for latexes made by miniemulsion polymerization. *Langmuir* 17:5775–5780.

Tissot, I., C. Novat, F. Lefebvre, and E. Bourgeat-Lami. 2001. Hybrid latex particles coated with silica. *Macromolecules* 34:5737–5739.

Tsuzuki, T., K. Pethick, and P. G. McCormick. 2000. Synthesis of CaCO$_3$ nanoparticles by mechanochemical processing. *Journal of Nanoparticle Research* 2:375–380.

Wang, C., Y. Sheng, H. Bala, X. Zhao, J. Zhao, X. Ma, and Z. Wang. 2007a. A novel aqueous-phase route to synthesize hydrophobic CaCO$_3$ particles in situ. *Materials Science and Engineering* 27:42–45.

Wang, C., Y. Sheng, X. Zhao, Y. Pan, H. Bala, and Z. Wang. 2006a. Synthesis of hydrophobic CaCO$_3$ nanoparticles. *Materials Letters* 60:854–857.

Wang, H., L. Tang, X. Wu, W. Dai, and Y. Qiu. 2007b. Fabrication and anti-frosting performance of super hydrophobic coating based on modified nanosized calcium carbonate and ordinary polyacrylate. *Applied Surface Science* 253:8818–8824.

Wang, C., P. Xiao, J. Zhao, X. Zhao, Y. Liu, and Z. Wang. 2006b. Biomimetic synthesis of hydrophobic calcium carbonate nanoparticles via a carbonation route. *Powder Technology* 170:31–35.

Wu, W., T. He, J. Chen, X. Zhang, and Y. Chen. 2006. Study on in situ preparation of nano calcium carbonate/PMMA composite particles. *Materials Letters* 60:2410–2415.

Xie, X., Q. Liu, R. Li, X. Zhou, Q. Zhang, Z. Yu, and Y. Mai. 2004. Rheological and mechanical properties of PVC/CaCO$_3$ nanocomposites prepared by in situ polymerization. *Polymer* 45:6665–6673.

Zhang, Z., C. Wang, Z. Yang, C. Chen, and K. Mai. 2008. Crystallization behavior and melting characteristics of PP nucleated by a novel supported -nucleating agent. *Polymer* 49:5137–5145.

Zhang, K., L. Zheng, X. Zhang, X. Chen, and B. Yang. 2006. Silica-PMMA core-shell and hollow nanospheres. *Colloids and Surfaces A: Physicochemical and Engineering Aspects* 277:145–150.

Index

Printed and bound by CPI Group (UK) Ltd, Croydon, CR0 4YY

18/10/2024

01776271-0008